STO

**ACPL ITEM
DISCARDED**

SOLUTIONS, MINERALS, AND EQUILIBRIA

HARPER'S GEOSCIENCE SERIES
Carey Croneis, Editor

Solutions, Minerals, and Equilibria

ROBERT M. GARRELS
Harvard University

CHARLES L. CHRIST
United States Geological Survey

HARPER & ROW
Publishers   New York

SOLUTIONS, MINERALS, AND EQUILIBRIA

Copyright © 1965 by Robert M. Garrels and Charles L. Christ. Printed in the United States of America. All rights reserved. No part of this book may be used or reproduced in any manner whatsoever without written permission except in the case of brief quotations embodied in critical articles and reviews. For information address Harper & Row. Publishers, Incorporated, 49 East 33rd Street, New York 10016, N.Y.

Library of Congress Catalog Card Number: 65-12674

Contents

Editor's Introduction	vii
Preface	xi
Acknowledgments	xii
1. Introduction	1
2. Activity-Concentration Relations	20
3. Carbonate Equilibria	74
4. Complex Ions	93
5. Measurement of Eh and pH	122
6. Partial Pressure Diagrams	144
7. Eh-pH Diagrams	172
8. Ion Exchange and Ion-Sensitive Electrodes	267
9. Effects of Temperature and Pressure Variations on Equilibria	306
10. Combination Diagrams	352
11. Some Geological Applications of Mineral Stability Diagrams	379
Appendix 1. Symbols and Physical Constants	401
Appendix 2. Tables of $\Delta H_f°$, $S°$, and $\Delta F_f°$ Values at 25 °C	403
Appendix 3. Crystal Radii of Some Ions	430
Appendix 4. Pressure of Saturated Water Vapour at Various Temperatures	431
Appendix 5. Reference Buffer Solutions	432
Appendix 6. Table of Atomic Weights—1961	433
Index	437

Editor's Introduction

The book *Solutions, Minerals, and Equilibria* by Robert Minard Garrels and Charles Louis Christ is, in many respects, a new text from title to index. Actually, however, it is a second edition based on the earlier book, *Mineral Equilibria*, written by Dr. Garrels and published in 1960 in the Harper's Geoscience Series. In the development of the original volume, author Garrels was greatly assisted by Dr. Christ —his close colleague at the United States Geological Survey. It is thus appropriate that the latter has now joined more formally with the senior author in the preparation of a revision and amplification of the earlier text, so complete and extensive that the new title is more than justified.

The central theme of the present volume is the application of thermodynamics to the many natural systems of geological interest and geochemical concern. The problem, as in the first edition, is to portray, by the simplest and most effective methods, simultaneous chemical reactions, particularly equilibrium relations among minerals. The need for such a book is obvious, for the significance of geochemical investigations is hard to overstate. Although the realization of this fact was somewhat slow in developing, geochemistry has now become so important a subdiscipline of the geosciences that its devotees in educational institutions, industries, surveys, and so forth, are beginning to number in the thousands rather than in the scores, as was the case as recently as a quarter century ago. The increase in the number of geochemists has been notable even in the brief period since the first edition of this volume was issued. There seems little question but that many, if not all, of the younger scientists, as well as most of the veteran geochemists, will appreciate the updating and expansion of the one book which is principally concerned with the geological applications of chemical thermodynamics. It will be a little surprising if they do not also find useful the modern treatment of the basic principles of their chosen field, which the new volume makes available.

The Garrels and Christ volume is divided into eleven chapters and a number of appendixes. The presentation is characterized by an economy of words and a direct no-nonsense approach to each of the subjects discussed. The "Introduction" and the section on activity-concentration relations are designed to introduce the reader to fundamental thermodynamic relationships in a style designed to make their value apparent in the solution of geochemical problems; and solid solutions are treated so as to emphasize not only their importance but also their scientific promise as a subject for additional experimental work.

The discussion of carbonate equilibria is of particular importance owing to the great significance which geologists properly attach to the carbonates of various types and of widely ranging ages. Their study by oceanographers, stratigraphers, paleontologists, and ecologists, as well as by geochemists, grows apace. Such high level investigations do not only have scientific import, but the industrial applications of the research have made carbonate studies sufficiently significant so as to engender large research funds. It is obvious that henceforth it will be imperative for all so-called "soft-rock geologists" to become at least reasonably literate in geochemical lore. If they will study and understand carbonate equilibria as presented by Garrels and Christ, such literacy can be readily achieved.

The material presented by the authors on complex ions and measurement of Eh-pH are really "state of the art" discussions, but, in addition, the research and interpretive difficulties which result from the complex ions involved in many systems are also outlined. Indeed, the authors attempt to initiate the student—as well as many of their professional colleagues—into the mysteries of relatively sophisticated inorganic chemistry and its impact on many geological and geochemical problems. Moreover, the applicability and the advantages of diagrammatic representations of a number of such problems are pointed out through examples.

The section on the effects of temperature and pressure variations on equilibria demonstrates the methods through which thermodynamic equations, which involve pressure and temperature as variables, can be effectively utilized, and the chapter on combination diagrams explains how such devices may be employed in arriving at relatively simple solutions to highly complex problems. The title of the final chapter of the book, "Some Geological Applications of Mineral Stability Diagrams", is self-explanatory; but, it should be observed that, in addition to the geological situations discussed in this section of the text, other applications of geochemistry to the solution of long standing geological problems appear throughout the entire volume.

The references and problems which are appended at the end of the various chapters should make *Solutions, Minerals, and Equilibria* particularly valuable as a text, and they in no way detract from its usefulness as a reference volume.

The authors of the present textbook have backgrounds which admirably complement each other. Robert Garrels grew up and was educated in the Middle West. He was graduated from the University of Michigan and received his doctorate in geology from Northwestern University and came into the geochemical field by way of a natural and appropriate approach, the study of mineralogy and economic geology. These investigations led to his assignment to the United States Geological Survey team of geochemists and the beginning of his collaboration with Dr. Charles Louis Christ. In 1955, Dr. Garrels joined the faculty of Harvard University, where he has continued his distinguished career as a leader among the growing group of scientists interested in, and effective with, chemical approaches to geological problems.

Dr. Christ, born on the East Coast and educated to the doctorate in chemistry at Johns Hopkins University, has more and more employed the theories and methodology of chemistry in the analysis of the same geological problems which have concerned Robert Garrels. Dr. Christ is a student of crystal structures with experience as a group leader in x-ray crystallography with the American Cyanamid Corporation. After teaching at Johns Hopkins and Wesleyan, he began his long association with the United States Geological Survey, which organization, as early as the turn of the century, had the wisdom and foresight to explore into the virtually unknown realm of geochemistry. F. W. Clarke, who inspired a high level of excellence in all of the early geochemical work done by the Survey, would, we think, have been pleased with the present progress of his special field and approve the work of two of his most successful latter-day disciples.

February, 1965 CAREY CRONEIS
Rice University
Houston, Texas

Preface

Mineral Equilibria, a book of seven chapters and 254 pages, was published in 1960. It covered a relatively restricted field of geochemistry, and was limited to consideration of stability relations at room temperatures and pressures near 1 atmosphere. It became apparent immediately that the coverage of the book should be increased generally, and that the restriction to low temperatures and pressures should be removed. It was also felt that it would be desirable to add illustrative problems at the ends of most of the chapters.

In *Solutions, Minerals, and Equilibria*, there are eleven chapters and 400 pages. We have added new chapters on complex ions; on cation exchange and cation electrodes; on "combination diagrams," diagrams plotted using a variety of activity variables and showing relations among silicates, carbonates, and other mineral species; and on the effects of changing temperature and pressure on equilibria. Most of the original chapters have been extensively rewritten—included among the major additions is a treatment of the effects of solution phenomena in solids and gases; generally, a more comprehensive treatment of activity-concentration relations is given. Problems and their answers are included at the ends of most of the chapters; the references cited are much more numerous.

The treatment of the material has, for the most part, been kept free of the necessity of a knowledge of the calculus. In some chapters, notably Chapter 9, which is on equilibria as a function of temperature and pressure, the calculus is used to derive various equations. However, a number of worked-out numerical examples are furnished there, and it should be possible to use the derived relations directly.

The growth of geochemistry has been so rapid that we feel that *Solutions, Minerals, and Equilibria* is as relatively incomplete today as was *Mineral Equilibria* in 1960, but we hope that its greatly expanded scope will increase its audience and its utility.

ROBERT M. GARRELS

February, 1965   CHARLES L. CHRIST

ACKNOWLEDGMENTS

Despite the fact that this new volume is so changed from *Mineral Equilibria* that it has been issued as a new book under a new title, our indebtedness to others goes back to the old book. Accordingly, we reprint the *Acknowledgments* from *Mineral Equilibria* as follows:

The material presented here has been accumulated over a number of years. Perhaps it is best to proceed historically in attempting to show the many people who are really responsible for this text. At Northwestern University I worked closely with John Castano and N. King Huber, who were graduate students at that time, and our interest was largely in relations among the iron minerals. At about this time discussion concerning field and laboratory aspects of sedimentary iron deposition with H. L. James, of the U.S. Geological Survey, were most invigorating. All my colleagues at Northwestern were helpful, but W. C. Krumbein was the person who opened my eyes to the possibilities of simultaneous representation of stability fields of minerals, and we collaborated in attempts to delineate the environments of chemical sediments. At the U.S. Geological Survey I became involved in studies of the geochemistry of the Colorado Plateau uranium deposits, and am indebted to so many of my colleagues there that it is hard to stop when I list the individuals whose help has been indispensable. C. L. Christ, A. D. Weeks, C. R. Naeser, H. T. Evans, E. S. Larsen, III, R. G. Coleman, V. E. McKelvey, and A. M. Pommer all have collaborated with me in various publications. The stimulus of daily contact and continual discussion with C. L. Christ, H. T. Evans, C. R. Naeser, and A. D. Weeks has provided me with a wealth of information and ideas. At Harvard University, H. E. McKinstry influenced me back into studies of relationships among sulfide minerals and the development of the partial pressure diagrams. I have drawn heavily on his syntheses of mineral relationships. My other colleagues have been helpful; I have had many discussions with J. B. Thompson, Jr., and R. Siever. C. C. Stephenson of Massachusetts Institute of Technology has made many helpful suggestions and has contributed free energy values. Students in my classes have not only provided many of the diagrams used as illustrations but also have been a constant source of fresh ideas and new techniques. I have used diagrams originally prepared by J. A. Silman, J. Smith, R. Natarajan, P. Hostetler, D. Emery, U. Peterson, P. Howard, J. Anthony, R. Honea, J. Anderson, R. Notkin, K. Linn, E. Hilton, W. McIntyre, I. Barnes, J. Reitzel, and E. Gaucher. The diagrams were drafted by C. Goodwin.

Ramifying all through this development is the influence of the work of M. J. N. Pourbaix, Director of the Belgian Institute for the Study of Corrosion. I have had access to the many publications of the Institute on Eh-pH diagrams, as well as to M. Pourbaix's classic book on the subject. My early papers are crude indeed when compared with the elegant methods Pourbaix and his coworkers have developed, methods I have tried to emulate.

ACKNOWLEDGMENTS

The sponsors of the work are implicit in the foregoing discussion: The Northwestern University Graduate School; the U.S. Geological Survey (largely in work done on behalf of the U.S. Atomic Energy Commission); and the Committee on Experimental Geology at Harvard University.

The length of this list, which is far from inclusive of all to whom I owe a debt, shows how much has come from others and to what a large extent this text is synthesized with their help. I hope that colleagues I have not indicated by name will know that I am nonetheless appreciative of their help.

In the preparation of the new volume, we have relied heavily again on our friends and colleagues, and especially want to thank for their suggestions and advice, R. E. Siever and J. B. Thompson, Jr., of Harvard University; C. C. Stephenson, of the Massachusetts Institute of Technology; Paul Mangelsdorf, Jr., of Swarthmore College; George Eisenman, of the University of Utah; Rolland Wollast, of the University of Brussels; Harold C. Helgeson, of Shell Development Co.; Paul Barton, Jr., John Hem, Julian Hemley, Motoaki Sato, and Alfred H. Truesdell, all of the U.S. Geological Survey; and Mary E. Thompson, of the graduate school of Harvard University, who collaborated with one of us (R. M. G.) for several years on problems of electrodes and complex ions.

Owen Bricker, of the graduate school at Harvard University, kindly made available to us free-energy values for manganese oxides; Patrick Butler and Donald Newberg, also of the graduate school at Harvard, were most helpful in checking and drawing diagrams, and searching out references.

Our deepest thanks go to Gertrude C. Christ for her extensive help in the preparation of the manuscript. The task would have been impossible without her.

We recognize the impossibility of complete formal acknowledgment, but wish to stress that we are fully aware of the many important contributions from others too numerous to cite.

CHAPTER I

Introduction

A great deal of the following discussion is devoted to a simple theme—the simultaneous representation of various chemical reactions. It is a little surprising, considering the importance always accorded to mineral associations, that few methods of portraying equilibrium relations among minerals are in general use. By far the largest percentage of diagrams portrays stability relations as functions of temperature and pressure, by using rectilinear graphs, or alternatively, shows compositional relations at constant temperature and pressure, by using triangular or tetrahedral axes. There are in addition numerous combinations, such as triangular composition diagrams with isotherms or isobars superimposed.

These methods are essentially as important as their widespread use would indicate, and in any comprehensive treatment of methods of portrayal of chemical relations would occupy a large part of the text. But because they are in such common use, and because adequate texts are available that discuss their construction and interpretation, they will be considered here only to a limited extent and in particular only in their relation to other types of diagrams. As a result, anyone who might use this book as a guide to all the methods available, treated at a length consistent with their importance, would find it lopsided. But it is chiefly a vehicle for making available a large number of diagrams of other types that should be useful to geologists.

EQUATIONS AS THE BASIS OF REPRESENTATION

All methods of representation must be synthesized from individual chemical reactions. Such a statement perhaps emphasizes the obvious, but there is so great a tendency to use pressure-temperature-composition diagrams to show fields of stability that the methods by which the relations are obtained are ofttimes neglected. If a return is made to individual equations as the building blocks, it can be seen that the graphical structures useful in portraying their interconnections are many, and that the ones chosen should depend upon the problem to be solved.

Because chemical terminology has become so complex, and various notations are used, it is advisable to devote a preliminary section to definitions of terms and symbols before attempting even the simpler chemical equations.

TERMS AND SYMBOLS

The formulas of compounds will be written in the standard manner, i.e., H_2O, CO_2. The subscripts generally will indicate only the atomic ratios of the elements, and will be coprime numbers. From place to place, multiple numbers may be used if knowledge of structural constitution is important. On the one hand, H_4SiO_4 indicates only that H, Si, and O are in the atomic ratios of 4/1/4, and carries no implication, unless stated otherwise, that the molecular species involved is not $H_8Si_2O_8$ or $H_{16}Si_4O_{16}$. On the other hand, if a formula is written with larger than minimum number subscripts, such as $V_3O_9^{3-}$, structural information has been taken into account.

To indicate polymerization of unknown degree, the minimum subscript formula may be enclosed in parentheses, and the whole given a subscript x, i.e., $(H_4SiO_4)_x$. To designate solid solution relations, fractional subscripts are used within parentheses. The formula of a certain magnesian calcite would be written $(Ca_{0.9}Mg_{0.1})CO_3$.

Lower case subscripts following a formula are used to designate the state of the substance, or to provide other descriptive information, as follows:

amorp	amorphous
gls	glass
c	crystalline
l	liquid
g	gas
aq	in aqueous solution
n atm	at n atmospheres pressure
T	temperature in degrees Kelvin
t	temperature in degrees centigrade

An additional subscript will be used as needed to indicate the polymorphic form of a crystalline solid.

Conversion of temperatures on the centigrade scale to those on the absolute scale is made by taking 0 °C = 273.150 °K. Thus, t = 25 °C is equivalent to T = 298.15 °K. When the temperature is given in degrees centigrade in the subscript, the fact is stated explicitly.

As an example, $O_{2\ g\ 298.15°\ 1\ atm}$ means molecular gaseous oxygen at 298.15 °K and 1 atmosphere pressure; an equivalent statement is

INTRODUCTION 3

$O_{2\ g\ 25\ °C\ 1\ atm}$. $S_{rhombic}$ means crystalline rhombic sulfur. The most commonly used subscripts are those listed above indicating state. Where other descriptive terms are used throughout the text, they will be spelled out.

A complete list of terms and their symbols and of the physical constants used in this book is given in Appendix 1.

CONCENTRATION UNITS

Many natural solutions are highly aqueous, and it is convenient to use the concentration units developed by chemists for water solutions, such as molality and molarity, as well as the mole fraction, which is perhaps more useful for solutions of all types.

Molality, m, is expressed as moles of solute per 1000 grams of water; *formality*,[1] f, is moles of solute per 1000 grams of solution, and *molarity*, M, is moles of solute per 1000 milliliters of solution. In the ensuing calculations molality will be used most frequently. For dilute aqueous solutions the differences between m, f, and M are so small that they can be neglected for many purposes. For example, in the calculation of free energy changes these differences are overshadowed by the errors in the free energy values themselves.

The definitions and interrelations of these three concentration units, where the weight is in grams and the volume in milliliters, are

$$\text{molality} = m = \frac{\text{weight solute} \times 1000}{\text{formula weight solute} \times \text{weight water}} \quad (1.1)$$

$$\text{formality} = f = \frac{\text{weight solute} \times 1000}{\text{formula weight solute} \times \text{weight solution}} \quad (1.2)$$

$$\text{molarity} = M = \frac{\text{weight solute} \times 1000}{\text{formula weight solute} \times \text{volume solution}}. \quad (1.3)$$

Dividing (1.1) by (1.2)

$$\frac{m}{f} = \frac{\text{weight solution}}{\text{weight water}}. \quad (1.4)$$

Since

$$\text{weight water} = \text{weight solution} - \text{total weight solutes}$$

$$m = f \frac{\text{weight solution}}{\text{weight solution} - \text{total weight solutes}}. \quad (1.5)$$

[1] There seems to be no consistent definition as used in the chemical literature. The definition used here is a commonly accepted one.

Dividing (1.3) by (1.2)

$$\frac{M}{f} = \frac{\text{weight solution}}{\text{volume solution}}. \tag{1.6}$$

Since

$$\text{weight solution} = \text{volume solution} \times \text{density of solution}$$

$$M = f \times \text{density of solution}. \tag{1.7}$$

From (1.4) and (1.7)

$$m = M \frac{\text{weight solution}}{(\text{weight solution} - \text{total weight solutes})} \times \frac{1}{\text{density}}. \tag{1.8}$$

As an example of the application of the relations, a calculation will be made of the formality, molality, and molarity of sodium ion in a brine with a density of 1.008 g cm^{-3} containing 10,000 ppm total dissolved solids, and in which the sodium ion is 1000 ppm. Inasmuch as the analysis is expressed in weight of solutes per weight of solution, the first step is to obtain the formality from equation (1.2)

$$f = \frac{\text{weight Na}^+ \times 1000}{\text{formula weight Na}^+ \times \text{weight solution}}$$

$$= \frac{1000 \times 1000}{23.0 \times 1,000,000}$$

$$= 0.0435.$$

Then, from equation (1.5)

$$m = 0.0435 \frac{\text{weight solution}}{\text{weight solution} - \text{total weight solutes}}$$

$$m = \frac{0.0435 \times 1,000,000}{1,000,000 - 10,000} = 0.0439.$$

From equation (1.7)

$$M = f \times \text{density of solution} = 0.0435 \times 1.008 = 0.0438.$$

Consequently, for such a brine, which is fairly characteristic of waters encountered underground, $f = 0.0435$, $m = 0.0439$, and $M = 0.0438$. These differences increase with increasing total solutes. However, a great many natural waters are so dilute that the differences can be disregarded.

The *mole fraction* of a given constituent in any solution (liquid, solid,

INTRODUCTION

or gas) is defined as the ratio of the number of moles of the given constituent to the total moles of all constituents

$$N_1 = \frac{\text{moles}_1}{\text{moles}_1 + \text{moles}_2 + \text{moles}_3 + \text{moles}_x}$$

$$= \frac{n_1}{n_1 + n_2 + n_3 + n_x}. \tag{1.9}$$

No distinction is made between solute and solvent; the mole fraction of NaCl in a 1.0 molal solution is

$$N_{\text{NaCl}} = \frac{1}{1 + 55.51} = 0.0177.$$

The value 55.51 is the number of moles of water in 1000 grams of water.

Concentrations expressed in molality, formality, or mole fraction have the advantage that their values are independent of temperature and pressure; the molarity of a solution changes with temperature and pressure.

Molality, partial pressure, and mole fraction are the concentration terms that will be used almost exclusively in this text. Molality of a given species will be designated either as a small m followed by a subscript indicating the species, as $m_{\text{Fe}^{++}}$, or the chemical symbol of the species will be enclosed in parentheses. Partial pressure is symbolized by a capital P followed by a subscript indicating the species, as P_{O_2}, P_{CO_2}. Mole fraction is symbolized by N followed by a subscript representing the species, as N_{PbS}, N_{FeO}.

ACTIVITY—STANDARD STATE

Most of the calculations in this book will involve the *activity* or *thermodynamic* concentration of a substance. The activity is a measure of the *effective* concentration of a reactant or product in a chemical reaction. A large portion of the efforts of investigators of chemical reactions has gone into deducing the relations between the concentrations of reactants or products as determined from the composition of a system, and their activities as determined by the thermodynamic behavior of the system.

A *standard state* is defined for each substance in terms of a set of reference conditions. Each pure substance in its standard state is assigned an activity of unity. The standard state chosen may vary, depending upon the problem to be solved.

The standard state of a *solid* element or compound is usually chosen as the pure substance under the standard conditions of 1 atmosphere pressure and a specified temperature. Thus $S_{\text{rhombic, 298.15°}}$ would have unit activity, as would Fe_2O_3 $_{\text{hematite, 400°}}$. Unless the pressure is explicitly stated, it is understood to be 1 atmosphere.

The standard state of a *liquid* element or compound is also chosen as the pure liquid under the standard conditions of 1 atmosphere pressure and a specified temperature. Thus, pure liquid water at a chosen temperature, and at 1 atmosphere total pressure, is in its standard state and has unit activity. Since 298.15 °K, equivalent to 25 °C, is a commonly used temperature, it will occasionally be called the *reference temperature*.

The standard state of a *gaseous* element or compound is not defined so simply as is that of a solid or liquid. The standard state of a perfect gas (which obeys the relation $PV = nRT$) is that at the stipulated temperature and 1 atmosphere pressure. The standard state of a real gas, at which its activity is unity, is defined in terms of the stipulated temperature and at that pressure at which the gas pressure would be 1 atmosphere, if the gas behaved ideally.

The standard state of a *dissolved* species, such as Fe^{++}_{aq}, or $H_4SiO_{4\ aq}$, is defined like that of a gas, by extrapolation to unit activity from behavior in infinitely dilute solutions, where behavior is ideal: i.e., where activity and concentration are numerically the same. A more detailed discussion of the definitions of the standard states for dissolved substances and for gases is given in Chapter 2.

Activities will be denoted by the symbol a, followed by a subscript indicating the species, as $a_{Fe^{++}}$, $a_{Fe_2O_3}$. Alternately, activity will be indicated by enclosing the species in brackets, as $[Fe^{++}]$, $[Fe_2O_3]$. Activity in the standard state (unit activity) is indicated by $a°$. Activities are dimensionless quantities.

THE LAW OF MASS ACTION AND THE EQUILIBRIUM CONSTANT

One of the cornerstones of the development of stability relations among minerals is the *Law of Mass Action*. It was discovered long ago that the driving force of a chemical reaction to right or to left could be related to the concentrations of the reactants and products. When a system is at equilibrium, the rates to the right and left are equal. After any temporary disturbance of the system, say a small change in temperature or pressure, the system will revert to its original condition if the temperature and pressure are restored to their original values. For the system at equilibrium, the relations of the concentrations of the reactants and products can be stated, "The product of the activities of the reaction products, each raised to the power indicated by its numerical coefficient,

INTRODUCTION

divided by the product of the activities of the reactants, each raised to a corresponding power, is a constant at a given temperature." This *thermodynamic equilibrium constant* varies with temperature, but for any given temperature is independent of the total pressure.

The basic simplicity of the law is obscured by the words necessary to state it. If the reaction of b moles of B with c moles of C has come to equilibrium with the products d moles of D and e moles of E,

$$bB + cC = dD + eE$$

and

$$\frac{a_D^d a_E^e}{a_B^b a_C^c} = K, \qquad (1.10)$$

where K is the thermodynamic equilibrium constant.

STANDARD FREE ENERGY OF FORMATION, ΔF_f°

When the interrelations of chemical reactions and their energy changes are considered, a new definition of equilibrium can be formulated. Any system not at equilibrium will change spontaneously with the release of energy. On that basis the relation

$$\Sigma \text{ free energy products} - \Sigma \text{ free energy reactants} = 0,$$

where the reactants and products are considered at the same temperature and pressure, defines an equilibrium condition.

The next step in this development is to relate free-energy changes of reactions to their equilibrium constants, and it becomes necessary to define the free-energy content of various types of substances. It will be necessary only to obtain the *standard free energy of formation*, the free energy of the reaction to form one mole of the substances in their standard states from the stable elements under standard conditions.

The standard free energy of formation of the *stable* configuration of an element in its standard state is taken to be zero by convention. Thus, the standard free energy of formation of $S_{\text{rhombic, 298.15}^\circ}$ is zero. Note that $S_{\text{monoclinic, 298.15}^\circ}$ has unit activity in its standard state, but its standard free energy of formation is not zero because it is not the stable composition of sulfur at that temperature.

Similarly, it proves to be convenient to take as zero the standard free energy of formation of hydrogen ion in aqueous solution.

The standard free energy of formation is symbolized ΔF_f°. The temperature will be indicated by an additional subscript, as $\Delta F_{f, T}^\circ$, or $\Delta F_{f, t}^\circ$. Generally, temperatures are given in °K; where given in °C, the fact will be explicitly noted, as $\Delta F_{f, 298.15^\circ}^\circ$, or $\Delta F_{f, 25^\circ C}^\circ$. Tables of

ΔF_f° values for substances of geologic interest are given in Appendix 2. Much more comprehensive lists are given in the references listed at the end of that Appendix.

STANDARD FREE ENERGY OF REACTION AND THE EQUILIBRIUM CONSTANT

The standard free-energy change of a reaction is the sum of the free energies of formation of the products in their standard states, minus the free energies of formation of the reactants in their standard states

$$\Delta F_r^\circ = \Sigma \Delta F_f^\circ \text{ products} - \Sigma \Delta F_f^\circ \text{ reactants}.$$

The standard free-energy change of reaction is related to the equilibrium constant by

$$\Delta F_r^\circ = -RT \ln K, \qquad (1.11)$$

where

$$K = \frac{a_D^d a_E^e}{a_B^b a_C^c},$$

R is the gas constant, and T the absolute temperature.

At 298.15°

$$\Delta F_r^\circ (\text{kcal}) = -0.001987 \text{ kcal/deg} \times 298.15° \times 2.303 \log K$$

$$\Delta F_r^\circ = -1.364 \log K. \qquad (1.12)$$

The relation between ΔF_r° and the equilibrium constant is the special case of the relation between the standard free-energy change of reaction and the activities of the products and reactants for a system at equilibrium.

More generally, for any chemical reaction:

$$\Delta F_r = \Delta F_r^\circ + RT \ln \frac{a_D^d a_E^e}{a_B^b a_C^c} \qquad (1.13)$$

where the activities of the reactants and products (measured at the same temperature) now refer to the initial and final activities of the substances involved in the chemical transformation. In accordance with the initial definition of ΔF_r°, when $a_B = a_C = a_D = a_E = 1$, then $\Delta F_r = \Delta F_r^\circ + RT \ln 1$, whence $\Delta F_r = \Delta F_r^\circ$. When for a given reaction, $\Delta F_r \neq 0$, the quotient

$$\frac{a_D^d a_E^e}{a_B^b a_C^c}$$

is called the *reaction quotient* Q to distinguish it from the equilibrium constant K. For a system at equilibrium $\Delta F_r = 0$ and $\Delta F_r^\circ = -RT \ln K$.

INTRODUCTION

REACTION TYPES

Reactions Involving Only Solids

Equilibrium is attained only when the free-energy change of a reaction is zero. If a reaction involves only pure solids at a given temperature and at 1 atmosphere pressure, the solids are in their standard states, and have unit activities. Under these conditions, for the reaction

$$A_c \rightleftharpoons B_c$$

$$\Delta F_r = \Delta F_r^\circ + RT \ln \frac{a_B}{a_A}$$

$$\Delta F_r = \Delta F_r^\circ + RT \ln \frac{1}{1}$$

$$\Delta F_r = \Delta F_r^\circ.$$

The condition that the two pure solids be in equilibrium is that $\Delta F_r = \Delta F_r^\circ = 0$. It will be only by chance that, at an arbitrary temperature, $\Sigma \Delta F_f^\circ$ of the products equals $\Sigma \Delta F_f^\circ$ of the reactants. The relation $\Delta F_r = \Delta F_r^\circ$ is a convenient one because it puts solid-solid reactions into a go-no-go category. Any proposed reaction can be quickly tested to see if the proposed reactant(s) or product(s) is the stable species at a given temperature and 1 atmosphere pressure (where the pure solids are in their standard states). If the ΔF_r° is negative, i.e., energy is released, then the product(s) is stable relative to the reactant(s); if ΔF_r° is positive, the reactant(s) is stable.

For example, there are two monoxides of lead—a red and a yellow variety. For the reaction

$$\text{PbO}_{c,\text{ red}} \rightleftharpoons \text{PbO}_{c,\text{ yellow}} \quad (25\ ^\circ\text{C, 1 atm})$$

$$\Delta F_f^\circ{}_{\text{PbO}_{\text{yellow}}} - \Delta F_f^\circ{}_{\text{PbO}_{\text{red}}} = \Delta F_r^\circ$$

$$-45.05 - (-45.25) = +0.20 \text{ kcal.}$$

Thus, PbO_{red} is the stable phase. The method is a convenient test for the relative stability of any polymorph under standard conditions.

For a reaction of geologic interest involving two reactants and two products:

$$\text{PbCO}_{3\ c} + \text{CaSO}_{4\ c} \rightleftharpoons \text{PbSO}_{4\ c} + \text{CaCO}_{3\ c,\text{ calcite}} \quad (25\ ^\circ\text{C, 1 atm})$$

$$\Delta F_f^\circ{}_{\text{PbSO}_4} + \Delta F_f^\circ{}_{\text{CaCO}_3} - \Delta F_f^\circ{}_{\text{PbCO}_3} - \Delta F_f^\circ{}_{\text{CaSO}_4} = \Delta F_r^\circ$$

$$(-193.89) + (-269.78) - (-149.7) - (-315.56) = +1.59 \text{ kcal.}$$

Therefore, cerussite and anhydrite are stable with respect to anglesite and calcite. If reactions go to completion in nature, anglesite and calcite should be an unknown association under the chosen standard conditions. Only by changing the temperature or the pressure, or both, could all four species be brought into equilibrium.

The effect of such changes of pressure and temperature may be considered explicitly, in a reaction involving only pure solids, by rewriting the general equation for the free-energy change

$$\Delta F_{r,\,T,\,P} = \Delta F^\circ_{r,\,T} + RT \ln \frac{a_D^d a_E^e}{a_B^b a_C^c}. \tag{1.14}$$

Here account is taken of the fact that the value of ΔF°_r depends upon the temperature, and that the relative activities of the solids involved in the reaction depend upon the total pressure. Thus, $\Delta F_{r,\,T,\,P}$ can be made to vanish, i.e., the system brought to equilibrium, by varying the temperature and pressure so that the equality

$$\Delta F^\circ_{r,\,T} = -RT \ln Q_P$$

is attained. Since the relations listed are exact ones, this process can always be carried out in principle; in actual practice the conditions necessary to achieve equilibrium for some chemical systems may be difficult to attain.

Reactions Involving Only Solids and Nearly Pure Water

Reactions that involve only solids and pure, or nearly pure, liquid water can be treated like solid-solid reactions, inasmuch as the activity of pure water is also fixed. This gives a test for such relations as stabilities of oxides versus hydroxides, or of oxides versus hydrates in dilute water solution. For example

$$Al_2O_3 \cdot H_2O_{\text{boehmite}} + 2H_2O_l = Al_2O_3 \cdot 3H_2O_{\text{gibbsite}} \quad (25\ °C, 1\ atm)$$

$$\Delta F^\circ_{f\ \text{gibbsite}} - \Delta F^\circ_{f\ \text{boehmite}} - 2\Delta F^\circ_{f\ H_2O} = \Delta F^\circ_r$$

$$(-554.6) - (-435.0) - (-113.4) = -6.2\ \text{kcal.}$$

Gibbsite is the stable phase relative to boehmite in dilute aqueous solution, at 25 °C and 1 atmosphere total pressure.

Reactions Involving a Gas Phase

If a reaction involves a gas phase, equilibrium can usually be achieved under standard conditions by varying the partial pressure[2] of the gas.

---

[2] We shall use partial pressure and activity synonymously for all ensuing calculations at 1 atmosphere total pressure; for systems at higher total pressures a distinction will be made.

INTRODUCTION

For example, if a single gas is involved, as in the reaction

$$FeO_c + CO_{2\,g} = FeCO_{3\,c},$$

the equilibrium constant is

$$K = \frac{[FeCO_3]}{[FeO]P_{CO_2}}.$$

Because the activities of the solids are unity at 1 atmosphere total pressure, the expression reduces to

$$K = \frac{1}{P_{CO_2}}.$$

Thus, a reaction involving pure solids and a single gas phase is at equilibrium under standard conditions at a single fixed partial pressure of the gas. In this instance

$$\Delta F^\circ_{f\ FeCO_3} - \Delta F^\circ_{f\ FeO} - \Delta F^\circ_{f\ CO_2} = \Delta F^\circ_r$$

$$(-161.06) - (-58.4) - (-94.26) = -8.40 \text{ kcal.}$$

Then

$$-1.364 \log K = -8.40$$

$$\log K = 6.1$$

and

$$P_{CO_2} = 10^{-6.1} \text{ atm.}$$

This indicates that, at the earth's surface, FeO is unstable relative to siderite, inasmuch as the partial pressure of CO_2 in the atmosphere is about $10^{-3.5}$ atmosphere. On the other hand, FeO may be unstable relative to other compounds as well, so that the reaction cited tests only the FeO-FeCO$_3$ pair. Yet it does give a specific answer to the specific question.

Many reactions can be expressed in terms of two or even more gas pressures. The determination of the equilibrium constant is similar to that for a single gas

$$Fe_c + CO_{2\,g} + \tfrac{1}{2}O_{2\,g} = FeCO_{3\,c}$$

$$\Delta F^\circ_{f\ FeCO_3} - \Delta F^\circ_{f\ Fe} - \Delta F^\circ_{f\ CO_2} - \tfrac{1}{2}\Delta F^\circ_{f\ O_2} = \Delta F^\circ_r$$

$$(-161.06) - (0) - (-94.26) - \tfrac{1}{2}(0) = -67.00 \text{ kcal}$$

$$-1.364 \log K = -67.00$$

$$\log K = 49.1$$

then
$$\log \frac{1}{P_{CO_2} P_{O_2}^{1/2}} = 49.1$$
$$P_{CO_2} P_{O_2}^{1/2} = 10^{-49.1}.$$

Reactions Involving Dissolved Species

As for gases, reactions involving dissolved species can be expressed in terms of the activities of the dissolved species. Although we are on the verge of having data available for a wide range of temperatures and pressures, abundant thermochemical data to handle this kind of reaction are currently available only for 298.15 °K and 1 atmosphere, and the following discussion of dissolved species refers to that temperature and pressure. In many ways a unique aspect of this text is the extensive use of data for dissolved species. Strangely enough, little use has been made of the thermochemical data by investigators interested in low temperature-low pressure systems of geological interest.

What happens when hematite is put into water? What is the activity of ferric ion? Questions of this kind can be answered, at least in part, by writing the reaction

$$Fe_2O_{3\ c} + 3H_2O_l = 2Fe^{3+}_{aq} + 6OH^-_{aq}.$$

The equilibrium constant for the reaction is

$$K = \frac{[Fe^{3+}]^2[OH^-]^6}{[Fe_2O_3][H_2O]^3}.$$

The activities of $Fe_2O_{3\ c}$ and H_2O are unity, so

$$K = [Fe^{3+}]^2[OH^-]^6.$$

The standard free energy of the reaction is

$$2\Delta F^\circ_{f\ Fe^{3+}} + 6\Delta F^\circ_{f\ OH^-} - \Delta F^\circ_{f\ Fe_2O_3} - 3\Delta F^\circ_{f\ H_2O} = \Delta F^\circ_r$$
$$(2 \times -2.53) + (6 \times -37.6) - (-177.1) - (3 \times -56.69)$$
$$= 116.51 \text{ kcal.}$$

Then
$$-1.364 \log K = 116.51$$
$$\log K = -85.4 = \log [Fe^{3+}]^2[OH^-]^6.$$

Rearranging and simplifying
$$\log [Fe^{3+}] = -42.7 - 3 \log [OH^-].$$

Therefore, the activity of the ferric ion is a function of the cube of the hydroxyl activity in the solution. The stability of Fe_2O_3 in water can

INTRODUCTION

be plotted as a linear equation if log [Fe^{3+}] and log [OH^-] are used as coordinates.

As compared to partial pressure diagrams, such plots can be termed activity-activity diagrams. Activity-activity diagrams can obviously be used to show stability relations using any activities of dissolved species as the axes. Because so many reactions of geologic interest involve the activity of hydrogen ions or can be rewritten to include activity of hydrogen ions, special attention is directed toward writing reactions so that the activity of hydrogen ions becomes one of the variables plotted.

THE ROLE OF pH. The activity of hydrogen ions is involved in, or can, by a little effort, be made to become involved in reactions with dissolved species. Because experimental techniques permit measurement of the activity of hydrogen ions, there has been a tendency to use a_{H^+} as a characterizing variable, whenever possible, so as to help provide a common reference activity for a variety of reactions. Also, because so many representations of activities are most conveniently shown as logarithmic functions, the term pH was developed long ago

$$\text{pH} = -\log a_{H^+}. \qquad (1.15)$$

If we reconsider the reaction discussed in the previous section

$$Fe_2O_{3\,c} + 3H_2O_l = 2Fe^{3+}_{aq} + 6OH^-_{aq},$$

we see that it can be rewritten by adding the reaction for the dissociation of water

$$Fe_2O_{3\,c} + 3H_2O_l = 2Fe^{3+}_{aq} + 6OH^-_{aq}$$
$$\underline{6OH^-_{aq} + 6H^+_{aq} = 6H_2O_l}$$
$$Fe_2O_{3\,c} + 6H^+_{aq} = 2Fe^{3+}_{aq} + 3H_2O_l$$

The only variable activities are those of Fe^{3+}_{aq} and H^+_{aq}. The equilibrium constant is

$$\log K = \log \frac{[Fe^{3+}]^2}{[H^+]^6}.$$

Rearranging

$$\log K = 2 \log [Fe^{3+}] - 6 \log [H^+].$$

But by substituting the relation

$$\text{pH} = -\log [H^+]$$

we obtain

$$\log K = 2 \log [Fe^{3+}] + 6\,\text{pH}.$$

The number of reactions that can be expressed with pH as a variable is remarkable, including not only reactions involving oxides, hydroxides, and basic salts, but also those involving carbonates, silicates, sulfides, and phosphates.

THE ROLE OF Eh. Just as pH can be developed as a variable in a great many reactions, so can the *oxidation potential* Eh be used to compare equilibrium conditions among diverse substances. Here it is convenient to develop the concept of Eh through those of pH and partial pressure of gases.

The first step in the development of Eh is the splitting of oxidation-reduction reactions into *half-cells* or *half-reactions*. For the oxidation of ferrous ions to ferric ions by reaction with hydrogen ions

$$Fe^{++}_{aq} + H^+_{aq} = Fe^{3+}_{aq} + \tfrac{1}{2}H_{2\,g}. \tag{1.16}$$

Ferrous ions are oxidized to ferric ions, and hydrogen ions are reduced to hydrogen. We can think of the overall reaction as the sum of two processes—oxidation of ferrous ion to ferric, with release of an electron

$$Fe^{++}_{aq} = Fe^{3+}_{aq} + e \tag{1.17}$$

and reduction of hydrogen ion to hydrogen, accompanied by the acceptance of an electron

$$H^+_{aq} + e = \tfrac{1}{2}H_{2\,g}. \tag{1.18}$$

Either of these reactions is called a half-cell, or half-reaction. Now consider the free-energy relations

$$\Delta F^\circ_{f\ Fe^{3+}} + \Delta F^\circ_{f\ e} - \Delta F^\circ_{f\ Fe^{++}} = \Delta F^\circ_{r\,(1.17)}$$

$$\tfrac{1}{2}\Delta F^\circ_{f\ H_2} - \Delta F^\circ_{f\ H^+} - \Delta F^\circ_{f\ e} = \Delta F^\circ_{r\,(1.18)}.$$

Upon adding these, it is seen that

$$\Delta F^\circ_{r\,(1.17)} + \Delta F^\circ_{r\,(1.18)} = \Delta F^\circ_{r\,(1.16)}.$$

That is, the sum of the free-energy changes of reaction of the half-cells is identical to that of the *whole* reaction. Note particularly that $\Delta F^\circ_{f\ e}$ disappears during the addition.

Therefore, the actual value for $\Delta F^\circ_{f\ e}$ is unimportant, for it cancels out in the addition of the free energies of any two half-cell reactions. Henceforth we can treat $\Delta F^\circ_{f\ e}$ as if it were zero. Then, since $\Delta F^\circ_{f\ H^+}$ also is zero by convention

$$H^+_{aq} + e = \tfrac{1}{2}H_{2\,g}$$

$$\tfrac{1}{2}\Delta F^\circ_{f\ H_2} - \Delta F^\circ_{f\ H^+} = \Delta F^\circ_r$$

$$0 - 0 = 0.$$

INTRODUCTION

The standard free energy of reaction of the hydrogen half-cell is zero. Consequently, any oxidation-reduction reaction can be written as two half-reactions with the hydrogen half-reaction as the reducing or oxidizing part of the two couples, and the standard free-energy change attributed entirely to the oxidation or reduction half-cells, ignoring the ΔF_f° of the electrons. A specific example will help to clarify this statement. For the reaction

$$Fe_{aq}^{++} + Cu_{aq}^{++} = Fe_{aq}^{3+} + Cu_{aq}^{+} \qquad (1.19)$$

we can write

$$Fe_{aq}^{++} = Fe_{aq}^{3+} + e \qquad (1.20)$$

$$Cu_{aq}^{++} + e = Cu_{aq}^{+} \qquad (1.21)$$

$$\Delta F_f^\circ{}_{Fe^{3+}} - \Delta F_f^\circ{}_{Fe^{++}} = \Delta F_r^\circ{}_{(1.20)}$$

$$\Delta F_f^\circ{}_{Cu^+} - \Delta F_f^\circ{}_{Cu^{++}} = \Delta F_r^\circ{}_{(1.21)}$$

$$\Delta F_r^\circ{}_{(1.20)} + \Delta F_r^\circ{}_{(1.21)} = \Delta F_r^\circ{}_{(1.19)}$$

where $\Delta F_r^\circ{}_{(1.20)}$ and $\Delta F_r^\circ{}_{(1.21)}$ are the standard free-energy changes of the half-reactions. Thus, if we have a series of standard free-energy changes for half-reactions available, we can take any pairs we choose to obtain the overall free-energy change of the oxidation-reduction reaction.

This leads to a relation between free energies of reactions and voltage measurements of galvanic cells. Let us suppose we have an electrode of pure copper dipping into a solution containing cupric ions at unit activity, and a hydrogen electrode consisting of an inert platinum electrode dipping into a solution at unit activity of hydrogen ions, and saturated with hydrogen gas at 1 atmosphere pressure (also unit activity).

If the electrodes are connected, electrons will flow through the circuit. At the copper electrode, copper will be deposited by discharge of cupric ions

$$Cu_{aq}^{++} + 2e = Cu_c$$

and at the hydrogen electrode, hydrogen will release electrons to become H^+

$$H_{2\,g} = 2H_{aq}^+ + 2e.$$

Thus, the overall reaction is

$$Cu_{aq}^{++} + H_{2\,g} = Cu_c + 2H_{aq}^+.$$

The voltage between the electrodes can be measured at the very beginning of the process, before any sensible change in activities has occurred because of the reactions, and will be found to be 0.337 volt. As the reaction proceeds toward equilibrium, the voltage will decrease, and at equilibrium will become zero.

Thus, in the beginning there is a voltage corresponding to a reaction under standard conditions, with all substances at unit activity; at the end, the voltage disappears when the free-energy change becomes zero. This suggests a relation between voltage and standard free-energy change, which is

$$\Delta F_r^\circ = n\mathrm{E}^\circ \mathscr{F} \qquad (1.22)$$

where E° is the voltage of the reaction when all substances involved are at unit activity; n is the number of electrons involved (two in the case cited for the copper-hydrogen cell), and $\mathscr{F}$ is a constant, the faraday. A faraday is 23.06 kcal per volt-gram equivalent—in the units necessary to express the faraday to have it be consistent with the other units used.

The equation relating E° directly to the equilibrium constant is then

$$\mathrm{E}^\circ = -\frac{RT}{n\mathscr{F}} \ln K. \qquad (1.23)$$

The standard free energy of the hydrogen half-cell reaction is zero; correspondingly, its voltage is also used as a zero reference. The voltage of any circuit, when a given standard half-cell is measured against the standard hydrogen half-cell, can be attributed entirely to the given standard half-cell and used to obtain its standard free energy of reaction.

Consider the standard copper-cupric ion electrode. Its voltage, measured against the standard hydrogen electrode, is 0.337 volt at 298.15°, and two electrons are involved. From equation (1.22)

$$\Delta F_r^\circ = n\mathrm{E}^\circ \mathscr{F}$$

$$\Delta F_r^\circ = 2 \times 0.337 \times 23.06 = 15.5_4 \text{ kcal.}$$

By writing the reaction and using ΔF_f° values at 298.15°

$$\mathrm{Cu}_c = \mathrm{Cu}_{aq}^{++} + 2e$$

$$\Delta F_{f\ \mathrm{Cu}^{++}}^\circ - \Delta F_{f\ \mathrm{Cu}}^\circ = \Delta F_r^\circ$$

$$15.53 - 0 = 15.5_3 \text{ kcal.}$$

More generally, if the equalities $\Delta F_r = n\mathrm{E}\mathscr{F}$, and $\Delta F_r^\circ = n\mathrm{E}^\circ \mathscr{F}$ are substituted into the equation

$$\Delta F_r = \Delta F_r^\circ + RT \ln \frac{a_D^d a_E^e}{a_B^b a_C^c} \qquad (1.13)$$

we obtain the relation:

$$\mathrm{E} = \mathrm{E}^\circ + \frac{RT}{n\mathscr{F}} \ln \frac{a_D^d a_E^e}{a_B^b a_C^c}. \qquad (1.24)$$

In this equation E is the EMF (electromotive force) or potential, i.e.,

INTRODUCTION

voltage measured in a reversible manner, for any *reversible* cell characterized by the chemical reaction

$$bB + cC = dD + eE.$$

E°, as defined previously, is the EMF of the cell when all substances involved are at unit activity; for a system at equilibrium, E = 0.

To achieve consistency in calculation, half-cell reactions are always written with the oxidation products and the electrons released on the right side of the chemical equation, e.g.,

$$V_2O_{3\,c} + H_2O_1 = V_2O_{4\,c} + 2H^+_{aq} + 2e$$

or more generally

$$bB + cC = dD + eE + ne.$$

The potential of a half-cell, measured against the standard hydrogen half-cell, for reactions written in this manner, is called the *oxidation potential* Eh. The same symbol E° is conventionally used for both the standard half-cell reaction and the standard full-cell reaction. Thus, for the half-cell reaction

$$bB + cC = dD + eE + ne$$

we write

$$\text{Eh} = \text{E}° + \frac{RT}{n\mathscr{F}} \ln \frac{a_D^d a_E^e}{a_B^b a_C^c}. \tag{1.24a}$$

For reactions at 25 °C

$$\frac{RT}{n\mathscr{F}} \ln Q = \frac{0.001987 \text{ kcal/deg} \times 298.15 \text{ deg} \times 2.303}{n \times 23.06 \text{ kcal/volt-gram equivalent}} \log Q$$

$$= \frac{0.05916}{n} \log Q.$$

For example, let us consider the situation when the potential of a copper electrode immersed in a solution of cupric ions is measured against a standard hydrogen electrode, but the activity of the cupric ions is not unity. For this example we can write

$$\text{Eh} = \text{E}° + \frac{RT}{n\mathscr{F}} \ln \frac{[\text{Cu}^{++}]}{[\text{Cu}]}. \tag{1.25}$$

At 25 °C this expression can be rewritten

$$\text{Eh} = \text{E}° + \frac{0.0592}{n} \log \frac{[\text{Cu}^{++}]}{[\text{Cu}]}. \tag{1.26}$$

If the measured potential is 0.100 volt, then remembering that the activity of Cu is unity and the number of electrons involved is 2, we have

$$0.100 = 0.337 + \frac{0.0592}{2} \log [Cu^{++}] \qquad (1.27)$$

$$[Cu^{++}] = 10^{-8.03}.$$

SUMMARY

The various terms and symbols to be used in subsequent chapters are defined. The molality, m, is chosen as the analytical concentration unit of a species dissolved in aqueous solution, the mole fraction, N, as the concentration unit for solid solutions, and partial pressure as the concentration unit for gases. The activity, a, is the thermodynamic concentration resulting from thermochemical calculations.

The basis for later development of stability relations is given, chiefly in the form of a few equations. These are

(1) The relation of the equilibrium constant to the standard free-energy change of a reaction

$$\Delta F_r^\circ = -RT \ln K = -RT \ln \frac{a_D^d a_E^e}{a_B^b a_C^c};$$

(2) The relation of the reaction quotient, the free-energy change of a reaction, and the standard free-energy change of the reaction

$$\Delta F_r = \Delta F_r^\circ + RT \ln Q;$$

(3) The relation between the standard free-energy change of a reaction and the potential of the corresponding cell

$$\Delta F_r^\circ = nE^\circ \mathscr{F};$$

(4) The relation between the reaction quotient, the free-energy change of a reaction, and the potential of the corresponding cell

$$\Delta F_r = nE\mathscr{F}$$

$$E = E^\circ + \frac{RT}{n\mathscr{F}} \ln Q$$

where E = Eh, the oxidation potential, for the half-cell measured against the standard hydrogen half-cell.

The use of these relations is demonstrated by a consideration of reactions of various types. The examples chosen for demonstration have been for reactions at 25 °C and 1 atmosphere, but the use of the equations is not restricted to these conditions, and calculations involving other temperatures and pressures will be illustrated in later chapters.

INTRODUCTION

SELECTED REFERENCES

Darken, L. S., and R. W. Gurry, *Physical Chemistry of Metals*. New York, McGraw-Hill, 1953, chap. 17.
    A concise summary of the basic theory of galvanic cells.

Glasstone, Samuel, *An Introduction to Electrochemistry*. New York, Van Nostrand, 1942.
    General reference for all aspects of theory and practice of electrochemistry; good for troubleshooting on specific problems related to details of calculations.

Kolthoff, I. M., and E. B. Sandell, *Textbook of Quantitative Inorganic Analysis*. New York, Macmillan, 1952, chaps. IV (Mass Action), V (Solubility and Solubility Product), X (Theory of Electroanalysis).
    Excellent introduction at the second year level to the basic relations of chemical equilibria.

Latimer, W. M., *Oxidation Potentials*, 2nd edition. Englewood Cliffs, N.J., Prentice-Hall, 1952, chap. I.
    Condensed discussion of units, conventions, and general methods of obtaining equilibrium constants and Eh equations from free energy data.

Pourbaix, M. J. N., *Thermodynamics of Dilute Aqueous Solutions*. London, E. Arnold, 1949.
    Detailed development of theory and practice of pH-oxidation potential relations.

PROBLEMS

1.1. A brine of density 1.030 g cm^{-3} contains 5.03 g K$^+$ and 15.00 g of other dissolved solids per 500 g *water*. Calculate the molality, formality, and molarity of K$^+$ in the brine.
    Ans. $m = 0.257$, $M = 0.255$, $f = 0.247$.

1.2. Which is stable at 298.15° and 1 atmosphere total pressure in the presence of nearly pure water, MnO or Mn(OH)$_2$?
    Ans. Mn(OH)$_2$ (-3.4 kcal).

1.3. Calculate K at 25 °C for the reaction $CO_{2\,g} + H_{2\,g} = CO_g + H_2O_g$.
    Ans. $K = 10^{-4.99}$.

1.4. For the reaction (at 25 °C)

$$SnO_{2\,c} + 4H^+_{aq} = Sn^{4+}_{aq} + 2H_2O_l,$$

what is [H$^+$] when [Sn^{4+}] $= 10^{-10}$? What is the pH of the solution? What can you conclude about the solubility of cassiterite in ordinary surface waters?
    Ans. [H$^+$] $= 10^{-0.40}$, pH $= 0.40$.

1.5. Calculate the EMF of the cell made up of a metallic Zn electrode in a solution with [Zn^{++}] $= 10^{-2.0}$, connected appropriately to an electrode of metallic Cu in a solution with [Cu^{++}] $= 10^{-3.0}$.
    Ans. E $= 1.07$ volts.

1.6. In problem 1.5, as the reaction in the cell proceeds, is [Cu^{++}] increasing or decreasing?

CHAPTER 2

Activity-Concentration Relations

The basic materials with which geochemists work are rocks and natural waters. With few exceptions, the chemical information available for these is expressed in terms of concentrations. There are tens of thousands of rock and mineral analyses available in the literature, and an even larger number of analyses of natural waters.

In calculating equilibria among minerals, or among minerals and natural waters, the data that are used in, or result from, the equations, are the activities. Thus, the handling of mineral relations through thermochemical calculations requires translation back and forth between concentrations and activities. In general, it can be said that the interrelations among activities as obtained by calculation are reliable, and that the values for concentrations obtained by various types of analytical techniques are reliable. The major source of error in treating natural systems lies in the translation process.

In any discussion of activity-concentration relations it is convenient to consider separately the several different kinds of solutions encountered in practice; this is done in the following by treating in sequence gases, nonelectrolyte liquid solutions, solid solutions and melts, and aqueous solutions of electrolytes.

Possibly the only broad generalization that can be made about all types of solutions is that "like dissolves like." Thus, liquids composed of neutral, nonpolar molecules such as benzene and toluene are freely soluble in one another; Br^- readily replaces some of the Cl^- in cerargyrite, $AgCl$, to form a solid solution; and all gases intermix in all proportions. Substances that are ionic in the solid state, like $NaCl$, tend to dissolve readily in water, a liquid having high dielectric constant. On the other hand, benzene is very sparingly soluble in water (and water in benzene); organic compounds do not form solid solutions with inorganic compounds, etc.

For the purpose of calculating activity-concentration relations in real systems, it is profitable to consider certain *model systems*. This is done in the present book by taking as models the *perfect gas*, the *perfect gas mixture*, and the *ideal solution*: gaseous, liquid, or solid. For such model systems, exact and simple relationships may be established. These relationships are obeyed approximately by real systems under many conditions, but importantly, are approached more and more nearly exactly under certain limiting conditions.

GASES

Pure Gases

ACTIVITY OF THE PERFECT GAS. Much experience has shown that the behavior of real gases may be approximated, at ordinary pressures and temperatures, by a limiting law, the *perfect gas law*, the mathematical expression for which is

$$PV = nRT. \tag{2.1}$$

In particular, at room temperatures, most gases of interest geologically behave nearly perfectly up to a pressure of at least 1 atmosphere. It is also experimentally true that as the pressure of a gas is lowered, or as its temperature is raised (in the absence of decomposition[1]), it becomes more nearly perfect, and in the limit as the pressure approaches zero, obeys the perfect gas law exactly. This effect is shown schematically in Figure 2.1. Here it is assumed that a given mass of gas of n moles, of total volume V^t, is allowed to expand, at constant temperature, in a series of discrete steps, and at each step the measured *molal volume*, $V_m = V_m^t/n$, and the measured pressure, P_m, are recorded. Then the corresponding calculated pressures, P_c, are obtained from the expression

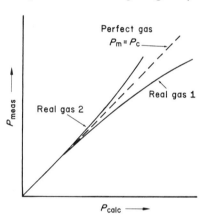

FIG. 2.1. Schematic representation of the behavior of two different real gases compared with that of the perfect gas, at constant temperature.

$$P_c = \frac{RT}{V_m}. \tag{2.2}$$

[1] If the molecules of a gas dissociate into smaller molecules, or into atoms, at an elevated temperature, and the resulting increase in the number of particles present in the system is taken into account in the quantity n, then the perfect gas law $PV = nRT$ will still hold. The gas system then becomes a *perfect gas mixture*, as discussed farther on in the text.

The results of plotting P_c vs. P_m are shown in Figure 2.1 for two different types of real gases. The dashed line with the slope of unity represents the relationship $P_c = P_m$, true for the perfect gas. The P_c vs. P_m curve for a real gas may lie above or below this line (or may, in fact, cross the line at higher pressures), but in either case the slope of the curve will approach that of the perfect gas as the pressure approaches zero.

The activity of the perfect gas is *defined* to be unity when the gas is in its standard state, i.e., at 1 atmosphere and some designated temperature. Further, for the perfect gas, the activity (a dimensionless quantity) is defined as numerically equal to the pressure, expressed in atmospheres, at all pressures. Thus,

$$a = |P(\text{atm})|. \tag{2.3}$$

ACTIVITY OF REAL GASES; THE ACTIVITY COEFFICIENT. As might be expected, for real gases equation (2.3) does not hold, in general. Accordingly, we introduce the useful relationship

$$a = \chi P(\text{atm}). \tag{2.4}$$

The quantity χ is called the *gas activity coefficient*. Its value for a given gas depends upon the temperature and pressure. Since a real gas approaches the perfect gas as its pressure is lowered, χ for any gas will approach unity at lower pressures. This effect is shown in Figure 2.2, where the activity coefficients are plotted for several gases at various temperatures and for pressures up to 1000 atmospheres. It can be seen from Figure 2.2 that, at pressures up to about 10 atmospheres, χ does not differ significantly from unity for any of the gases considered. This fact is of considerable practical importance, as it enables us to substitute numerical values of pressures for activities in thermochemical calculations involving gases at not too high pressures.

The pressure calculated through the use of the perfect gas law, as expressed in equation (2.2),

$$P_c = \frac{RT}{V_m},$$

should not be confused with the activity, as is often done. In fact, a very useful relationship between P_c and a, which holds for moderate pressures, is given by the expression

$$a = \frac{P_m^2}{P_c}. \tag{2.5}$$

Equation (2.5) shows that a and P_c lie on opposite sides of P_m; for

ACTIVITY-CONCENTRATION RELATIONS

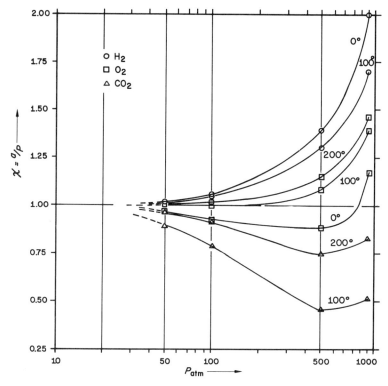

FIG. 2.2. Activity coefficients of some important gases, as a function of temperature and pressure. The pressure, in atmospheres, is plotted on a logarithmic scale. [Data from L. S. Darken and R. W. Gurry, *Physical Chemistry of Metals*. New York, McGraw-Hill, 1953, p. 210.]

example, if for a certain gas P_m is 10 atmospheres and P_c is 10.5 atmospheres,

$$a = \frac{(10)^2}{10.5} = \frac{100}{10.5} = 9.5.$$

As a result, although both a and P_c approach P_m as P_m approaches zero, they do so from opposite sides.

The relationship given by equation (2.5) holds quite accurately for gases like oxygen, up to pressures of the order of 100 atmospheres, and within an error of about 4 percent even for such imperfect gases as CO_2 at pressures of 50 atmospheres (see Figure 2.2). Thus, if both the pressure and the molal volume of a gas are known, its activity may be easily calculated.

A useful universal chart of the activity coefficients of gases, as a function of temperature and pressure, can be made through the use of "reduced" temperatures and pressures. Such a chart is given in Figure 2.3. The "reduced" temperature is the ratio of the temperature of interest to the critical temperature of the gas, the temperatures being expressed in °K; the "reduced" pressure is the ratio of the pressure of interest to the critical pressure of the gas. The necessary critical constants for a number of substances are given in Table 2.1.

TABLE 2.1. Critical Temperatures and Pressures for Some Gases

Gas	Critical Temperature (°C)	Critical Temperature (°K)	Critical Pressure (atmospheres)	Density (g cm^{-3})
CO_2	31.1	304.3	73.0	0.460
CS_2	273	546	76	—
CO	−139	134	35	0.311
COS	105	378	61	—
Cl_2	144.0	417.2	76.1	0.573
He	−267.9	5.3	2.26	0.0693
H_2	−239.9	33.3	12.8	0.0310
HBr	90	363	84	—
HCl	51.4	324.6	81.6	0.42
HI	151	424	82	—
H_2Se	138	411	88	—
H_2S	100.4	373.6	88.9	—
Hg	>1550	>1823	>200	4–5
CH_4	−82.5	190.7	45.8	0.162
N_2	−147.1	126.1	33.5	0.311
O_2	−118.8	154.4	49.7	0.430
SiF_4	−1.5	271.7	50	—
SiH_4	−3.5	269.7	48	—
$SnCl_4$	318.7	591.9	37.0	0.742
S	1040	1313	—	—
SO_2	157.2	430.4	77.7	0.52
SO_3	218.3	491.5	83.6	0.630
H_2O	374.0	647.2	217.7	0.4

SOURCE: *Handbook of Chemistry and Physics*, 41st edition. Cleveland, Ohio, Chemical Rubber Publishing Co., 1959, pp. 2303–5.

As an example of the use of the chart in Figure 2.3 we calculate the activity of oxygen gas at a temperature of 473 °K and a pressure of 1000 atmospheres. The reduced temperature is

$$\frac{T}{T_{cr}} = \frac{473}{154} = 3.1,$$

and the reduced pressure is

$$\frac{P}{P_{cr}} = \frac{1000}{49.7} = 20.$$

With these numbers the chart yields a value of χ_{O_2} of about 1.5—in excellent agreement with the more accurate value of 1.46, found from the χ values for O_2 in Figure 2.2.

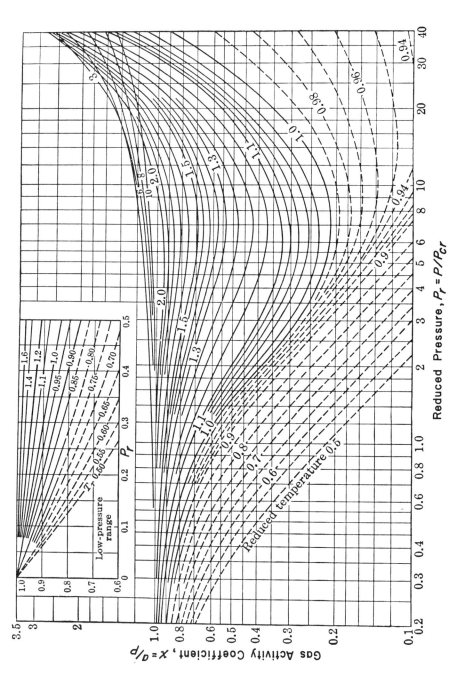

FIG. 2.3. Universal gas activity coefficient chart. [Reproduced (with slight modification of ordinate notation) by permission from O. A. Hougen and K. M. Watson, *Chemical Process Principles, Part II, Thermodynamics*, copyright 1947 by John Wiley & Sons, Inc.; based on data taken from B. W. Gamson and K. M. Watson, *National Petroleum News, Tech. Section*, 36, R623 (Sept. 6, 1944).]

The accuracy of the χ values obtained from the universal activity coefficient chart of Figure 2.3 will depend upon the reduced temperature and reduced pressure of interest, and should be assessed separately for each gas.

Mixtures of Gases

The molecular picture implied for a perfect single gas, one that obeys the relation $PV = nRT$, is that the gas consists of freely moving particles of negligible volume and having no significant interaction. This picture is in agreement with the observation that a real gas approaches the perfect gas as the pressure is lowered. For, as the pressure of a gas diminishes, the number of molecules in a given containing volume becomes fewer, so that the total volume of the particles themselves, relative to that of the containing volume, becomes less important; also, since the molecules will on the average be farther apart, the forces between them will decrease.

ACTIVITY OF THE PERFECT GAS MIXTURE. We may also consider a *perfect gas mixture*, for which the same molecular considerations given above for a single perfect gas apply. A perfect gas mixture is defined as one that obeys the perfect gas law, $PV = nRT$, where n now represents the total number of moles of all the different kinds of gases present in the mixture. The perfect gas mixture law is given by the expression

$$PV = RT \sum_j n_j, \qquad (2.6)$$

where

$$n = \sum_j n_j = n(N_1 + N_2 + \cdots + N_j). \qquad (2.7)$$

The activity of each of the components in a perfect gas mixture is dependent only upon the relative number of molecules of that component present in the mixture, hence upon its mole fraction. Thus, consistently with equation (2.3), we write

$$a_j = |PN_j|, \qquad (2.8)$$

i.e., the activity of the jth component of a perfect gas mixture is *defined* as the numerical value of the *total* pressure (in atmospheres) times the mole fraction of component j.

Another definition, *valid for any gas*, relates the partial pressure of each component to the total pressure, and is given by equation (2.9)

$$P_j = PN_j. \qquad (2.9)$$

Since, by equation (2.9)

$$\sum_j P_j = PN_1 + PN_2 + \cdots + PN_j = P(N_1 + N_2 + \cdots + N_j),$$

and

$$N_1 + N_2 + \cdots + N_j = 1,$$

then

$$\sum_j P_j = P, \qquad (2.10)$$

i.e., the sum of the partial pressures of all the gases present in a mixture is equal to the total pressure.

Combining equations (2.8) and (2.9), we see that for a perfect gas mixture

$$a_j = |P_j|. \qquad (2.11)$$

Equation (2.11) does not hold, in general, for real gas mixtures.

ACTIVITY OF THE IDEAL GAS SOLUTION. So far as is known, all gases are completely miscible with each other in all proportions. Mixtures of real gases deviate from perfect behavior, just as real single gases do. However, little is known, for most gases, about the effect of one chemical species on the activity of another chemical species present in a gas solution. Hence, it is usually assumed that real gas solutions are *ideal solutions*. This assumption means essentially that each gas present in a gas solution behaves independently of the other gases present, and thus that each gas carries with it, into the mixture, the same departure from perfection that it had as a pure gas.[2] Accordingly, for an ideal gas solution, we *define* the activity of the gas component j as

$$a_j = \chi_j P N_j \qquad (2.12)$$

where P is the total pressure (in atmospheres), χ_j is the activity coefficient for the given gas species, *at the total pressure P* (and the given temperature), and N_j is the mole fraction.

As a specific example, let us calculate the activity of oxygen in a mixture of 2 moles of oxygen and 6 moles of nitrogen at a total pressure of 500 atmospheres and at a temperature of 120 °C. The mole fraction of oxygen is

$$N_{O_2} = \frac{2}{2 + 6} = 0.25.$$

[2] Much more testing of this assumption is needed before confidence can be placed in the results of calculations based upon it; however, for most gases of geologic interest, the work done to date indicates that the assumption is most seriously in error when the gas density is in the neighborhood of the critical density of a component, i.e., about 0.4 g cm^{-3}. At densities higher or lower than the critical density, ideal gas solution behavior may more safely be assumed.

The value of the activity coefficient of oxygen is that of pure oxygen at the total pressure of the mixture. From Table 2.1, the reduced temperature is

$$\frac{T}{T_{cr}} = \frac{393}{154} = 2.55,$$

and the reduced pressure is

$$\frac{P}{P_{cr}} = \frac{500}{49.7} = 10.1.$$

From Figure 2.3, χ_{O_2} is 1.1. Substituting these values into equation (2.12), we obtain

$$a_{O_2} = 1.1 \times 500 \times 0.25 = 138.$$

Only a knowledge of the chemistry of a gaseous mixture will permit decisions as to whether the mixture can be treated as an ideal solution. At one end of the spectrum are mixtures, such as that of oxygen and argon, that can be treated with confidence; at the other end, are mixtures like sulfur and oxygen. Not only does sulfur gas contain a variety of sulfur molecules whose proportions change with the temperature and pressure, but these react with oxygen to form molecules, such as SO, SO_2, or SO_3.

SUMMARY—GASES

At ordinary temperatures, pure gases tend to follow the *perfect gas law*, $PV = nRT$, up to pressures of at least 1 atmosphere. Under such conditions the activity of the gas may be taken as numerically equal to the pressure in atmospheres. In the same way, gas mixtures at most temperatures of interest, and in the pressure range 0 to 1 atmosphere, behave as *perfect gas mixtures*. For such mixtures the activity of each gas present in the mixture may be taken as numerically equal to its partial pressure.

At higher pressures, gas solutions are considered to be ideal ones, when it is known that the gases involved do not interact chemically. For such solutions the activity of any component of the solution can be found by various analytical and graphical devices; here a chart is given (Figure 2.3) that provides activity coefficients in terms of the reduced temperatures and pressures of the gases. The activity of a given component in an ideal gas solution is given by the product of three quantities: the activity coefficient of the corresponding pure gas at the total pressure of the mixture, the total pressure, in atmospheres, and the mole fraction of the component.

NONELECTROLYTE SOLUTIONS

Nonelectrolyte solutions are by definition solutions consisting of uncharged particles. Typical binary nonelectrolyte solutions are those formed by various pairs of organic liquids, such as the benzene-toluene pair. The treatment of the behavior of nonelectrolyte solutions, which is uncomplicated and easily visualized in terms of simple kinetic pictures, leads directly to certain fundamental ideas, including an exact definition of the activity. The theory of nonelectrolyte solutions, developed initially for treatment of molecular solutions, has been found to apply to metallic solutions. Moreover, the same theory describes the behavior of solid and molten salt solutions, even though they contain ions. Consequently, for all solution phenomena in the realm of geology, except aqueous solutions of electrolytes, the nonelectrolyte solution theory is used. Included in the scope of this theory are solid solutions in minerals, cation exchangers, processes in magmas, and organic materials, both solid and liquid.

Ideal Solution

RAOULT'S LAW. It has been found that many binary liquid nonelectrolyte solutions obey, with varying degree of accuracy, a simple relationship between the vapor pressure exerted by each component of the solution and the mole fraction of that component in the solution. This relationship, known as *Raoult's law*, is given by equation (2.13), as follows

$$P_j = P_j^* N_j. \tag{2.13}$$

Here P_j is the equilibrium vapor pressure of component j of the solution, N_j is the mole fraction, in solution, of that component, and P_j^* is the equilibrium vapor pressure of the pure substance j. Since the values of the vapor pressures vary with changes in temperature and *total* pressure, Raoult's law applies only for conditions of constant temperature and constant total pressure.

Equation (2.13) may be applied to a solution of any number of components. For a binary solution, we may write the pair of simultaneous equations

$$P_1 = P_1^* N_1 \tag{2.14a}$$

$$P_2 = P_2^* N_2. \tag{2.14b}$$

The graphical representation of this pair of equations is shown in Figure 2.4.

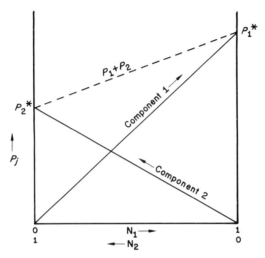

FIG. 2.4. Plot of the vapor pressures P_j vs. mole fractions N_j for a binary system obeying Raoult's law. P_1^* and P_2^* represent the vapor pressures of the two pure substances. The dashed line gives the total pressure over the solution due to both vapors, as a function of the composition of the solution.

The more nearly alike all the physical properties of the components of a solution, the more likely is the solution to obey Raoult's law. Thus, solutions of liquid organic optical isomers, and solutions of isotopes obey

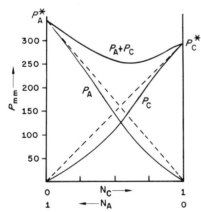

FIG. 2.5. Partial pressures vs. composition in the system chloroform-acetone, at 35.2 °C. The solid lines marked P_A and P_C denote the vapor-pressure curves for acetone and chloroform, respectively; the dashed lines indicate the corresponding Raoult's law behavior. [Data from J. v. Zawidski, Zeit. f. Physikal. Chemie, 35, 129 (1900).]

Raoult's law nearly exactly. Most solutions, however, show departures, sometimes marked ones, from Raoult's law. An example of the vapor pressure-composition relations of an actual solution is shown in Figure 2.5 for the system acetone-chloroform. It will be seen that for this system there is negative departure from Raoult's law. Both positive and negative departures are observed in various systems.

A solution obeying Raoult's law, and for which the vapors behave as perfect gases, is *defined* as an *ideal solution*. Thus, for an ideal solution

$$P'_j = P^{*\prime}_j N_j, \tag{2.15}$$

where the primed quantities indicate that the vapors obey the perfect gas law.

THE FUGACITY. The pressure of a perfect gas plays a central role in thermochemical considerations, and has been given a special name—it is called the *fugacity*, and is written f_j. Using the above notation

$$P'_j = f_j. \tag{2.16}$$

Combining equations (2.15) and (2.16), we have

$$f_j = f_j^* N_j, \tag{2.17}$$

for the ideal solution.

Most solutions, at ordinary temperatures and total pressures, have relatively low vapor pressures. At these low vapor pressures, the vapors approach perfect gas behavior. Under these conditions, fugacities and measured vapor pressures have, for all practical purposes, identical values, i.e.,

$$P_j \doteq f_j \quad (\doteq \text{means nearly equal}). \tag{2.17a}$$

For conditions of temperature and pressure where the vapor does not behave as a perfect gas, the value of the fugacity will differ from that of the measured vapor pressure; however, it is always feasible to determine one from the other; the fugacity may be thought of as a corrected pressure.

It should be clearly understood that the fugacity of a component dissolved in any condensed phase (liquid or solid) is exactly equal to the fugacity of that component in the vapor phase in equilibrium with the condensed phase. Although the concept of fugacity has been introduced here through considerations of the ideal solution and the perfect gas, the fugacity of a dissolved component in any solution, however nonideal, is a definite and generally measurable quantity.

ACTIVITY-FUGACITY RELATIONS. With these facts in mind, it is now possible to give the fundamental definition of the activity of a dissolved component of *any* solution. The activity a_j is *defined* as

$$a_j = \frac{f_j}{f_j^\circ}, \tag{2.18}$$

where f_j is the fugacity of the dissolved component j, and f_j° is its fugacity in the standard state. The standard state of a liquid or solid is usually taken as the pure liquid or solid at 1 atmosphere total pressure and at a specified temperature. For these standard conditions, the standard-state fugacity is the fugacity of the pure substance, i.e.,

$$f_j^\circ = f_j^* \tag{2.19}$$

and equation (2.18) becomes

$$a_j = \frac{f_j}{f_j^*}. \tag{2.20}$$

In these latter equations, it is to be remembered that the asterisk denotes a pure substance, that fugacities are measured in pressure units (atmospheres, bars, mm Hg, etc.), and that since activity is a ratio of pressures, and the pressure units cancel out, activity is a pure number. It will also be noted that for a pure substance

$$f_j = f_j^*,$$

and by virtue of equation (2.20), therefore

$$a_j^* = 1,$$

i.e., the activity of a pure substance is unity, under the standard state conditions given above.

Equation (2.20) holds for any solution. For the ideal solution we may combine (2.20) and (2.17) to obtain the important result

$$a_j = N_j. \tag{2.21}$$

For an ideal binary solution, the pair of simultaneous equations

$$a_1 = N_1 \tag{2.22a}$$

$$a_2 = N_2 \tag{2.22b}$$

will hold. The graphical representation of (2.22a,b) is shown in Figure 2.6.

It is seen from equation (2.20) that the activity of any dissolved component may be obtained by determining the equilibrium vapor pressure of that component over the solution. Where the vapor behaves as a perfect gas, the fugacity is equal to the partial pressure; where the vapor does not follow the perfect gas law, the fugacity may be calculated from the vapor pressure, as will be discussed shortly. Hence, vapor pressure determinations furnish a practical way for determining activities of substances having appreciable vapor pressures, such as the chloroform-acetone system discussed previously, or the water in aqueous solutions of all kinds. For less volatile components, such as $MgCO_3$ dissolved in

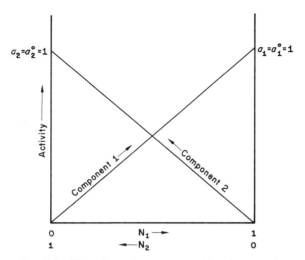

FIG. 2.6. Plot of activity vs. composition (expressed as mole fraction) for an ideal binary solution, at 1 atmosphere total pressure and a specified temperature.

$CaCO_3$ at room temperature, a system for which the vapor pressure of either of the two components is negligible, equation (2.20) has no practical operational meaning, and other, more indirect, methods for determining activities must be resorted to.

Equation (2.18) is perfectly general and applies to all states of matter—solid, liquid, or gaseous. For a gas, the standard state of which is taken as the pure gas at unit fugacity, equation (2.18) yields

$$a_{\text{gas}} = \frac{f_{\text{gas}}}{f^\circ_{\text{gas}}} = \frac{f_{\text{gas}}}{1} = |f_{\text{gas}}|, \tag{2.23}$$

i.e., the activity of a gas is numerically equal to the fugacity of the gas (the pressure units of the fugacities canceling out).

In the previous section on the activity-concentration relations for gases, several methods for relating activities to pressure and composition were developed. Since by equation (2.23) the activity and the fugacity of a gas are numerically the same, it follows that we may substitute fugacity everywhere for activity in these gas equations [(2.3, 2.5, 2.8, 2.11, 2.12)], remembering only that the fugacity must be expressed in the same units as the pressure (atmospheres), in the given equation. It is to be emphasized that it is only for a gas that activity and fugacity are numerically the same; by equations (2.18) and (2.20), the activity for a liquid or solid is always given by a ratio of fugacities, where the denominator of the ratio is not unity (except by chance). Since it is always possible to derive the fugacity of a vapor from its pressure, this completes the operational definition of the activity given by equation (2.18).

Let us use Figure 2.5 to show the significance of the various quantities that have been introduced. The measured vapor pressure of pure acetone, P_a^* (at 35.2 °C) is 344 mm Hg; that of pure chloroform, P_c^*, is 293 mm Hg.[3] At these low pressures it is reasonable to assume that the pure vapors behave as perfect gases. Consequently, the pressure is equal to the fugacity. For pure chloroform, by (2.17a)

$$P_{cg}^* \doteq f_{cg}^* = f_{cl}^* = \frac{293}{760} = 0.386 \text{ atm.}$$

The activity of chloroform vapor in equilibrium with pure liquid chloroform is, according to (2.23),

$$a_{cg}^* = \frac{f_{cg}^*}{f_{cg}^\circ} = \frac{0.386}{1} = 0.386.$$

The activity of pure liquid chloroform is, from (2.18),

$$a_{cl}^* = \frac{f_{cl}^*}{f_{cl}^\circ} = \frac{0.386}{0.386} = 1.$$

When the mole fraction of acetone in the solution is 0.419, the total measured vapor pressure is 248 mm. If the acetone-chloroform solution obeyed Raoult's law (and the partial vapor pressures were additive), the total vapor pressure at this composition would be, from (2.13),

$$P_{total} = P_a + P_c = P_a^* N_a + P_c^* N_c$$
$$= 344 \times 0.419 + 293 \times 0.581 = 314 \text{ mm.}$$

Clearly, this solution has negative deviation from Raoult's law.

For the acetone-chloroform solution of composition $N_a = 0.419$, $N_c = 0.581$, let us now calculate the fugacity and activity of the acetone. Again we start with equation (2.17a)

$$P_a \doteq f_a.$$

The measured vapor pressure of the acetone, over a solution of the stated composition, is 108 mm. The fugacity is

$$P_{ag} \doteq f_{ag} = f_{al} = \frac{108}{760} = 0.142 \text{ atm.}$$

The activity of the acetone in the vapor phase is, from (2.23),

$$a_{ag} = \frac{f_{ag}}{f_{ag}^\circ} = \frac{0.142}{1} = 0.142;$$

[3] Numerical data for the acetone-chloroform system taken from J. v. Zawidski, *Zeit. f. Physikal. Chemie*, 35, 129 (1900).

ACTIVITY-CONCENTRATION RELATIONS

the activity of the acetone in the liquid phase is, by (2.18),

$$a_{a\,1} = \frac{f_{a\,1}}{f^{\circ}_{a\,1}} = \frac{0.142}{344/760} = 0.313.$$

If the acetone-chloroform pair had formed an ideal solution, the activity of acetone in solution would have been given by (2.21)

$$a_{a\,1} = N_{a\,1} = 0.419.$$

THE EFFECT OF PRESSURE ON THE FUGACITY AND ACTIVITY. Because so many geologically important physical and chemical transformations take place at pressures greatly exceeding 1 atmosphere, it is necessary that we consider how the activity changes with *total* pressure.[4] For pure gases and gas solutions there is no problem, since the equations for activity involve the total pressure explicitly. For liquid and solid solutions the problem is more involved.

In our fundamental definition of activity, the relation (2.18)

$$a_j = \frac{f_j}{f^{\circ}_j},$$

was defined for all temperatures and pressures. Here the quantity f°_j is the fugacity of component j in its standard state. The standard state is defined as the pure liquid or solid at the specified temperature and at 1 atmosphere total pressure. Hence, f°_j does not change with change in total pressure. However, f_j, and consequently a_j, does change with pressure, in the manner given by equation (2.24)

$$\left(\frac{\partial \ln f_j}{\partial P}\right)_T = \left(\frac{\partial \ln a_j}{\partial P}\right)_T = \frac{\bar{V}_j}{RT}. \qquad (2.24)$$

Equation (2.24) states that the rate of change of the $\ln a_j$ (or $\ln f_j$) with change in total pressure, at constant temperature, is equal to the partial molal volume of component j divided by RT.

As an example of the use of equation (2.24) we may consider pure water. For this we write

$$\left(\frac{\partial \ln f_{H_2O}}{\partial P}\right)_T = \frac{V_{H_2O}}{RT}, \qquad (2.25)$$

---

[4] In a beaker or flask open to the atmosphere, a liquid or solid is subjected to a pressure of approximately 1 atmosphere (depending on the barometric pressure at the time). Increased pressure may be applied to the liquid or solid by means of a piston, in which case no vapor exists. However, a *fictive* vapor pressure (measured by the force of bombardment of molecules against the piston) may be calculated. Alternatively, increased pressure on the liquid or solid may be obtained by introducing an inert gas, at any desired pressure, into a closed space over the liquid or solid.

where V_{H_2O} is the volume of a mole of pure water at the specified temperature T and at each pressure considered. The expression given in (2.25) may be partially integrated to give

$$RT[\ln f_{H_2O}(\text{at } P_2) - \ln f_{H_2O}(\text{at } P_1)] = \int_{P_1}^{P_2} V_{H_2O}\, dP. \qquad (2.26)$$

The right-hand member of (2.26) is simply the area under the curve obtained by plotting V_{H_2O} against P. Since values of V_{H_2O} at various pressures have been experimentally determined, it is possible to evaluate (2.26) and obtain the fugacity and the activity of water at each of these pressures. Such a list of the activities and fugacities of water, at 298 °K, is given in Table 2.2. Note that in Table 2.2, the activity is given at

TABLE 2.2. The Fugacity and Activity of Liquid Water at 298 °K for Various Total Pressures

Pressures (atmospheres)	Fugacity (atmospheres)	Activity
1	0.03125	1
100	0.03362	1.0757
200	0.03618	1.1576
300	0.03892	1.2454
400	0.04186	1.3394
500	0.04501	1.4402
600	0.04838	1.5481
700	0.05199	1.6637
800	0.05586	1.7874
900	0.06000	1.9200
1000	0.06443	2.0618

SOURCE: Adapted from M. Randall and B. Sosnick, *J. Am. Chem. Soc.*, 50, 967 (1928).

each pressure by dividing the fugacity at that pressure by the fugacity at 1 atmosphere, in accordance with equation (2.18). Thus, at 500 atmospheres total pressure

$$a = \frac{0.04501}{0.03125} = 1.4402.$$

Similar calculations can be made for any system for which the data on $\bar{V}_j$ are available.

Since, as shown above, f_j^* changes with pressure, and f_j° does not change, the relationship given by equation (2.19) is valid only at a pressure of 1 atmosphere. Consequently, equations (2.20) and (2.21) hold only at 1 atmosphere. At pressures other than 1 atmosphere we may use only equation (2.17)

$$f_j = f_j^* N_j,$$

to define an ideal solution.

ACTIVITY-CONCENTRATION RELATIONS

The expression for the activity of an ideal solution at pressures other than 1 atmosphere is obtained by combining equations (2.17) and (2.18), as follows

$$a_j = \frac{f_j}{f_j^\circ} = \frac{f_j^* N_j}{f_j^\circ}. \quad (2.27)$$

As an example of the application of (2.27) consider a hypothetical ideal aqueous solution at 1000 atmospheres pressure and 298 °K. From Table 2.2, at $N_{H_2O} = 1$, $f_{H_2O}^* = 0.06443$, and since $f_{H_2O}^\circ = 0.03125$, we have for this ideal solution, according to (2.27),

$$a_{H_2O} = \frac{0.0644}{0.0312} N_{H_2O} = 2.06 N_{H_2O}. \quad (2.27a)$$

A plot of a_{H_2O} vs. N_{H_2O} for this system is shown in Figure 2.7.

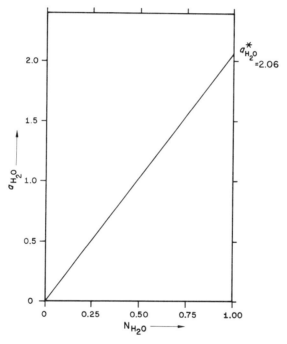

FIG. 2.7. The activity-composition relations for water in an hypothetical ideal aqueous nonelectrolyte solution at 1000 atmospheres pressure, and 298 °K.

It will be noted that we have not discussed the change of fugacity or activity with temperature. Although it is possible to write an equation for this change, the variables involved are, in general, more difficult of measurement than the fugacity or activity themselves. Hence, we shall adopt the attitude that the fugacity or activity are known or can be measured at each specified temperature.

Nonideal Behavior

RATIONAL ACTIVITY COEFFICIENT. Most actual solutions do not exhibit ideal behavior. Nevertheless, the concept of ideal behavior furnishes a useful yardstick against which actual behavior may be discussed.

As an example of the activity-composition relations in an actual solution, we show in Figure 2.8 the results obtained for the liquid solution cadmium-lead at 500 °C and 1 atmosphere. It will be noted that the activities of Cd at low concentrations are much larger than the corresponding mole fractions—there is large positive departure from Raoult's law. As the mole fraction of Cd approaches unity, the activity also approaches unity.

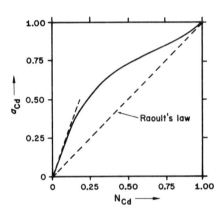

FIG. 2.8. Activity of cadmium in the system Cd-Pb at 500 °C, and 1 atmosphere. [Data of J. F. Elliott and J. Chipman, *Trans. Faraday Soc.*, **47**, 138 (1951).]

To preserve the form of the ideal solution law for nonideal systems, it is convenient to introduce again the activity coefficient. For nonelectrolyte solutions, we define λ_j, the *rational activity coefficient*, for all temperatures and pressures, as

$$\lambda_j = \frac{a_j}{N_j}. \tag{2.28}$$

A plot of the activity coefficient against mole fraction for Cd in the Cd-Pb system can be obtained simply by dividing each value of the activity by the corresponding value of the mole fraction. Thus, from Figure 2.8, for $N_{Cd} = 0.50$, $a_{Cd} = 0.72$, and therefore

$$\lambda_{Cd} = \frac{0.72}{0.50} = 1.44.$$

A complete plot of λ_{Cd} vs. N_{Cd} is given in Figure 2.9. It will be noted that since a_{Cd} approaches unity as N_{Cd} approaches unity (Figure 2.8), λ_{Cd} must similarly approach unity.

For the ideal solution, at 1 atmosphere pressure and the specified temperature, $\lambda_j = 1$. However, at pressures other than 1 atmosphere, for the ideal solution, $\lambda_j \neq 1$. According to (2.27)

$$a_j = \frac{f_j^*}{f_j^\circ} N_j,$$

and since, generally, from (2.28)

$$a_j = \lambda_j N_j,$$

we have

$$\lambda_j = \frac{f_j^*}{f_j^\circ}. \tag{2.29}$$

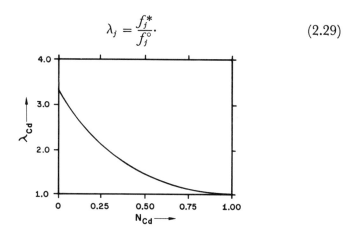

FIG. 2.9. Activity coefficients of cadmium in the Cd-Pb system at 500 °C, and 1 atmosphere pressure.

Since f_j° has the same value at all pressures, and f_j^* is a constant for the pressure under consideration, λ_j will have a different constant value at each pressure, for the ideal solution. This value will be unity at a pressure of 1 atmosphere, and different from unity at pressures other than 1 atmosphere.

Thus, for the example we considered previously, that of a hypothetical ideal aqueous solution at 298 °K and 1000 atmospheres pressure [equation (2.27a)], λ_{H_2O} would have the constant value 2.06.

HENRY'S LAW. So far in our discussion, there has been no necessity to make any distinction between the solute (the component being dissolved), and the solvent (the component doing the dissolving). The distinction has meaning only when that component called the solute is present in the solution in relatively small proportion, i.e., when we are dealing with dilute solutions.

It has been found, without exception, that for all actual nonelectrolyte solutions, the fugacity of the dissolved component, f_2, becomes more nearly proportional to the mole fraction N_2 as the solution becomes more and more dilute, i.e., as N_2 approaches zero, at constant temperature and total pressure. The equation expressing this result is written

$$f_2 = k''N_2 \quad \text{as } N_2 \to 0. \tag{2.30}$$

Equation (2.30) is known as Henry's law.

The difference between the expressions for the ideal solution law and Henry's law is due to the difference in the proportionality constants of these expressions, as may be seen by comparing equations (2.17) and (2.30)

$$\text{ideal solution} \quad f_2 = f_2^* N_2 \tag{2.17}$$

$$\text{Henry's law} \quad f_2 = k'' N_2. \tag{2.30}$$

This distinction is made clear in the schematic drawing shown in Figure 2.10. The value of the proportionality constant, k″, and the range of composition over which Henry's law is obeyed depend upon the particular solute and solvent and upon the temperature and total pressure being considered.

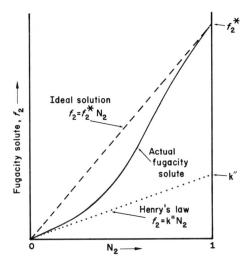

FIG. 2.10. Schematic diagram showing the distinction between Henry's law and the ideal law for an actual binary solution.

If both sides of equation (2.30) are divided by f_2°, the standard-state fugacity of the solute, we obtain

$$\frac{f_2}{f_2^\circ} = \frac{k''N_2}{f_2^\circ} \tag{2.31}$$

or

$$\frac{f_2}{f_2^\circ} = k'N_2 \tag{2.32}$$

ACTIVITY-CONCENTRATION RELATIONS

Since the activity of any component of a solution is always, by definition,

$$a_j = \frac{f_j}{f_j^\circ}, \tag{2.18}$$

it follows that equation (2.32) may be written

$$a_2 = k'N_2, \tag{2.33}$$

for the infinitely dilute solution. Moreover, since

$$k' = \frac{k''}{f_2^\circ}, \tag{2.34}$$

it is obvious that the value of k' will depend upon the standard state chosen for the solute. This choice depends upon the particular problem to be solved and will be considered further in connection with specific problems.

By equation (2.28), the activity coefficient, λ_2, of the solute is defined

$$\lambda_2 = \frac{a_2}{N_2}, \tag{2.28}$$

and since by Henry's law

$$a_2 = k'N_2 \qquad \text{as } N_2 \to 0, \tag{2.33}$$

then

$$\lambda_2 = k' \qquad \text{as } N_2 \to 0. \tag{2.35}$$

Equation (2.35) tells us that in the infinitely dilute solution the value of λ_2 will be the same as that of k'. This result is illustrated in Figures 2.8 and 2.9. The dashed line in Figure 2.8 represents Henry's law behavior; the slope of the line, k', has a value of about 3.3, and in Figure 2.9 it is seen that λ_{Cd} has the limiting value 3.3 at $N_{Cd} = 0$. It will also be noted from these illustrations that as N_{Cd} approaches 1, the ideal solution law is approached, and consequently λ_{Cd} approaches unity. It can be shown that for any solution for which Henry's law is obeyed by all the solutes, the solvent will obey the ideal solution law.

PRACTICAL ACTIVITY COEFFICIENT. In the dilute solution range where Henry's law holds, we may make a further simplification in expressing the composition of the solution. Since, for a binary solution, the mole fraction of the solute is given by

$$N_2 = \frac{n_2}{n_1 + n_2},$$

and since $n_1 + n_2 \cong n_1$ for small values of n_2, we may write, for dilute solutions,

$$N_2 \cong \frac{n_2}{n_1}. \qquad (2.36)$$

By definition, the molality of a solution, m, is given by the number of the moles of solute per 1000 grams of solvent

$$m = \frac{n_2}{\text{grams solvent}} \times 1000,$$

or

$$m = \frac{n_2}{n_1} \times \frac{1000}{\text{formula wt. solvent}} = \frac{n_2}{n_1} \times \text{const.} \qquad (2.37)$$

Comparing (2.36) and (2.37), we see that

$$m \cong \text{const.} \times N_2. \qquad (2.38)$$

If this result is put into equation (2.33), we have,

$$a_2 = km. \qquad (2.39)$$

When Henry's law holds at finite concentrations of the solute, equation (2.39) will be an approximation; in the infinitely dilute solution, the activities given by (2.33) and (2.39) will converge on the same value. For a solution containing more than one dissolved component, the approximations made in arriving at equation (2.39) will hold, and for each dissolved component j we may write the equation

$$a_j = k_j m_j. \qquad (2.40)$$

It is useful to introduce an activity coefficient based on the molal scale. This is called the *practical activity coefficient*, γ_j, and is given by the equation

$$a_j = \gamma_j m_j. \qquad (2.41)$$

The value of γ_j varies with m_j (and is dependent on the temperature and total pressure), so that equation (2.41) holds at all concentrations, regardless of whether or not Henry's law is obeyed. In the infinitely dilute solution, where Henry's law behavior is approached by the solute, γ_j approaches k_j, as seen by comparison of equations (2.40) and (2.41).

Solid Solutions and Melts

Minerals are invariably solid solutions: the chemical analysis of a mineral almost always shows deviations from that corresponding to the pure substance. Thus, rhodochrosite, $MnCO_3$, usually contains dissolved calcium, and may also contain ferrous iron, magnesium, and zinc.

The extent and kind of solid solution varies, depending upon the composition and crystal structure of the mineral acting as solvent. Most minerals are salts, that is, crystalline compounds whose structures are made up of cations, positively charged ions, and anions, negatively charged ions. Typical examples are rhodochrosite, cited above, consisting of Mn^{++} and CO_3^{--} ions, and olivine, consisting of Mg^{++} and SiO_4^{4-} ions. Solid solution takes place through the substitution of an ion for a like ion in the host structure. In rhodochrosite, Ca^{++} or Fe^{++} ions may replace Mn^{++}. If, for example x gram-atoms of Fe^{++} replace the corresponding quantity of Mn^{++}, the formula of the resulting solid solution is $(Fe_xMn_{1-x})CO_3$. Substitution of one anion by another may also take place. Thus, the molybdate ion, MoO_4^{--}, substitutes readily for the chromate ion, CrO_4^{--}, in some compounds. Other more complex types of substitution are possible.

When a solid solution is formed by substituting one cation for another, the extent to which solution takes place and the stability of the resulting solution depend upon how nearly alike are the two cations. If the substituent cation has the same charge and nearly the same ionic radius as the dispossessed cation, stable solution takes place readily. Thus, Fe^{++}, ionic radius 0.75 Å (Appendix 3), and Mn^{++}, ionic radius 0.80 Å, replace one another freely in many minerals, whereas Ba^{++}, ionic radius 1.35 Å, for example, is not found to replace Fe^{++} or Mn^{++}.

At elevated temperatures, the mutual solubility of minerals becomes considerably enhanced. This is exactly what is to be expected on a crystal-structure basis. With increasing temperature, each ion in the structure has increased thermal motion, and thus has increased effective ionic radius. Since the lighter ions, having smaller ionic radii, gain relatively larger thermal motion, the overall effect is to make the effective ionic radii of all the elements more nearly the same with increasing temperature. Thus, although K^+ ($r_{K^+} = 1.33$ Å) replaces Na^+ ($r_{Na^+} = 0.95$ Å) in NaCl to a negligible extent at 25 °C, KCl and NaCl are mutually soluble in all proportions above 500 °C.

It is important to note that the reported chemical analyses of many minerals represent solid solutions that are thermodynamically unstable at room temperatures. As pointed out above, solubility is greatly increased with increasing temperature, and when a mineral solution formed at elevated temperatures is cooled, the sluggishness of the unmixing reaction often prevents the system from reaching equilibrium. The fact that a mineral commonly represents an unstable system must be taken into account in thermochemical considerations.

The activity of a pure solid at 1 atmosphere pressure and the specified temperature is taken as unity by definition. However, in the case of a solid solution under these conditions, the solvent crystal does not usually have an activity of 1. Thus a_{MnCO_3} in $(Fe_{0.2}Mn_{0.8})CO_3$ is not 1. If the solid solution is an ideal one, equation (2.21) is valid, i.e.,

$$a_j = N_j.$$

Thus, in the above example, if $MnCO_3$ performs ideally in the range of composition around that given by $(Fe_{0.2}Mn_{0.8})CO_3$ (and we do not know this to be true), then a_{MnCO_3} would be given by

$$a_{MnCO_3} = N_{MnCO_3} = \frac{0.8}{0.2 + 0.8} = 0.8.$$

In all cases, equation (2.28) is valid, i.e.,

$$a_j = \lambda_j N_j.$$

REGULAR SOLUTION. A close approximation to the behavior of many binary nonelectrolyte solutions of various kinds is given by the equation

$$\ln \lambda_1 = \frac{B}{RT} N_2^2, \tag{2.42}$$

where B is a constant, independent of composition. Equation (2.42) may be rewritten as

$$\log \lambda_1 = \frac{B}{2.303 RT} N_2^2, \tag{2.43a}$$

or

$$\log \lambda_1 = B'N_2^2 = B'(1 - N_1)^2. \tag{2.43b}$$

Solutions obeying equation (2.42) are called *regular solutions*.[5]

Inspection of equation (2.43b) shows that as N_j approaches unity, λ_j also approaches unity, and therefore a_j approaches unity. This result corresponds to a choice of the pure component j as the standard state of that component. Hence, implicit in the formulation of the regular solution equation is the fact that the standard states of the components of the binary solution are the pure components.

In equation (2.43a), the quantities $\ln \lambda$ and N are pure numbers, and hence B has the same units as RT. If R is expressed in kilocalories per degree mole, then kcal/deg mole × deg = kcal/mole, the units of B. At 298.15 °K (25 °C), equation (2.43a) becomes

$$\log \lambda_1 = \frac{B}{2.303 \times 0.001987 \times 298.15} N_2^2$$

$$\log \lambda_1 = \frac{B(\text{kcal/mole})}{1.364} N_2^2. \tag{2.43c}$$

---

[5] The term "regular solution," as used here, is not meant to imply more than is stated in the text, and is not a universal designation. The expression "symmetrical solution" is used instead by some authors. (See, e.g., J. H. Hildebrand and R. L. Scott, *Regular Solutions*. Englewood Cliffs, N.J., Prentice-Hall, 1962.)

ACTIVITY-CONCENTRATION RELATIONS

Although there are relatively few data available on the behavior of solid solutions and melts of geological interest, it appears that of those studied, most behave nearly as regular solutions. It has been estimated that for most systems of geological interest, the value of the constant B' ranges from zero (ideal solution) to ± 5 (C. C. Stephenson, Massachusetts Institute of Technology; personal communication).

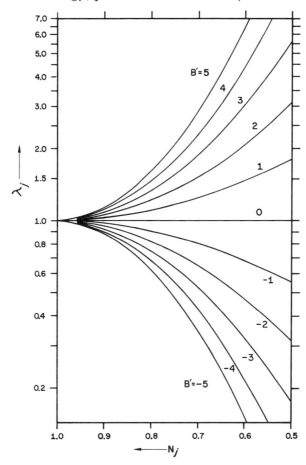

FIG. 2.11. Plot of activity coefficient λ_j (log scale) vs. the mole fraction N_j, for the function $\log \lambda_j = B'(1 - N_j)^2$, for values of B' from -5 to $+5$.

In Figure 2.11 is plotted the family of curves relating the activity coefficient, λ_j, and the mole fraction, N_j, for the component j of a binary solution, for values of B' ranging from -5 to $+5$. An important result to be gained from Figure 2.11 is that for any of the values of B' shown, the activity coefficient approaches unity rapidly as the mole fraction approaches unity. Hence, even for solutions exhibiting large departures from ideal behavior over much of the composition range, the solvent may be considered to behave ideally as the solution becomes dilute. Specifically, for minor substitution in a mineral, it may be assumed without important error that the host mineral behaves ideally.

Our knowledge of the behavior of the solute, the minor constituent of a solution, is much more uncertain. From Figure 2.11, it can be seen that as N_j becomes smaller, the value of λ_j for a given N_j depends more and more strongly on the value of B' chosen. To sum up these conclusions, one can say that it is reasonably safe to assume ideal behavior for the solution component having mole fraction in the range of about $N_j \geq 0.9$, but that in the absence of specific and accurate knowledge of the value of B' (assuming regular solution) no prediction can be made about the behavior of component j for $N_j < 0.9$.

SOME EXPERIMENTAL RESULTS ON REGULAR SOLUTIONS. In the study of the activity of silver bromide in solution in the molten bromides of lithium, sodium, potassium, and rubidium, at 500 to 600 °C, Salstrom

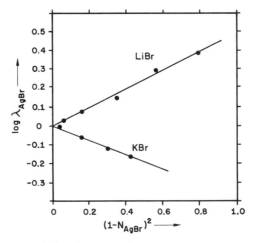

FIG. 2.12. Plot of the log λ_{AgBr} vs. $(1 - N_{AgBr})^2$ for AgBr dissolved in LiBr and in KBr, respectively, at 550 °C. [Data derived from Salstrom and Hildebrand, op. cit.]

and Hildebrand[6] have shown that each system behaves essentially as a regular solution. In Figure 2.12 are plotted the results obtained by these authors for AgBr dissolved in LiBr, and AgBr dissolved in KBr, at 550 °C. In this figure we have plotted log λ_{AgBr} vs. $(1 - N_{AgBr})^2$,

[6] E. J. Salstrom and J. H. Hildebrand, *J. Am. Chem. Soc.*, 52, 4650 (1930). E. J. Salstrom, *ibid*, 53, 1794, 3385 (1931); 54, 4252 (1932).

which according to equation (2.43b) should yield a straight line of slope B' for each of the solutions, provided that it is a regular solution. From the results shown, it is seen that the solutions do indeed approximate regular solutions. For AgBr dissolved in LiBr, the constant B' of equation (2.43b) is approximately 0.50, and for AgBr dissolved in KBr, B' is approximately -0.38.

Other salt pairs and certain alloy systems have been investigated experimentally and found to behave as regular solutions, at least approximately. The results obtained in some of these studies are discussed in Darken and Gurry, chapter 10 (reference list, end of Chapter 1).

CORRECTIONS TO THE ACTIVITIES OF SOLIDS BECAUSE OF COMPOSITIONAL VARIATIONS. Minerals are generally not pure substances, and this must be taken into account in thermochemical calculations involving mineral activities. Currently, because of the lack of knowledge of the deviation of solid solutions from ideality, the procedure is to assume ideal behavior and to estimate the range of compositions within which this assumption is a reasonable one.

One kind of question that arises in attempting to make calculations is as follows: thermochemical data are available to calculate the equilibrium product of the activities of calcium and carbonate ions in aqueous solution in equilibrium with pure calcite, but how much is this product changed if the calcite contains 5 mole percent $MnCO_3$? To answer this question, we first write the equilibrium constant for the dissociation of calcite at 25 °C and 1 atmosphere total pressure. For the reaction under these conditions

$$CaCO_{3\ calcite} = Ca^{++}_{aq} + CO^{--}_{3\ aq}$$

$$\Delta F^\circ_r = \Delta F^\circ_{f\ Ca^{++}} + \Delta F^\circ_{f\ CO_3^{--}} - \Delta F^\circ_{f\ CaCO_3}$$

$$11.38 \text{ kcal} = -132.18 + (-126.22) - (-269.78).$$

Since by equation (1.12)

$$\Delta F^\circ_r = -1.364 \log K$$

then

$$11.38 = -1.364 \log \frac{[Ca^{++}][CO_3^{--}]}{[CaCO_3]}$$

and

$$\frac{[Ca^{++}][CO_3^{--}]}{[CaCO_3]} = 10^{-8.34}. \qquad (2.44)$$

For pure calcite $[CaCO_3]$ is unity, so the ion activity product is $10^{-8.34}$.

But what is the activity of $CaCO_3$ in $(Ca_{0.95}Mn_{0.05})CO_3$? On the assumption of ideal solution

$$[CaCO_3] = N_{CaCO_3} = 0.95.$$

Substituting this result in (2.44)

$$[Ca^{++}][CO_3^{--}] = 10^{-8.34} \times 0.95 = 10^{-8.36}.$$

If the $(Ca,Mn)CO_3$ is not an ideal solution, but it is assumed that it is a regular solution, reference to Figure 2.11 shows that λ_{CaCO_3} is between about 0.98 and 1.02 (assuming that B' is a minimum or a maximum). Therefore, the range of activity of $CaCO_3$ is, according to equation (2.28),

$$[CaCO_3] = \lambda_{CaCO_3} N_{CaCO_3}$$
$$0.93 = 0.98 \times 0.95$$
$$0.97 = 1.02 \times 0.95.$$

From these relations, the activity products of calcium and carbonate ions, in equilibrium with $(Ca,Mn)CO_3$, are calculated and summarized in Table 2.3.

It is seen from Table 2.3 that the assumption of ideal solution does not lead to significant error in the activity product for solid solutions of small impurity content. In fact, from Figure 2.11 it is seen that dilution of a pure phase by impurities of mole fraction values up to $N_{impurity} = 0.1$ will not lead to serious error if ideal solution is assumed.

TABLE 2.3. Activity Product $[Ca^{++}][CO_3^{--}]$ for Calcite in Equilibrium with Its Aqueous Solution at 25 °C, 1 atmosphere

Solid Phase	$[CaCO_3]$	$[Ca^{++}][CO_3^{--}]$
Pure calcite	1	$10^{-8.34}$
$(Ca_{0.95}Mn_{0.05})CO_3$	0.95 (ideal solution)	$10^{-8.36}$
$(Ca_{0.95}Mn_{0.05})CO_3$	0.97 ($B' = +5$)	$10^{-8.35}$
$(Ca_{0.95}Mn_{0.05})CO_3$	0.93 ($B' = -5$)	$10^{-8.37}$

On the other hand, no conclusions can be drawn from these considerations concerning the activity of the component present in minor amount. In the case under discussion, the activity of $MnCO_3$ in $(Ca_{0.95}Mn_{0.05})CO_3$ must be obtained from experiment.

Phase equilibria studies show that in many natural systems, solid solubility is limited. In these cases we are dealing with compositions not far removed from pure substances, and can, as shown above, make fairly safe estimates of the activities of major components. Where solubility is more extensive, or even complete over the entire range of composition, in the absence of evidence to the contrary, again it can be assumed only that the major component will behave ideally for compositions approaching the pure components. An example of this is given by our previous discussion of the completely miscible molten salt pairs AgBr-LiBr and AgBr-KBr. It was shown that in these systems (Figure 2.12), λ_{AgBr} approaches unity only as N_{AgBr} approaches unity, and that over the entire composition range the solutions obey regular solution theory, with B' significantly different from zero.

Values of activity coefficients can be obtained from data on equilibrium compositions. For example, it is known[7] that at 25 °C, the equilibrium composition of calcite and siderite approximate $(Ca_{0.95}Fe_{0.05})CO_3$ and $(Fe_{0.95}Ca_{0.05})CO_3$, respectively. When these two phases are in equilibrium, the activity of $CaCO_3$ in calcite must be equal to that of $CaCO_3$ in siderite (the standard state being taken as pure calcite at 1 atmosphere total pressure and the temperature of interest for both phases). Thus,

$$a_{CaCO_3 \text{ (calcite)}} = a_{CaCO_3 \text{ (siderite)}}.$$

Because the calcite is essentially pure

$$a_{CaCO_3 \text{ (calcite)}} = a' = \lambda'N' \cong 1 \times 1 \cong 1.$$

Therefore,

$$a_{CaCO_3 \text{ (siderite)}} = a'' = \lambda''N'' \cong 1$$

and since

$$N'' = 0.05,$$

then

$$\lambda'' \cong \frac{1}{0.05} \cong 20.$$

Where data similar to those for calcite and siderite are available, activity coefficients can be obtained through similar reasoning.

SUMMARY—NONELECTROLYTE SOLUTIONS

The relations developed by chemists for solutions of nonelectrolytes have been found useful for the geologist in describing a wide range of solution phenomena in nature, including processes in magmas, solid solution relations in minerals, cation exchange phenomena, and the behavior of natural organic substances. Among the key relations is Raoult's law, $P_j = P_j^* N_j$, which can be applied directly to a system having a volatile component, and which, importantly, leads to the concepts of the fugacity f_j and the activity a_j. The development of a relationship between the fugacity and the activity of a component

[7] P. E. Rosenberg, *Am. J. Sci.*, *261*, 689 (1963).

permits the calculation of the activity of a volatile component in aqueous, or other medium, from measurement of its fugacity. Henry's law, $f_2 = k''N_2$, describes the limiting dilute-solution behavior of the solute in all nonelectrolyte solutions. For nonideal solution, over all composition ranges, the concept of the activity coefficient λ_j is introduced and used to relate activity to mole fraction, i.e., $a_j = \lambda_j N_j$. It has been found that many binary systems, including minerals, are "regular" solutions, i.e., the activity coefficient of each component is given by an equation of the type $\log \lambda_1 = B'(1 - N_1)^2$. Sample calculations show the use of the theory of nonelectrolytes in typical problems.

AQUEOUS SOLUTIONS OF ELECTROLYTES

An *electrolyte solution* is one in which the solute exists partially or completely in the form of ions. In the treatment given in the following, liquid water is considered the solvent,[8] although with appropriate modifications the treatment would apply for any solvent.

The actual nature of the solute as a kinetic entity in an aqueous solution is a complicated affair, and depends upon the particular solute, the concentration of the solution, and the presence or absence of other chemical species. For purposes of convenience of discussion we divide electrolytes into two classes: *nonassociated electrolytes* and *associated electrolytes*.[9]

By "nonassociated electrolyte" is meant a solute that exists in solution wholly in the form of simple cation and anion; these may be hydrated. Such salts as NaCl and KCl are believed to be nonassociated, or "strong electrolytes." However, as Robinson and Stokes (*op. cit.*) point out, "the chief criterion for placing an electrolyte in this class is the absence of valid evidence for any form of association."

Associated electrolytes are divided into two classes: *weak electrolytes* and *ion pairs*. By "weak electrolyte" is meant a solute that can exist as undissociated covalent molecules as well as in the form of ions. Acids, some bases, and a very few inorganic salts fall into this class. Solutions of carbon dioxide, for example, contain undissociated H_2CO_3 molecules, as well as the ions HCO_3^- and CO_3^{--}; hydrogen chloride solutions, at high concentrations, contain a high proportion of HCl molecules as well as H^+ and Cl^- ions.

[8] Other solvents, such as the fluids carbon dioxide, ammonia, or hydrogen sulfide, are possibly important in subsurface geologic processes; however, data for these as ionizing solvents are meager.

[9] The treatment of nonassociated and associated electrolytes given here follows that of R. A. Robinson and R. H. Stokes, *Electrolyte Solutions*. New York, Academic, 1959, chap. 3.

"Ion-pairing" is the term used to describe the association of oppositely charged ions resulting from electrostatic interaction only, no covalent (electron-pair) bond formation being involved, as in the case of weak electrolytes. For example, in solutions of calcium sulfate, as in other bivalent metal sulfate solutions, a certain proportion of the Ca^{++} and SO_4^{--} ions associate to form neutral $CaSO_4^o$ particles; in solutions of calcium hydroxide, some of the solute exists in the form of the ion pair $Ca(OH)^+$.

It is not always possible to decide whether a given dissolved species that clearly exists as an associated electrolyte should be called an ion pair or a weak electrolyte, so that the classification rests on grounds of convenience rather than of rigor.

"Complex ion" is the term used to designate all ions other than monatomic ones. The ions $Ca(OH)^+$, CO_3^{--}, HCO_3^-, HS^-, $ZnCl_4^{--}$, and $B_3O_3(OH)_4^-$ are all complex ions; no account need be taken here of the nature of the forces binding the atoms together.

It is to be emphasized that the particular state of the solute depends markedly upon the concentration. With increasing concentration, the solute particles will on the average be closer together, so that interaction, whether by covalent or electrostatic bond formation, takes place more readily. Thus, the formation of neutral molecules, or of complex ions, is favored by increasing concentration. Conversely, separation of more complex entities into simple hydrated ions is favored by dilution.

Activity and Activity Coefficients

The activity of an electrolyte solute is found to obey Henry's law when the dissociation of the electrolyte is taken into account. For a uni-univalent solute like HCl, which dissociates into H^+ and Cl^- ions, the expression for Henry's law contains the square of the molality. In the limit of infinite dilution, Henry's law has the form

$$a_2 = km_2^2. \tag{2.45}$$

All uni-univalent electrolyte solutes, such as NaCl and LiF, and all symmetrical salts in which anion and cation have the same valence, such as $MgSO_4$ and $CaCO_3$, obey equation (2.45). In the case of a ternary salt like $BaCl_2$, for which each mole of salt dissociates into a total of 3 moles of ions, Henry's law has the form

$$a_2 = km_2^3. \tag{2.46}$$

In general, for an electrolyte dissociating into $\nu = \nu_+ + \nu_-$ ions according to the equation

$$C_{\nu_+}A_{\nu_-} = \nu_+ C + \nu_- A,$$

where C stands for cation (of appropriate charge) and A stands for anion, Henry's law is

$$a_2 = km_2^\nu. \tag{2.47}$$

SYMMETRICAL ELECTROLYTES. The value of the proportionality constant k depends upon the standard state chosen for the solute. For an electrolyte solute, the standard state is not taken as the pure solid or liquid. In the case of a symmetrical compound, for which equation (2.45) holds, we choose the standard state in the manner illustrated in Figure 2.13. It will be seen from this drawing that the standard state

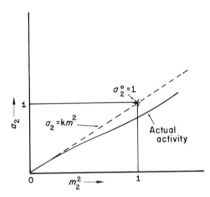

FIG. 2.13. Schematic representation of the activity-molality plot for a symmetrical electrolyte solute, at a given temperature and total pressure. The plot is constructed in the following way: the measured activity is plotted on an arbitrary numerical scale against m_2^2, as the solid line. The limiting slope of this line, as m_2^2 approaches zero, is established, and is represented by the dashed line; this dashed line has the equation $a_2 = km_2^2$. The asterisk on the dashed line, at $m_2^2 = 1$, represents the standard state of the solute, for which $a_2 = a_2^0 = 1$; and thus establishes the numerical scale of the activity. With the choice of this standard state, we have $k = 1$, and hence $a_2 = m_2^2$, as the limiting law.

is a hypothetical one established by the behavior of the solution at infinite dilution, and that for this choice of standard state of a symmetrical electrolyte

$$a_2 = m_2^2, \quad \text{as } m_2 \to 0. \tag{2.48}$$

As the solution approaches infinite dilution, the interaction between the cations and anions will vanish, and for the symmetrical electrolyte, we may assume that the molalities of the cations and anions will be the same and equal to the molality of the dissolved solute, i.e.,

$$m_2 = m_+ = m_-. \tag{2.49}$$

Substitution of this result into equation (2.48) gives us

$$a_2 = (m_+)(m_-). \tag{2.50}$$

ACTIVITY-CONCENTRATION RELATIONS

The individual ion activities of the cation and anion, designated a_+ and a_-, respectively, *are defined* so that

$$a_+ = m_+, \quad \text{as } m_2 \to 0 \tag{2.51a}$$

$$a_- = m_-, \quad \text{as } m_2 \to 0. \tag{2.51b}$$

Since $m_+ = m_-$, in the infinitely dilute solution, then $a_+ = a_-$ under this condition.

Combining equations (2.50) and (2.51a, b), we have

$$a_2 = (a_+)(a_-). \tag{2.52}$$

Although equation (2.52) was derived for the infinitely dilute solution, it is a useful relationship to have at all concentrations, and so is taken *by definition* to hold at all concentrations.

The mean ionic activity of a symmetrical electrolyte, $a_\pm$, *is defined* as

$$a_\pm = (a_+ a_-)^{1/2} = a_2^{1/2}. \tag{2.53}$$

We define the *individual ion activity coefficients* as

$$\gamma_+ = \frac{a_+}{m_+} \tag{2.54a}$$

$$\gamma_- = \frac{a_-}{m_-}. \tag{2.54b}$$

Since the ratios a_+/m_+ and a_-/m_- approach unity as m_+ and m_- approach zero, that is at infinite dilution, γ_+ and γ_- approach unity at infinite dilution.

The *mean ionic activity coefficient* is defined as

$$\gamma_\pm = (\gamma_+ \gamma_-)^{1/2}. \tag{2.55}$$

By equations (2.54a, b)

$$\gamma_\pm = \left(\frac{a_+}{m_+} \frac{a_-}{m_-}\right)^{1/2}. \tag{2.56}$$

By equation (2.53), $(a_+ a_-)^{1/2} = a_\pm = a_2^{1/2}$. If we define the *mean ionic molality* of a symmetrical electrolyte as

$$m_\pm = (m_+ m_-)^{1/2}, \tag{2.57}$$

and retain the *formal* relationship $m_2 = m_+ = m_-$ for *all concentrations*, we have

$$m_\pm = m_2,$$

and equation (2.56) reduces to

$$\gamma_\pm = \frac{a_\pm}{m_\pm} = \frac{a_2^{1/2}}{m_2}. \tag{2.58}$$

Since by equation (2.48) the ratio $a_2^{1/2}/m_2$ approaches unity as m_2 approaches zero, then $\gamma_\pm$ approaches unity at infinite dilution.

UNSYMMETRICAL ELECTROLYTES. The development of the relationships between the activity and concentration of an unsymmetrical electrolyte is carried out by procedures analogous to that used for the symmetrical electrolyte. For an electrolyte $C_{\nu_+}A_{\nu_-}$ dissociating into ν_+ positive ions and ν_- negative ions, Henry's law is

$$a_2 = km_2^{(\nu_+ + \nu_-)} = km_2^\nu. \qquad (2.59)$$

The individual ion activities are related to the activity of the solute, a_2, by

$$a_2 = (a_+)^{\nu_+}(a_-)^{\nu_-}; \qquad (2.60)$$

the mean activity, $a_\pm$, is given by

$$a_\pm = [(a_+)^{\nu_+}(a_-)^{\nu_-}]^{1/\nu}; \qquad (2.61)$$

the mean ionic molality, $m_\pm$, by

$$m_\pm = m_2[(\nu_+)^{\nu_+}(\nu_-)^{\nu_-}]^{1/\nu}; \qquad (2.62)$$

and the mean activity coefficient, $\gamma_\pm$, as

$$\gamma_\pm = \frac{a_\pm}{m_\pm} = [(\gamma_+)^{\nu_+}(\gamma_-)^{\nu_-}]^{1/\nu}. \qquad (2.63)$$

The activity coefficient $\gamma_\pm$ is so defined, for the general case, that it approaches unity in the infinitely dilute solution. The definitions as given by equations (2.54a, b), for the individual ion activity coefficients for the symmetrical electrolyte, are retained for the general case.

Although individual ion activity coefficients are not directly measurable, the mean ion activity coefficients can be obtained accurately. A variety of methods is available, and values for a wide range of concentrations are available in standard chemical reference books.[10] From the measured mean ion activity coefficients it is possible to derive semi-empirical values for individual ion activity coefficients in various ways.

Solubility-Activity Relations

Before taking up the methods used for obtaining values of individual ion activity coefficients, we wish to consider one of the major needs for these values—the calculation of solubilities.

[10] I. M. Klotz, *Chemical Thermodynamics*. Englewood Cliffs, N.J., Prentice-Hall, 1950, chap. 21; W. M. Latimer, *Oxidation Potentials*. New York, Prentice-Hall, 1952, pp. 349–358; H. S. Harned and B. B. Owen, *The Physical Chemistry of Electrolytic Solutions*, 3rd ed. (ACS Monograph 137). New York, Reinhold, 1958; R. A. Robinson and R. H. Stokes, *Electrolyte Solutions*. New York, Academic, 1959, appendices; Roger Parsons, *Handbook of Electrochemical Constants*. London, Butterworth, 1959; B. E. Conway, *Electrochemical Data*. New York, Elsevier, 1952.

The geologist wants to be able to calculate solubilities of substances in various aqueous environments. This requires development of a relation between the activities of dissolved species of a given element and total concentration of the element as determined by chemical analysis.

It must be recognized from the beginning that calculation of solubilities is still more of an art than a science, and that any values determined entirely theoretically are suspect. The solubility of a given element is the sum of the stoichiometric concentrations of all dissolved species containing the element. Thus, accurate calculation of solubility requires knowledge of all species contributing to the result of chemical analysis, plus a relation between concentration of a given species and its activity.

Therefore, to make solubility calculations, the geochemist needs detailed knowledge of the chemistry of the element in question; it might almost be said that the result, to be trusted, has to be known in advance from experimental work. Yet it is possible in many instances to calculate a *minimum solubility* that is nearly as useful as the true solubility in answering geochemical questions.

The concentration of a given element in aqueous solution as determined by quantitative analysis can be described as

$$m_A = \Sigma\, m_s \qquad (2.64)$$

where m_A is the total molality defined as

m_A = moles per 1000 grams of water

$$= \frac{\text{weight of element obtained by analysis} \times 1000}{\text{atomic weight of element} \times \text{weight of containing water}}.$$

Then $\Sigma\, m_s$ is the sum of the molalities of all species containing the element, where the molality of a given species is defined similarly to m_A, on the assumption that the species could be isolated and determined quantitatively.

To relate m_A to calculated activities of individual species, a connection between activity and molality of individual species is required; thus,

$$a_s = \gamma_s m_s; \quad m_s = \frac{a_s}{\gamma_s}, \qquad (2.65)$$

and

$$m_A = \Sigma\, m_s = \Sigma \frac{a_s}{\gamma_s}. \qquad (2.66)$$

As an example, assume that water is saturated with carbon dioxide under a given partial pressure, and that the amount dissolved is determined by loss of weight of the gas from the source container. Then

$$m_{A(CO_2)} = \Sigma\, m_s.$$

In this instance, knowledge of the chemistry tells us that the important dissolved species are $CO_{2\ aq}$, H_2CO_3, HCO_3^-, and CO_3^{--}. Then

$$m_{A(CO_2)} = m_{CO_2\ aq} + m_{H_2CO_3} + m_{HCO_3^-} + m_{CO_3^{--}} + m_x + m_y + \cdots$$

The terms m_x, m_y, etc., are used to denote those species undoubtedly present of which we have no current knowledge. Thus, if only $CO_{2\ aq}$, H_2CO_3, HCO_3^-, and CO_3^{--} are considered, the value of $m_{A(CO_2)}$ obtained by summing them up will be less than the analytical value by the sum of the molalities of these unknowns. If their sum is negligible in terms of analytical error, then the approximation

$$m_{A(CO_2)} = m_{CO_2\ aq} + m_{H_2CO_3} + m_{HCO_3^-} + m_{CO_3^{--}} \qquad (2.67)$$

is satisfactory. But at any rate, $m_{A(CO_2)}$ will not be less than their sum.

If we wish to calculate $m_{A(CO_2)}$ from thermochemical considerations, we can, knowing the partial pressure of CO_2 used, calculate values for $a_{CO_2\ aq}$, $a_{H_2CO_3}$, $a_{HCO_3^-}$, and $a_{CO_3^{--}}$. If we also know values for $\gamma_{CO_2\ aq}$, $\gamma_{H_2CO_3}$, $\gamma_{HCO_3^-}$, and $\gamma_{CO_3^{--}}$, we can write

$$m_{A(CO_2)} = \frac{a_{CO_2\ aq}}{\gamma_{CO_2\ aq}} + \frac{a_{H_2CO_3}}{\gamma_{H_2CO_3}} + \frac{a_{HCO_3^-}}{\gamma_{HCO_3^-}} + \frac{a_{CO_3^{--}}}{\gamma_{CO_3^{--}}}. \qquad (2.68)$$

This approximation is subject to the same limitations as equation (2.67), plus any uncertainties that may arise specifically in the determination of activities and activity coefficients.

In summary, an approximation of the solubility of a given element can often be obtained by thermochemical calculations if detailed information is available on all dissolved species contributing to the solubility, and if accurate values of their activity coefficients can be obtained. When anything less than this knowledge is available, the calculated value is less than the observed one,[11] and underestimation may be serious indeed.

Ionic Strength

An extremely useful generalization for assessing the combined effects of the activities of several electrolytes in a solution on a given electrolyte is embodied in the Lewis and Randall concept of the *ionic strength*.[12,13] In the words of Lewis and Randall: *in dilute solutions, the activity coefficient of a given strong electrolyte is the same in all solutions of the same ionic strength*.[13] The ionic strength, I, is defined as

$$I = \tfrac{1}{2} \sum m_i z_i^2, \qquad (2.69)$$

[11] In highly concentrated solutions, a_s may exceed m_s, and the calculated value may exceed the observed one. This effect is rare at concentrations less than 1 molal.
[12] G. N. Lewis and Merle Randall, *J. Am. Chem. Soc.*, **43**, 1112 (1921).
[13] *Idem, Thermodynamics*. New York, McGraw-Hill, 1923, chap. 28.

ACTIVITY-CONCENTRATION RELATIONS

where m_i is the molality and z_i is the charge of the ith ion in the solution, the summation being taken over all ions, positive and negative. For example, the ionic strength of a 1 molal solution of $CaCl_2$ is

$$I = \tfrac{1}{2}(m_{Ca^{++}} \times 2^2 + m_{Cl^-} \times 1^2) = \tfrac{1}{2}(1 \times 4 + 2 \times 1) = 3.$$

That of a 1 molal NaCl solution is

$$I = \tfrac{1}{2}(1 \times 1^2 + 1 \times 1^2) = 1.$$

The ionic strength of natural waters can be calculated from chemical analyses, inasmuch as the analyses are usually expressed in terms of major ionic species or in terms of dissolved salts.

Analysis of a water from the Madison sand of Mississippian Age from the Cut Bank oil field in Montana provides the values given in Table 2.4.[14]

TABLE 2.4. Analysis of Water from a Mississippian Sand (ppm)

Na^+	Ca^{++}	Mg^{++}	SO_4^-	Cl^-	CO_3^-	HCO_3^-
2187	39	57	232	1680	84	2850

The first step is conversion from parts per million to molality (Table 2.5), from the relation[15]

$$\text{molality} = \frac{\text{parts per million}}{\text{gram formula weight}} \times 10^{-3}.$$

TABLE 2.5. Analysis of Water from a Mississippian Sand (molality)

Na^+	Ca^{++}	Mg^{++}	SO_4^-	Cl^-	CO_3^-	HCO_3^-
0.0951	0.00097	0.0023	0.00242	0.0474	0.0014	0.0467

Taking the ions as they appear from left to right in the table,

$$I = \tfrac{1}{2}(0.0951 \times 1^2 + 0.00097 \times 2^2 + 0.0023 \times 2^2 + 0.0024 \times 2^2$$
$$+ 0.0474 \times 1^2 + 0.0014 \times 2^2 + 0.0467 \times 1^2)$$
$$= 0.109.$$

This value of ionic strength is about average for the waters in rocks. Stream and lake waters commonly run about one-tenth of this value, whereas the oceans are about ten times higher.

Ionic strength is a relation useful in comparing solutions of diverse compositions because the specific electrical effects of the interactions of

---

[14] Taken from James G. Crawford, Water analysis, in *Subsurface Geologic Methods*. Colorado School of Mines, 1951, p. 193.

[15] A correction should be made here because the analysis is expressed in terms of grams per 10^6 grams solution, and not grams per 10^6 grams H_2O. The error involved is, however, very small.

the variously charged ions present are taken into consideration. Because electrical effects are functions of the squares of the charges on the ions, use of ionic strength gives a more useful criterion of the behavior of a solution than does concentration. It is not a cure-all, but it does remove one complexity that otherwise tends to obscure specific compositional effects. Determination of ionic strength is, like determination of solubility, a process requiring successive approximations. To calculate an ionic strength, we must know the major ions present. The process consists, as in the example cited, of a first consideration of known major species. If the presence of other species is deduced, as a result of theory or experiment, the ionic strength must be changed by taking cognizance of them. In general, the major species of natural waters are sufficiently well known so that the presentation of ions in typical chemical analyses provides a value of ionic strength that is little changed by more detailed study. Undoubtedly the water from the Madison sand contains traces of boron and iodine, as well as small amounts of potassium, but their total contribution would be unlikely to change the ionic strength determined by more than a percent or two.

In Figure 2.14 are illustrated typical curves resulting from plots of experimentally measured mean activity coefficients against ionic strength. The values illustrate well the marked effect of valence on the values of the activity coefficient. KCl, a uni-univalent salt, has the highest values; Na_2SO_4 and $CuCl_2$, which are combinations of a divalent ion and two univalent ions, correspond well below ionic strengths of about 0.1; $CuSO_4$, a di-divalent salt, exhibits the lowest activity coefficients.

Individual Ion Activity Coefficients

As was mentioned previously, although it is generally possible to obtain accurate experimental values of mean ion activity coefficients, it is not possible to measure directly individual ion activity coefficients. A way of circumventing this difficulty is to make the assumption that $\gamma_+ = \gamma_-$ for a standard uni-univalent electrolyte, over the ionic strength range of interest. The behavior of KCl in solution is the standard basis for obtaining individual ion activity coefficients.[16] Various lines of evidence indicate that γ_{K^+} and γ_{Cl^-} have similar values;[17] hence, as a reasonable approximation,

$$\gamma_{\pm KCl} = [(\gamma_{K^+})(\gamma_{Cl^-})]^{1/2} = \gamma_{K^+} = \gamma_{Cl^-}. \qquad (2.70)$$

---

[16] G. N. Lewis and Merle Randall, *Thermodynamics.* New York, McGraw-Hill, 1923, p. 381.

[17] The suggestion that γ_{K^+} and γ_{Cl^-} are equal in a pure solution of KCl is due to MacInnes: D. A. MacInnes, *J. Chem. Soc., 41,* 1086 (1919); this assumption is often called the *MacInnes assumption.*

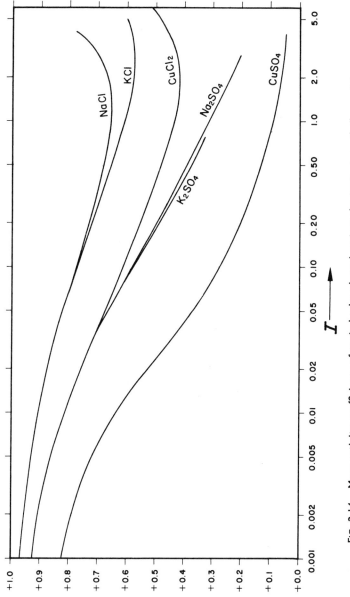

FIG. 2.14. Mean activity coefficients of typical salts plotted against the ionic strength (log scale). Note the similarity of the behavior of salts of the same valence type at ionic strengths below 0.1. [Data from W. M. Latimer, op. cit., p. 354.]

Using this relationship, a table of values for other ions can be built up from the appropriate mean ion activity coefficients.

Thus, for a monovalent chloride

$$\gamma_{\pm MCl} = [(\gamma_{M^+})(\gamma_{Cl^-})]^{1/2} = [(\gamma_{M^+})(\gamma_{\pm KCl})]^{1/2},$$

$$\gamma_{M^+} = \frac{\gamma_{\pm MCl}^2}{\gamma_{\pm KCl}}. \tag{2.71}$$

For a divalent chloride

$$\gamma_{\pm MCl_2} = [(\gamma_{M^{++}})(\gamma_{Cl^-})^2]^{1/3} = [(\gamma_{M^{++}})(\gamma_{\pm KCl})^2]^{1/3},$$

$$\gamma_{M^{++}} = \frac{\gamma_{\pm MCl_2}^3}{\gamma_{\pm KCl}^2}. \tag{2.72}$$

For salts of potassium other than the chloride, the reverse relation can be used to obtain activity coefficients for anions. For example

$$\gamma_{\pm K_2SO_4} = [(\gamma_{K^+})^2(\gamma_{SO_4^{--}})]^{1/3} = [(\gamma_{\pm KCl})^2(\gamma_{SO_4^{--}})]^{1/3},$$

$$\gamma_{SO_4^{--}} = \frac{\gamma_{\pm K_2SO_4}^3}{\gamma_{\pm KCl}^2}. \tag{2.73}$$

For a salt like $CuSO_4$, a double bridge must be used to obtain $\gamma_{Cu^{++}}$, since

$$\gamma_{\pm CuSO_4} = [(\gamma_{Cu^{++}})(\gamma_{SO_4^{--}})]^{1/2}, \tag{2.74}$$

substituting the value for $\gamma_{SO_4^{--}}$ from equation (2.73) and rearranging

$$\gamma_{Cu^{++}} = \frac{\gamma_{\pm CuSO_4}^2 \gamma_{\pm KCl}^2}{\gamma_{\pm K_2SO_4}^3}. \tag{2.75}$$

In the following, we shall call this method of calculating individual ion activity coefficients from mean activity coefficients the *mean salt method*.

The errors involved in the mean salt method are difficult to estimate and become greater as the bridge becomes longer. The best check is always against observational data. Thus, a valid set of single ion activity coefficients will have values that, taken in appropriate combinations, lead to mean activity coefficients that agree with experiment. Since the mean activity coefficient of an electrolyte is a function of the ionic strength, the individual ion activity coefficient similarly depends on the ionic strength, and *in dilute solutions of mixed electrolytes, we assume as an approximation that the activity coefficient of a given ion depends only on the ionic strength.*

Debye-Hückel Theory of Activity Coefficients

Even in very dilute solutions of electrolytes the charged ions contained in these solutions exert long-range electrostatic forces upon one another with the result that the values of the activity coefficients are lowered. This effect has been evaluated in the Debye-Hückel theory and several very useful equations have been derived on the basis of that theory, which are of great value in handling experimental data.[18] In the following discussion we consider two of these equations.

In dilute solutions, the individual ion activity coefficient is given by the Debye-Hückel expression

$$-\log \gamma_i = \frac{A z_i^2 \sqrt{I}}{1 + \mathring{a}_i B \sqrt{I}}. \tag{2.76}$$

Here, z_i and I have the meanings previously ascribed in the definition of ionic strength, and A and B are constants characteristic of the solvent (here considered to be water), at the specified temperature and pressure. Values of A and B as a function of temperature (at 1 atmosphere) are given in Table 2.6. The quantity $\mathring{a}_i$ has a value dependent upon the

TABLE 2.6. Values of Constants for Use in Debye-Hückel Equation (aqueous solution)

Temperature, °C	A	B ($\times 10^{-8}$)
0	0.4883	0.3241
5	0.4921	0.3249
10	0.4960	0.3258
15	0.5000	0.3262
20	0.5042	0.3273
25	0.5085	0.3281
30	0.5130	0.3290
35	0.5175	0.3297
40	0.5221	0.3305
45	0.5271	0.3314
50	0.5319	0.3321
55	0.5371	0.3329
60	0.5425	0.3338

SOURCE: G. G. Manov, R. G. Bates, W. J. Hamer, S. F. Acree, J. Am. Chem. Soc., 65, 1765 (1943).

"effective diameter" of the ion in solution, and is determined largely from experiment; values of $\mathring{a}_i$ are listed in Table 2.7. The physical

---

[18] See, e.g., I. M. Klotz, *op. cit.*, pp. 328–336; G. N. Lewis and Merle Randall, *Thermodynamics*, 2nd ed., revised by K. S. Pitzer and Leo Brewer. New York, McGraw-Hill, 1961, chap. 23.

TABLE 2.7. Values of $å_i$ for Some Individual Ions in Aqueous Solutions

$å_i \times 10^8$	Ion
2.5	Rb^+, Cs^+, NH_4^+, Tl^+, Ag^+
3.0	K^+, Cl^-, Br^-, I^-, NO_3^-
3.5	OH^-, F^-, HS^-, BrO_3^-, IO_4^-, MnO_4^-
4.0–4.5	Na^+, HCO_3^-, $H_2PO_4^-$, HSO_3^-, Hg_2^{2+}, SO_4^{--}, SeO_4^{--}, CrO_4^{--}, HPO_4^{--}, PO_4^{3-}
4.5	Pb^{++}, CO_3^{--}, SO_3^{--}, MoO_4^{--}
5.0	Sr^{++}, Ba^{++}, Ra^{++}, Cd^{++}, Hg^{++}, S^{--}, WO_4^{--}
6	Li^+, Ca^{++}, Cu^{++}, Zn^{++}, Sn^{++}, Mn^{++}, Fe^{++}, Ni^{++}, Co^{++}
8	Mg^{++}, Be^{++}
9	H^+, Al^{3+}, Cr^{3+}, trivalent rare earths
11	Th^{4+}, Zr^{4+}, Ce^{4+}, Sn^{4+}

SOURCE: Adapted from I. M. Klotz, *Chemical Thermodynamics*. Englewood Cliffs, N.J., Prentice-Hall, 1950, p. 331.

significance of $å_i$, which is commonly related to the diameter of the ion in solution, merits a brief digression. Values of $å_i$ are ordinarily larger than values of ionic diameters given for ions in crystals. This difference presumably stems from the envelope of water molecules that surround the ions in aqueous solution. Some attempts have been made to interpret $å_i$ values structurally, but a clear-cut picture of the coordination of water molecules around the charged ions has not yet emerged. Detailed discussion of the hydration problem is given by Robinson and Stokes.[19]

The mean salt method for obtaining individual ion activities is the better method for use in solutions of ionic strengths higher than about 0.05. In solutions of ionic strength below this value, the mean salt method and the results of equation (2.76) generally agree fairly well; moreover, the Debye-Hückel equation provides a method of extrapolating into the range of infinite dilution. When the ionic strength becomes very small, equation (2.76) reduces to the limiting expression

$$-\log \gamma_i = Az_i^2 \sqrt{I}. \qquad (2.77)$$

The expression similar to (2.77) for the mean activity coefficient of an electrolyte composed of *two different ions only*, in the limiting case of the infinitely dilute solution, is

$$-\log \gamma_\pm = A|z_+ z_-|\sqrt{I}, \qquad (2.78)$$

(where the symbol | | denotes absolute value).

[19] R. A. Robinson and R. H. Stokes, *Molecular Interaction*, Ann. N.Y. Acad. Sci., *51*, 593 (1949).

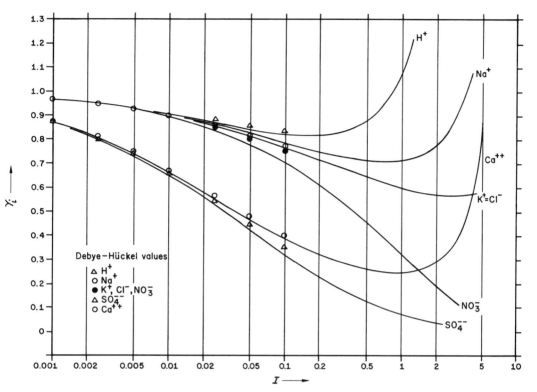

FIG. 2.15. Single ion activity coefficients vs. ionic strength for some common ions. Solid lines represent the values calculated by the mean salt method. Debye-Hückel values were calculated using equation (2.76), with $10^8/\mathring{a}_i = 9$ for H^+; 4 for Na^+; 3 for K^+, Cl^-, NO_3^-; 6 for Ca^{++}; and 4 for SO_4^{--}. The Debye-Hückel Y_i values for the monovalent ions converge, within experimental error, for $I < 0.01$.

Some Ionic Activity Coefficients Determined by Mean Salt and Debye-Hückel Methods

In Figure 2.15 are plotted the individual activity coefficients for some common ions, as determined by the mean salt method and by the Debye-Hückel method. For the calculations by the mean salt procedure, the methods outlined previously were used. It was assumed that $\gamma_{K^+} = \gamma_{Cl^-} = \gamma_{\pm KCl}$; values of γ_{H^+}, γ_{Na^+}, and $\gamma_{Ca^{++}}$ were then calculated, using values of the mean activity coefficients of the respective chlorides in equation (2.71) or in equation (2.72). Values of $\gamma_{NO_3^-}$ were calculated from the relation

$$\gamma_{NO_3^-} = \frac{\gamma^2_{\pm KNO_3}}{\gamma_{\pm KCl}}, \qquad (2.79)$$

and values of $\gamma_{SO_4^{--}}$ by using equation (2.73).

The Debye-Hückel values shown in Figure 2.15 are those calculated from equation (2.76), using the appropriate values for the constants A, B, and $å_i$ as given, for example, in Tables 2.6 and 2.7. Such calculations for ions in water at 25 °C (and 1 atmosphere) have been made and are listed in convenient tabular form in Klotz.[20]

It will be noted in Figure 2.15 that the γ_i values for all the univalent ions converge rapidly upon one another, and upon the Debye-Hückel limiting values, at ionic strengths below about 0.02. Except for the nitrate ion, the D-H and mean salt values for each univalent ion agree fairly well up to $I = 0.1$. The ion activity coefficient values of the two divalent ions plotted do not converge so rapidly upon each other or upon the D-H values. However, the agreement between the mean salt values and the D-H values, for all practical purposes, is fairly good up to ionic strengths of about 0.05.

Activity of Water in Electrolyte Solutions

It will be recalled that the activity of a pure liquid at 1 atmosphere total pressure and a specified temperature is defined to be unity. Pure water, then, under these conditions, will have unit activity. Further, the activity of any substance is related to its fugacity by equation (2.18), written here for water, i.e.,

$$a_{H_2O} = \frac{f_{H_2O}}{f^*_{H_2O}},$$

where $f^*_{H_2O}$ is the fugacity of pure water. Water vapor at moderate pressures behaves sufficiently ideally that the fugacity may be replaced by partial pressure, and we have

$$a_{H_2O} = \frac{P_{H_2O}}{P^*_{H_2O}}. \tag{2.80}$$

Hence, the activity of water in any electrolyte solution can be determined simply by measuring the vapor pressure of water over the solution, at a constant, known temperature, and dividing that value by the value of the vapor pressure of pure water at the same temperature. Values of the vapor pressure of water at various temperatures are listed in Appendix 4.

Tables of values of the activity of water in several electrolyte solutions are given in Robinson and Stokes.[21] In Figure 2.16 are plotted data

---

[20] I. M. Klotz, *op. cit.*, p. 332.
[21] R. A. Robinson and R. H. Stokes, *op. cit.*, appendices 8.3, 8.5.

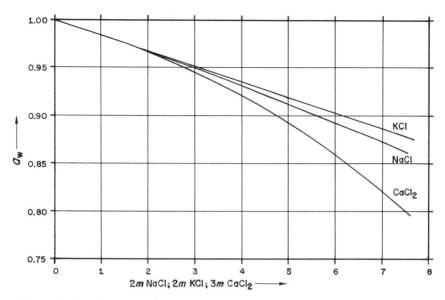

FIG. 2.16. Plot of activity of water, a_w vs. νm for several electrolyte solutions, at 25 °C. [Data from R. A. Robinson and R. H. Stokes, *Electrolyte Solutions*. New York, Academic, 1959, Appendices 8.3, 8.5.]

taken from these tables for NaCl, KCl, and $CaCl_2$ solutions, at 25 °C. Here the water activities are plotted against concentration units that take into account the total number of solute particles. Thus, a 1 molal solution of NaCl or KCl is considered to have a concentration of 2 molal total solute, a 1 molal $CaCl_2$ solution to have a concentration of 3 molal total solute, etc. It is seen that the activity curves for the three salt solutions, when plotted this way, converge on a straight line in the dilute solution range.

This behavior may be explained in the following way. Assuming that Raoult's law, as given by equation (2.13), may be applied here, we have

$$P_1 = P_1^* N_1 = P_1^* (1 - N_2), \qquad (2.81)$$

where P_1 is the vapor pressure of water over the electrolyte solution, P_1^* is the vapor pressure of pure water, N_1 the mole fraction of water, and N_2 the mole fraction of total solute. For dilute solutions, by equation (2.38),

$$N_2 = k\nu m, \qquad (2.82)$$

where ν is the number of ions into which the electrolyte dissociates. Substituting (2.82) into (2.81), and rearranging, we have

$$a_{H_2O} = \frac{P_1}{P_1^*} = 1 - k_w \nu m. \qquad (2.83)$$

By the definition of k [equation (2.38)] the value for k_w is given by

$$k_w = \frac{\text{formula wt. of water}}{1000} = 0.0180.$$

The limiting slope of the curves shown in Figure 2.16 has the concordant value 0.017.

The relationship expressed by equation (2.83) is a limiting one that can be expected to hold only in the dilute solution range. For accurate values of water activities at higher concentrations, recourse must be made to the experimental values. Data for the vapor pressure of water, or for the lowering of the vapor pressure of water, over various electrolyte solutions, as a function of temperature and concentration, are collected in several places.[22]

Equations (2.81) and (2.82) can be combined and rearranged to yield the result

$$\Delta P_1 = P_1^* - P_1 = P_1^* k_w \nu m; \qquad (2.84)$$

i.e., in dilute solutions the lowering of the vapor pressure of water, ΔP_1, over an electrolyte solution varies only with the quantity νm, at a constant temperature. Thus, all electrolytes of given formula type, if completely dissociated, should yield the same vapor pressure lowering at a specified (low) concentration. This means that ν would have the value 2 for such salts as KCl, $BaSO_4$, and others of the formula of type AB, the value 3 for compounds like $CaCl_2$, 4 for $AlCl_3$, etc. A perusal of the available ΔP_1 values for various electrolytes shows that this is indeed approximately true for many substances. There are, however, marked exceptions. For example, although for KCl, $\Delta P_1 = 12.2$ mm at $0.5\ m$, the ΔP_1 value for $MgSO_4$ is 6.5 mm at $0.5\ m$, both ΔP_1 values being measured at 100 °C. This result indicates that $MgSO_4$ exists to a large extent as the ion pair $MgSO_4^\circ$, in moderately concentrated solutions. Similar ΔP_1 deviations, indicating complex ion formation or ion-solvent interaction, occur. The important subject of complex ions in aqueous solutions will be taken up in some detail in Chapter 4.

The rate of change of the activity of water with temperature, over the range 0° to 100 °C, is inappreciable up to moderate concentrations of the dissolved electrolyte. With increasing concentrations the temperature dependence becomes larger. A good example of this is shown by the data for H_2SO_4 solutions given in Harned and Owen.[23] The plot of a_{H_2O} vs. $m_{H_2SO_4}^{1/2}$ at 0°, 25°, and 60 °C, given by these authors, is illustrative of the points we wish to make here.

---

[22] *International Critical Tables*, Vol. III. New York, McGraw-Hill, 1928; *Handbook of Chemistry and Physics*, 43rd ed. Cleveland, Ohio, Chemical Rubber, 1961, pp. 2467–2468.

[23] H. S. Harned and B. B. Owen, *op. cit.*, pp. 574–575.

Thus, at higher concentrations, an activity value of water is accurate only at the temperature at which it was measured; at lower concentrations, the statement of the temperature becomes less important. In particular, the ΔP_1 values listed in the Handbook of Chemistry and Physics (op. cit.), were measured at 100 °C. Strictly speaking, then, the activity values derived from these are appropriate only to that temperature. However, as indicated, at lower concentrations the activities may be considered to be 25 °C ones, without much loss of accuracy.

Activity Coefficients of Neutral Molecules in Electrolyte Solutions

The reaction for a gas in equilibrium with an aqueous electrolyte solution, at a constant temperature, may be written as follows, using oxygen as an example,

$$O_{2\,g} = O_{2\,aq}. \tag{2.85}$$

The equilibrium constant K is related to the activities of oxygen in the two phases by the expression

$$K = \frac{a_{O_2\,aq}}{a_{O_2\,g}} = \frac{a_{O_2\,aq}}{|f_{O_2\,g}|} = \frac{a_{O_2\,aq}}{|P_{O_2\,g}|}, \tag{2.86}$$

where $f_{O_2\,g}$ is the fugacity of oxygen gas (in atmospheres) and it is assumed that the partial pressure P_{O_2} (also in atmospheres) is low enough so that it equals the fugacity.

Rewriting equation (2.86) for any gas with activity a_2, and dropping the symbol $|\;|$ (but remembering that P_2 has numerical value only), we have

$$a_{2\,aq} = KP_{2\,g}. \tag{2.87}$$

Since $a_2 = \gamma_2 m_2$

$$\gamma_2 m_2 = KP_{2\,g}. \tag{2.88}$$

By convention, the activity coefficient of a gas dissolved in water containing no other solute is taken to be unity. Designating the molality in such a solution as m_0, and that in a solution of finite electrolyte concentration as m_s, we write

$$\gamma_s m_s = KP_s \tag{2.89a}$$

$$m_0 = KP_0. \tag{2.89b}$$

Combining equations (2.89a, b) yields a relationship between the activity coefficient of the dissolved gas in an electrolyte solution, its solubility in water, and in the electrolyte solution, and the equilibrium partial pressures, i.e.,

$$\gamma_s = \frac{m_0}{m_s} \times \frac{P_s}{P_0}. \tag{2.90}$$

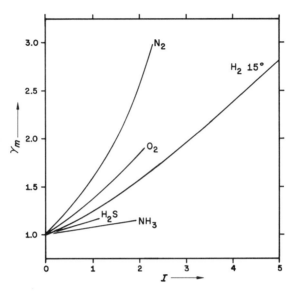

FIG. 2.17. Activity coefficients, γ_m, for various gases dissolved in NaCl solution at 25 °C (H_2 at 15 °C). [Data from Merle Randall and C. F. Failey, Chem. Rev., 4, 271 (1927).]

Equation (2.90), or similar relations, provide an experimental basis for the determination of γ_s values.

For a constant value of P_s we rewrite equation (2.89a) as

$$\gamma_s = \frac{KP_s}{m_s} = \frac{K'}{m_s}. \tag{2.91}$$

Experimentally, it is observed that as the concentration of the dissolved electrolyte increases, the value of m_s generally decreases, at constant P_s. Hence, γ_s will usually change in value from unity at zero value of dissolved electrolyte to a number greater than 1 with increasing ionic strength. The effect of the dissolved electrolyte on decreasing the solubility is called the *salting-out-effect*.

In Figure 2.17 are shown the activity coefficients, γ_s, for several gases dissolved in NaCl solution, plotted as a function of ionic strength of the electrolyte solution. It is seen from this plot that the rate of change of the activity coefficient with concentration of electrolyte depends markedly upon the particular gas. The data used in the plot were taken from Randall and Failey,[24] who list extensive tables of γ_s values for a number of different gases dissolved in many different electrolyte solutions. Examination of these data shows that for a given gas, the trend of the γ_s values varies significantly from electrolyte to electrolyte.

[24] Merle Randall and C. F. Failey, *Chem. Rev.*, **4**, 271 (1927).

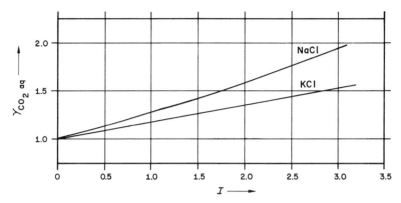

FIG. 2.18. Activity coefficients of CO_2 dissolved in NaCl and KCl solutions, at 25 °C. [Data from A. A. Markam and K. A. Kobe, *J. Am. Chem. Soc.*, 63, 449 (1941).]

Activity coefficient values for CO_2 dissolved in KCl or NaCl solutions at 25 °C are shown in Figure 2.18. The data for this plot are calculated from the solubility determinations of Markam and Kobe,[25] by means of equation (2.90). These authors give solubility data for CO_2 in solutions of KCl, NaCl, KNO_3, $NaNO_3$, $Mg(NO_3)$, $MgSO_4$, and Na_2SO_4, at temperatures of 0.2°, 25°, and 40 °C. They list similar data for nitrous oxide.

Values of the activity coefficient of the undissociated part of weak acids, such as acetic acid, in various electrolyte solutions, have also been determined.[26] This coefficient relates the activity of the undissociated, neutral, molecules to the concentration of those molecules only, i.e., $a_u = \gamma_u m_u$, where u = undissociated part. The coefficient γ_u is to be contrasted with the ordinary mean activity coefficient, $\gamma_\pm$, for which $a_2^{1/2} = \gamma_\pm m_2$, where $a_2^{1/2}$ and m_2 relate to the total solute.

A linear relationship that serves as a good first approximation for representing the experimental data is given by the empirical equation

$$\log \gamma_s = k_m I. \tag{2.92}$$

Here k_m is called the *salting coefficient* (on the molal scale), and I is the ionic strength of the solution. Such k_m values are listed by Randall and Failey[27] and by Harned and Owen[28] for various gases, for neutral organic molecules, and for the undissociated part of several weak organic acids, in a large variety of electrolyte solutions.

It would be satisfying to be able to assign accurate values to the activity coefficients of various neutral ion pairs in aqueous solution, such

---

[25] A. A. Markam and K. A. Kobe, *J. Am. Chem. Soc.*, 63, 449 (1941).
[26] Merle Randall and C. F. Failey, *op. cit.*, p. 291.
[27] Merle Randall and C. F. Failey, *op. cit.*, pp. 271, 285, 291.
[28] H. S. Harned and B. B. Owen, *op. cit.*, pp. 532–536, 736, 737.

as $MgSO_4^\circ{}_{aq}$, as a function of concentration. In the present state of the art of the treatment of ion pairs, however, this does not seem possible. Usually, such ion pairs are considered in dilute to only moderately concentrated solutions, and the activity coefficient of the ion pair is taken to be unity. In view of the various approximations made in calculating dissociation constants for ion pairs, this assumption probably does not greatly impair the accuracy of the final results; this subject will be discussed further in Chapter 4.

Dissociation Constant of Water

One of the most important reactions we must consider is the ionization of water. The equilibrium constant K_w for the reaction

$$H_2O_l = H_{aq}^+ + OH_{aq}^-, \qquad (2.93)$$

where

$$K_w = \frac{[H^+][OH^-]}{[H_2O]}, \qquad (2.94)$$

has been precisely measured by several different methods. Values for K_w for temperatures from 0° to 60 °C are listed in Table 2.8. These

TABLE 2.8. Ionization Constant of Water, K_w

t °C	$K_w \times 10^{14}$
0	0.1139
5	0.1846
10	0.2920
15	0.4505
20	0.6809
25	1.008
30	1.469
35	2.089
40	2.919
45	4.018
50	5.474
55	7.297
60	9.614

SOURCE: From H. S. Harned and B. B. Owen, op. cit., p. 638.

values were obtained from EMF measurements of cells without liquid junction.

It is noted that the activity of water appears in the denominator of equation (2.94). Often, this equation is written without $[H_2O]$ appearing at all, i.e., the implicit assumption is made that $[H_2O] = 1$. It should be clear from our previous discussion on the lowering of the activity of water by dissolved solutes that while this is a reasonable assumption for dilute solutions, it does not hold for more concentrated solutions. If, for example, we wished to calculate [OH$^-$] in a moderately concentrated solution, using equation (2.94), we would need accurate values of both [H$^+$] and [H$_2$O] for that solution.

SUMMARY—AQUEOUS SOLUTIONS OF ELECTROLYTES

In electrolyte solutions, interaction between solvent and solute is so important that dissolved substances are largely dissociated into charged particles. The nature of the interaction of the dissolved ions with each other and with the solvent changes in a complex way with concentration. For an electrolyte solution, Henry's law is found to be obeyed, provided that the dissociation of the solute is taken into account. This fact provides a ready basis for the assignment of a logical scheme of relationships among activities and molal activity coefficients, γ, and standard states for every kind of polyelectrolyte. In the dilute solution range, activity coefficients are given by the Debye-Hückel equation

$$-\log \gamma_i = \frac{A z_i^2 \sqrt{I}}{1 + \mathring{a}_i B \sqrt{I}}.$$

In more concentrated solutions the mean salt method is used. This method, based on assumptions concerning the complete dissociation of KCl and the equality of the activity coefficients of K$^+$ and Cl$^-$, and the concept of the ionic strength, is used to obtain individual ion activity coefficients of cations and anions from measurable mean activity coefficients. The growing recognition of the importance of neutral and charged ion pairs, even in moderately concentrated natural waters, while it has complicated calculations of the chemical relations in natural waters, has aided in our understanding of their behavior. Examples are given for the calculation of ionic strength, of individual ion activity coefficients for various ionic species, and for the effect of electrolytes on dissolved gases. The effect of dissolved electrolytes on the activity of solvent water is discussed.

SELECTED REFERENCES

Darken, L. S., and R. W. Gurry, *Physical Chemistry of Metals*. New York, McGraw-Hill, 1953.
    Especially useful for the very readable treatment of activity-concentration relations of nonelectrolyte systems, and treatment of the standard state problem.

Denbigh, K. G., *The Principles of Chemical Equilibrium*. New York, Cambridge, 1957.
> An excellent textbook; more advanced and more concise than that of Klotz listed below.

Harned, H. S., and B. B. Owen, *The Physical Chemistry of Electrolytic Solutions*, 3rd ed. (ACS Monograph 137). New York, Reinhold, 1958.
> Poorly organized, but encyclopedic coverage of subject matter.

Klotz, I. M., *Chemical Thermodynamics*. Englewood Cliffs, N.J., Prentice-Hall, 1950.
> Excellent elementary treatment of the thermodynamic basis of matters treated in chap. 2.

Latimer, W. M., *Oxidation Potentials*. New York, Prentice-Hall, 1952.
> Source of ΔF_f° and $\gamma_\pm$ data.

Lewis, G. N., and Merle Randall, *Thermodynamics*. New York, McGraw-Hill, 1923.
> This classic work, now out of print, is well worth study for the insight it provides into the ideas leading up to the modern treatment of practical free-energy calculations. See especially chaps. 25 to 28, incl.

Lewis, G. N., and Merle Randall, *Thermodynamics*, 2nd ed., revised by K. S. Pitzer and Leo Brewer. New York, McGraw-Hill, 1961.
> Lewis and Randall brought up to date, and enlarged.

Robinson, R. A., and R. H. Stokes, *Electrolyte Solutions*. New York, Academic, 1959.
> Comprehensive treatment of electrolyte solutions.

PROBLEMS

2.1. A 1000 ml container holds 0.05 mole of a gas at 25 °C. The pressure (measured) of the gas is 1.02 atmospheres.
   a. What is the calculated pressure if the gas obeys the perfect gas law?
   b. What is the activity coefficient (X) for this gas?
   Ans. (a) $P_{calc} = 1.22$ atm; (b) $X = 0.83$.

2.2. What is the activity coefficient of $CO_{2\,g}$ at 500 atmospheres pressure and a temperature of 500 °K? What is its activity? (Use Figure 2.3 and Table 2.1.) *Ans.* $X_{CO_2} = 0.78$; $a_{CO_2} = 390$.

2.3. What is the activity of $CO_{2\,g}$ in a mixture of 3 moles of CO_2 and 3 moles of O_2 at a total pressure of 700 atmospheres and a temperature of 300 °C?
Ans. $a_{CO_2} = 329$.

2.4. Calculate the fugacity and activity of chloroform in an acetone-chloroform solution where the mole fraction of chloroform is 0.25 (Figure 2.5).
Ans. $f_{CHCl_3} \cong 0.06$ atm; $a_{CHCl_3} \cong 0.16$.

2.5. Strontium is found substituting for calcium in aragonite. A particular chemical analysis gives the following results:

CaO	0.05 mole
SrO	0.005 mole
CO_2	0.055 mole

a. What is the mole fraction of $CaCO_3$? Of $SrCO_3$?
b. If when Sr substitutes for Ca it forms an ideal solution, what is the activity of $CaCO_{3\ arag}$?
c. Under the conditions of (b) what is the *ion product* of $CaCO_{3\ arag}$? (Compare with that of pure $CaCO_{3\ calcite}$.)

Ans. (a) $N_{CaCO_3} = 0.91$; $N_{SrCO_3} = 0.09$; (b) $a_{CaCO_3} = 0.91$; (c) ion product $= 10^{-8.20}$.

2.6. Calculate $\gamma_{Ca^{++}}$ at 25 °C in a solution $0.10m$ in $CaCl_2$, using the Debye-Hückel equation. *Ans.* $\gamma_{Ca^{++}} = 0.29$.

2.7. The mean activity coefficient of $CaCl_2$ in $0.10m$ $CaCl_2$ solution is 0.52. The mean activity coefficient of KCl in $0.30m$ KCl solution is 0.70. Calculate $\gamma_{Ca^{++}}$ using the mean salt method. *Ans.* $\gamma_{Ca^{++}} = 0.29$.

2.8. The chemical analysis of the major components of average surface sea water is given below (solutes in terms of mg/kg solution)

Cl	18,980	Ca	400
Na	10,561	K	380
Mg	1,272	Br	65
S	884	C	28

Density 1.024 gm/cm³
[Data from H. U. Sverdrup, M. W. Johnson, and R. H. Fleming, *The Oceans*. Englewood Cliffs, N.J., Prentice-Hall, 1942, Table 36, p. 176.]

a. Calculate the molalities of the species listed.
b. Calculate the ionic strength of sea water. S is present as SO_4^{--}, C as HCO_3^-. *Ans.* (b) $I = 0.7$.

2.9. At a constant CO_2 gas pressure, the solubility of CO_2 in water is reduced 10 percent by addition of a neutral salt. What is the activity coefficient of the dissolved CO_2 in the salt solution? *Ans.* $\gamma_{m\ CO_2} = 1.1$.

Carbonate Equilibria

INTRODUCTION

To many geologists, the carbonate rocks are the most important rock type in providing information about the geologic past. In addition to the abundant faunal and floral evidence they may contain, the fact that they are precipitates from their ancient environment means that if we are sufficiently astute we should be able to read from their minerals the conditions of their deposition and the events to which they have been subjected subsequently. A mere listing of the bibliography of investigations to this end would provide a larger volume than this one. Here the emphasis is entirely on the mechanics of handling several typical situations that may occur geologically to show the equilibria involved and the procedures used in obtaining the activities of metal ions in equilibrium with solid carbonates. Also, the question of the calculation of the actual solubility of the metal cation is discussed.

The solubility values for carbonate minerals in the literature are many, conflicting, and confusing. One of the reasons for this is the high degree of accuracy that is now expected from such work. To know the solubility of a mineral such as powellite, $CaMoO_4$, within an order of magnitude is most gratifying, but great labor has been expended to determine the difference in solubility of calcite and aragonite, in which the difference in free energy of formation of the dimorphs is somewhere between 0.2 and 0.5 kilocalorie. Another reason is one that will be discussed here, namely, that so many variables have to be specified to fix the solubility of a carbonate mineral that there is a tendency to compare values obtained under apparently similar but actually different conditions. In addition, mineral carbonates have highly variable compositions. The following discussion will treat the solubility of the pure end member calcite as a function of other variables, before considering the possible effects of compositional variation.

Five cases will be discussed. These five sets of conditions cover many of the situations of geological interest, and most others can be stipulated as combinations or permutations of the five. For convenience all calculations will be referred to calcite, with the recognition that the same procedures can be used for any other metal carbonate. The various sets of conditions are:

1. The reactions involved in placing pure calcite in pure water, with negligible gas phase present.
2. The reaction of calcite in pure water, but with the system open to CO_2; i.e., in contact with a reservoir, such as the atmosphere, of fixed partial pressure of CO_2.
3. Equilibrium relations in a system with a fixed quantity of dissolved carbonate species, but with pH arbitrarily fixed, i.e., controlled by other reactions in the system.
4. Equilibrium in a system connected to an external reservoir of fixed partial pressure of CO_2, but with pH arbitrarily determined.
5. Equilibrium resulting from addition of $CaCO_3$ to a system originally open to a CO_2 reservoir, but closed to that reservoir before addition of $CaCO_3$.

Case 1 can be regarded as that used for determining the solubility of calcite—it is the closed system $CaCO_3$-H_2O. It is, in fact, of little geologic utility, inasmuch as such a restricted composition is almost never encountered.

Case 2 is of considerable geological importance—it represents relations in lakes and streams and other dilute natural waters in intimate contact with the atmosphere, in which the pH of the system is controlled entirely by the carbonate equilibria.

Case 3 can be represented by the questions raised by the analysis of an underground water. The analysis commonly gives total dissolved carbonate, $HCO_3^- + CO_3^{2-} + H_2CO_3$, calcium concentration, and pH. Is such a solution in equilibrium with calcite? Is it in equilibrium with the atmosphere? If no analysis of calcium is given, but if the system is known to be in equilibrium with calcite, can calcium ion be calculated?

Case 4 is also of great geological importance. Given knowledge that a system is in equilibrium with a given partial pressure of CO_2 and that it has a given pH, whatever the control may be, what is the activity of calcium ion in equilibrium with calcite? What is it necessary to know to approximate the solubility of calcite in such a system?

Case 5 can be illustrated by the situation in which rain water in equilibrium with the atmosphere sinks into the ground, perhaps through an unreactive soil, and comes into contact with calcite. What is the activity and concentration of calcium ion in the resultant solution at equilibrium? What is the equilibrium pH of the water?

The calculations given in this chapter have direct application to waters of low ionic strength (≤ 0.1), such as those encountered in lakes, streams, and in potable underground waters, in which little complexing of ions occurs. For the effects of complexes in more concentrated solutions, see Chapter 4.

GENERAL PATTERN OF SOLUTION

In all five cases considered, the fundamental pattern of solution of the carbonate equilibrium problem is the same. There are at most seven variables involved at constant temperature and at constant total pressure, assuming knowledge of the activity coefficients for all species—they are P_{CO_2}, $a_{H_2CO_3}$, $a_{HCO_3^-}$, $a_{CO_3^{--}}$, a_{H^+}, a_{OH^-}, $a_{Ca^{++}}$. In every case the equilibrium constant for $CaCO_3$, the first and second dissociation constants for carbonic acid, and the equilibrium constant for water are applicable, inasmuch as equilibrium with $CaCO_3$ is always stipulated, and internal equilibrium of all ionic species in the solution assumed. Thus, the following five equations apply, where the numerical values of the equilibrium constants are for 25 °C:

$$\frac{[Ca^{++}][CO_3^{--}]}{[CaCO_{3\ c}]} = K_{CaCO_3} = 10^{-8.3} \tag{3.1}$$

$$\frac{[H^+][HCO_3^-]}{[H_2CO_3]} = K_{H_2CO_3} = 10^{-6.4} \tag{3.2}[1]$$

$$\frac{[H^+][CO_3^{--}]}{[HCO_3^-]} = K_{HCO_3^-} = 10^{-10.3} \tag{3.3}$$

$$\frac{[H^+][OH^-]}{[H_2O]} = K_{H_2O} = 10^{-14.0} \tag{3.4}$$

$$\frac{[H_2CO_3]}{P_{CO_2\ g}} = K_{CO_2} = 10^{-1.47}. \tag{3.5}$$

[1] Aqueous solutions of carbon dioxide contain CO_2 molecules as well as carbonic acid molecules, H_2CO_3. In fact, by rate studies, the equilibria involving the two species have been investigated and equilibrium constants determined from which the ratio $[CO_2]/[H_2CO_3]$ is shown to be approximately 386. Regardless of assumptions made about the activity coefficients appropriate to the two different species, CO_2 and H_2CO_3, the ratio of CO_2 molecules to H_2CO_3 molecules is considerably greater than unity in a solution of carbon dioxide dissolved in pure water. Despite these findings it has been customary to represent all aqueous carbon dioxide as H_2CO_3, and to use equation (3.2) for the first dissociation constant; we shall follow this practice here. No loss of generality results, since the hydration state of a dissolved species does not need to be stated for thermodynamic purposes. These matters are discussed in detail by Kern. [D. B. Kern, The hydration of carbon dioxide: *J. Chem. Ed.*, *37*, 14 (1960).]

In the above equations all of the species are to be taken as dissolved in water, except those designated otherwise, i.e., $CaCO_{3\,c}$ and $CO_{2\,g}$. Accurate values for the equilibrium constants are listed in Table 3.2 for temperatures from 0° to 80 °C, except for K_{H_2O}, which is discussed in Chapter 2. Because of the various approximations made in solving the equations in the following cases, the degree of precision given in the numerical values of the equilibrium constants for equations (3.1) to (3.5) is sufficient.

Equations (3.1) to (3.5) provide, then, the backbone of each situation described under the five cases listed. The necessary additional equations required are provided by the special circumstances chosen.

Case I. $CaCO_3$ in Pure Water

When an excess of pure $CaCO_3$ is placed in pure water, i.e., water from which reactive gases such as CO_2 have been swept out by bubbling through an inert gas such as nitrogen or helium, the ensuing process can be visualized. The calcium and carbonate ions of the calcite spread through the aqueous solution. Some hydration of the Ca^{++} takes place, which is of no importance to us here; on the other hand, carbonate ions react appreciably with water, in a succession of reactions

$$CO_{3\,aq}^{--} + H_2O_l = HCO_{3\,aq}^{-} + OH_{aq}^{-} \qquad (3.6a)$$

$$HCO_{3\,aq}^{-} + H_2O_l = H_2CO_{3\,aq} + OH_{aq}^{-}. \qquad (3.6b)$$

The solution becomes alkaline because of the release of OH^-. A third reaction of general importance is

$$H_2CO_{3\,aq} = CO_{2\,g} + H_2O_l. \qquad (3.7)$$

However, for Case 1 we specify that the gas phase present shall be negligible[2]; hence, equation (3.7) does not apply to this case.

Therefore, in this system the species of importance are Ca_{aq}^{++}, $CO_{3\,aq}^{--}$, $HCO_{3\,aq}^{-}$, $H_2CO_{3\,aq}$, OH_{aq}^{-}, H_{aq}^{+}, $CaCO_{3\,c}$, and H_2O_l. The activity of $CaCO_{3\,c}$ is unity, and that of H_2O_l also can be considered unity, inasmuch as the solution is a dilute one. This leaves six variables. Equations (3.1) to (3.4) all hold for the system, so two more equations must be found, to obtain solutions for the six variables.

Because the original water is devoid of carbonate species, it follows that every species in solution containing a carbon atom must be matched by a calcium ion, for the source of such species is calcite, which cannot dissociate to free calcium in excess of carbons, or vice versa, without building up an electric charge. Then we can write

$$m_{Ca^{++}} = m_{CO_3^{--}} + m_{HCO_3^{-}} + m_{H_2CO_3}. \qquad (3.8)$$

---

[2] Experimentally, this corresponds to a saturated $CaCO_3$ solution in contact with excess calcite and completely filling the container.

Furthermore, the solution must remain electrically neutral, so that the sum of the positive charges on cations must equal that of the negative charges on anions. Remembering that each calcium and each carbonate ion is doubly charged, we can write an equation for electric neutrality

$$2m_{Ca^{++}} + m_{H^+} = 2m_{CO_3^-} + m_{HCO_3^-} + m_{OH^-}. \quad (3.9)$$

Equations (3.8) and (3.9), in conjunction with equations (3.1) to (3.4), are then sufficient for solution of the problem, for there are six equations and six unknowns, with one stipulation: *because equations (3.1) to (3.4) are written in terms of activities, and (3.8) and (3.9) in terms of molalities, a solution of the problem is valid only if $m_i \cong a_i$, i.e., if $\gamma_i = 1$.*

Many solutions of the problem in the literature ignore this qualification, so before proceeding it is well to look into the basis for making the assumption originally and also to see the procedure required for a rigorous solution. It appears at first as if the dilemma is hopeless, for γ_i values cannot be obtained until the ionic strength is determined, and the ionic strength cannot be determined until the problem is solved. This difficulty can be circumvented by a series of successive approximations, in which the problem is first solved, assuming activity coefficients are unity and determining an ionic strength. Then activity coefficients are determined, and the problem is solved again to obtain a new ionic strength, until successive approximations show no appreciable differences.

Consequently, the next step is to solve the problem, assuming that $m_i = a_i$ and it is not possible to be more rigorous at the outset. Because back-substitution into six equations is a chore, it is also convenient in a first approximation to see if there are other methods of simplification. Equations (3.6a, b) show that the solution will be more alkaline than the original water, so that the final pH will be above 7. If it is 8, for example, then from equation (3.2),

$$\frac{[H^+][HCO_3^-]}{[H_2CO_3]} = 10^{-6.4}$$

$$\frac{[HCO_3^-]}{[H_2CO_3]} = \frac{10^{-6.4}}{10^{-8}} = 10^{1.6}.$$

Thus, $[H_2CO_3]$ is small relative to $[HCO_3^-]$ and can be neglected in a first approximation. Also, if pH $\geq$ 8, $[H^+] \leq 10^{-8}$, and can be temporarily ignored. This leaves

$$[Ca^{++}][CO_3^{--}] = 10^{-8.3} \quad (3.1a)$$

$$\frac{[H^+][HCO_3^-]}{[H_2CO_3]} = 10^{-6.4} \quad (3.2)$$

CARBONATE EQUILIBRIA

$$\frac{[H^+][CO_3^{--}]}{[HCO_3^-]} = 10^{-10.3} \tag{3.3}$$

$$[H^+][OH^-] = 10^{-14.0} \tag{3.4a}$$

$$[Ca^{++}] = [CO_3^{--}] + [HCO_3^-] \quad \text{(3.8, modified)}$$

$$2[Ca^{++}] = 2[CO_3^{--}] + [HCO_3^-] + [OH^-]. \tag{3.9, modified}$$

Multiplying (3.8$_{\text{mod.}}$) by 2 and subtracting from (3.9$_{\text{mod.}}$)

$$[HCO_3^-] = [OH^-]. \tag{3.10}$$

The easiest path of substitution perhaps is to express all equilibria in terms of $[H^+]$ and substitute into (3.8$_{\text{mod.}}$). From (3.4a)

$$[H^+][OH^-] = 10^{-14}$$

$$[OH^-] = \frac{10^{-14}}{[H^+]}.$$

Substituting from (3.10)

$$[HCO_3^-] = \frac{10^{-14}}{[H^+]}. \tag{3.11}$$

Substituting (3.11) into (3.3)

$$[CO_3^{--}] = \frac{10^{-24.3}}{[H^+]^2}. \tag{3.12}$$

Substituting (3.12) in (3.1a)

$$[Ca^{++}] = 10^{16}[H^+]^2. \tag{3.13}$$

Finally, by using (3.11), (3.12), and (3.13) in (3.8$_{\text{mod.}}$)

$$10^{16}[H^+]^2 = \frac{10^{-24.3}}{[H^+]^2} + \frac{10^{-14}}{[H^+]}. \tag{3.14}$$

Multiplying through by $[H^+]^2$ and rearranging

$$10^{16}[H^+]^4 - 10^{-14}[H^+] = 10^{-24.3}. \tag{3.15}$$

The easiest method of solution is by trial and error, and yields $[H^+] = 10^{-9.9_5}$.

Back-substitution into (3.13), (3.12), (3.11), and (3.10) yields

$$[Ca^{++}] = 10^{-3.9}$$
$$[CO_3^{--}] = 10^{-4.4}$$
$$[HCO_3^-] = 10^{-4.05}$$
$$[OH^-] = 10^{-4.05}.$$

It is clear that ignoring [H$^+$] in the electrical balance equation (3.9) has not led to appreciable error, and that ignoring [H$_2$CO$_3$] in equation (3.8) is equally justified ([H$_2$CO$_3$] from (3.2) = $10^{-7.6}$).

Now consider the error involved in assuming $\gamma_i = 1$. The ionic strength of the solution is[3]

$$I = \tfrac{1}{2}([\text{Ca}^{++}] \times 2^2 + [\text{H}^+] \times 1^2 + [\text{CO}_3^{--}] \times 2^2 + [\text{HCO}_3^-] \times 1^2 + [\text{OH}^-] \times 1^2)$$

$$I = \tfrac{1}{2}(5.0 \times 10^{-4} + 1.1 \times 10^{-10} + 1.6 \times 10^{-4} + 8.9 \times 10^{-5} + 8.9 \times 10^{-5})$$

$$I = 4.2 \times 10^{-4}.$$

This ionic strength is clearly in the range where the Debye-Hückel equation (2.76) can be used. For $\gamma_{\text{Ca}^{++}}$, with values from Tables 2.6 and 2.7,

$$-\log \gamma_{\text{Ca}^{++}} = \frac{0.5085 \times 4\sqrt{I}}{1 + 0.3281 \times 6\sqrt{I}} = +0.040$$

$$\gamma_{\text{Ca}^{++}} = 0.91.$$

Similarly, $\gamma_{\text{CO}_3^{--}} = 0.91$, $\gamma_{\text{HCO}_3^-} = 0.98$, $\gamma_{\text{OH}^-} = 0.98$, $\gamma_{\text{H}^+} = 0.98$, $\gamma_{\text{H}_2\text{CO}_3} = 1.0$ (molecular species). These coefficients are sufficiently close to unity so that the error resulting from neglecting them is small. But for the sake of completeness in this first case, the problem will be solved by using them.

Rewriting all equations in terms of molalities and activity coefficients:

$$m_{\text{Ca}^{++}}\gamma_{\text{Ca}^{++}} m_{\text{CO}_3^{--}}\gamma_{\text{CO}_3^{--}} = 10^{-8.3} \qquad (3.1\text{a})$$

$$m_{\text{Ca}^{++}} m_{\text{CO}_3^{--}} = \frac{10^{-8.3}}{\gamma_{\text{Ca}^{++}}\gamma_{\text{CO}_3^{--}}} = 10^{-8.3} \times 10^{0.1} = 10^{-8.2*}$$

$$\frac{m_{\text{H}^+}\gamma_{\text{H}^+} m_{\text{HCO}_3^-}\gamma_{\text{HCO}_3^-}}{m_{\text{H}_2\text{CO}_3}\gamma_{\text{H}_2\text{CO}_3}} = 10^{-6.4} \qquad (3.2)$$

$$\frac{m_{\text{H}^+} m_{\text{HCO}_3^-}}{m_{\text{H}_2\text{CO}_3}} = \frac{10^{-6.4}\gamma_{\text{H}_2\text{CO}_3}}{\gamma_{\text{H}^+}\gamma_{\text{HCO}_3^-}} = 10^{-6.4} \times 10^{0.02} = 10^{-6.4}$$

$$\frac{m_{\text{H}^+}\gamma_{\text{H}^+} m_{\text{CO}_3^{--}}\gamma_{\text{CO}_3^{--}}}{m_{\text{HCO}_3^-}\gamma_{\text{HCO}_3^-}} = 10^{-10.3} \qquad (3.3)$$

[3] As explained above, the ionic strength must of necessity be calculated using activities, rather than molalities, at this stage. After finding the values of the γ_i, as shown in the ensuing calculations, tentative values of m_i can be calculated, a new value of I found by using these m_i, and the whole process repeated. However, at the low value of the ionic strength involved, the γ_i do not change significantly, so that this sort of calculation is not worth while.

CARBONATE EQUILIBRIA

$$\frac{[H^+][CO_3^{--}]}{[HCO_3^-]} = 10^{-10.3} \tag{3.3}$$

$$[H^+][OH^-] = 10^{-14.0} \tag{3.4a}$$

$$[Ca^{++}] = [CO_3^{--}] + [HCO_3^-] \tag{3.8, modified}$$

$$2[Ca^{++}] = 2[CO_3^{--}] + [HCO_3^-] + [OH^-]. \tag{3.9, modified}$$

Multiplying (3.8$_{mod.}$) by 2 and subtracting from (3.9$_{mod.}$)

$$[HCO_3^-] = [OH^-]. \tag{3.10}$$

The easiest path of substitution perhaps is to express all equilibria in terms of $[H^+]$ and substitute into (3.8$_{mod.}$). From (3.4a)

$$[H^+][OH^-] = 10^{-14}$$

$$[OH^-] = \frac{10^{-14}}{[H^+]}.$$

Substituting from (3.10)

$$[HCO_3^-] = \frac{10^{-14}}{[H^+]}. \tag{3.11}$$

Substituting (3.11) into (3.3)

$$[CO_3^{--}] = \frac{10^{-24.3}}{[H^+]^2}. \tag{3.12}$$

Substituting (3.12) in (3.1a)

$$[Ca^{++}] = 10^{16}[H^+]^2. \tag{3.13}$$

Finally, by using (3.11), (3.12), and (3.13) in (3.8$_{mod.}$)

$$10^{16}[H^+]^2 = \frac{10^{-24.3}}{[H^+]^2} + \frac{10^{-14}}{[H^+]}. \tag{3.14}$$

Multiplying through by $[H^+]^2$ and rearranging

$$10^{16}[H^+]^4 - 10^{-14}[H^+] = 10^{-24.3}. \tag{3.15}$$

The easiest method of solution is by trial and error, and yields $[H^+] = 10^{-9.9_5}$.

Back-substitution into (3.13), (3.12), (3.11), and (3.10) yields

$$[Ca^{++}] = 10^{-3.9}$$
$$[CO_3^{--}] = 10^{-4.4}$$
$$[HCO_3^-] = 10^{-4.05}$$
$$[OH^-] = 10^{-4.05}.$$

It is clear that ignoring $[H^+]$ in the electrical balance equation (3.9) has not led to appreciable error, and that ignoring $[H_2CO_3]$ in equation (3.8) is equally justified ($[H_2CO_3]$ from (3.2) = $10^{-7.6}$).

Now consider the error involved in assuming $\gamma_i = 1$. The ionic strength of the solution is[3]

$$I = \tfrac{1}{2}([Ca^{++}] \times 2^2 + [H^+] \times 1^2 + [CO_3^{--}] \times 2^2 + [HCO_3^-] \times 1^2 + [OH^-] \times 1^2)$$

$$I = \tfrac{1}{2}(5.0 \times 10^{-4} + 1.1 \times 10^{-10} + 1.6 \times 10^{-4} + 8.9 \times 10^{-5} + 8.9 \times 10^{-5})$$

$$I = 4.2 \times 10^{-4}.$$

This ionic strength is clearly in the range where the Debye-Hückel equation (2.76) can be used. For $\gamma_{Ca^{++}}$, with values from Tables 2.6 and 2.7,

$$-\log \gamma_{Ca^{++}} = \frac{0.5085 \times 4\sqrt{I}}{1 + 0.3281 \times 6\sqrt{I}} = +0.040$$

$$\gamma_{Ca^{++}} = 0.91.$$

Similarly, $\gamma_{CO_3^{--}} = 0.91$, $\gamma_{HCO_3^-} = 0.98$, $\gamma_{OH^-} = 0.98$, $\gamma_{H^+} = 0.98$, $\gamma_{H_2CO_3} = 1.0$ (molecular species). These coefficients are sufficiently close to unity so that the error resulting from neglecting them is small. But for the sake of completeness in this first case, the problem will be solved by using them.

Rewriting all equations in terms of molalities and activity coefficients:

$$m_{Ca^{++}}\gamma_{Ca^{++}}m_{CO_3^{--}}\gamma_{CO_3^{--}} = 10^{-8.3} \tag{3.1a}$$

$$m_{Ca^{++}}m_{CO_3^{--}} = \frac{10^{-8.3}}{\gamma_{Ca^{++}}\gamma_{CO_3^{--}}} = 10^{-8.3} \times 10^{0.1} = 10^{-8.2*}$$

$$\frac{m_{H^+}\gamma_{H^+}m_{HCO_3^-}\gamma_{HCO_3^-}}{m_{H_2CO_3}\gamma_{H_2CO_3}} = 10^{-6.4} \tag{3.2}$$

$$\frac{m_{H^+}m_{HCO_3^-}}{m_{H_2CO_3}} = \frac{10^{-6.4}\gamma_{H_2CO_3}}{\gamma_{H^+}\gamma_{HCO_3^-}} = 10^{-6.4} \times 10^{0.02} = 10^{-6.4}$$

$$\frac{m_{H^+}\gamma_{H^+}m_{CO_3^{--}}\gamma_{CO_3^{--}}}{m_{HCO_3^-}\gamma_{HCO_3^-}} = 10^{-10.3} \tag{3.3}$$

---

[3] As explained above, the ionic strength must of necessity be calculated using activities, rather than molalities, at this stage. After finding the values of the γ_i, as shown in the ensuing calculations, tentative values of m_i can be calculated, a new value of I found by using these m_i, and the whole process repeated. However, at the low value of the ionic strength involved, the γ_i do not change significantly, so that this sort of calculation is not worth while.

$$\frac{m_{H^+} m_{CO_3^-}}{m_{HCO_3^-}} = \frac{10^{-10.3} \gamma_{HCO_3^-}}{\gamma_{H^+} \gamma_{CO_3^-}} = 10^{-10.3} \times 10^{0.04} = 10^{-10.3}$$

$$m_{H^+} \gamma_{H^+} m_{OH^-} \gamma_{OH^-} = 10^{-14.0} \tag{3.4a}$$

$$m_{H^+} m_{OH^-} = \frac{10^{-14}}{\gamma_{H^+} \gamma_{OH^-}} = 10^{-14.0} \times 10^{0.02} = 10^{-14.0}$$

$$m_{Ca^{++}} = m_{HCO_3^-} + m_{CO_3^-} \tag{3.8a}$$

$$2m_{Ca^{++}} = 2m_{CO_3^-} + m_{HCO_3^-} + m_{OH^-} \tag{3.9a}$$

Therefore, the only significant change, within the precision of the calculation, is in the constant of equation (3.1a), marked with an asterisk. As a result, substitution in the various equilibria would yield only one change, and the equation parallel to (3.15) would be

$$10^{16.1}[H^+]^4 - 10^{-14}[H^+] = 10^{-24.3}.$$

This change makes the equilibrium pH slightly lower—close to 9.9—but the whole operation is essentially within the limits of the errors involved. On the other hand, the procedure used contains all the steps necessary to correct calculations for other carbonates if the activity coefficients are sufficiently small to require consideration.

In summary of Case 1, it can be said that the equilibrium pH lies between 9.9 and 10, that $m_{Ca^{++}} \cong 10^{-3.9}$; $m_{CO_3^-} \cong 10^{-4.4}$; $m_{HCO_3^-} \cong 10^{-4.05}$; $m_{H_2CO_3} \cong 10^{-7.6}$. These theoretical values check well against experiment; in three runs of the equilibrium pH of powdered calcite in distilled, deaerated water, pH values of 9.88, 9.92, and 9.96 were obtained.[4] As previously indicated, this system has little geological interest because it is too simple. This point is illustrated by the "defining equations"; few natural systems can be assumed to have carbonate species derived only from a single solid carbonate, nor to have their ionic relations described only in terms of calcium, hydrogen, carbonate, bicarbonate, and hydroxyl ion activities.

Case 2. Calcium Carbonate-Water, With Externally Fixed Pressure of CO_2

The second case, in which calcite is in equilibrium with water and with a given partial pressure of CO_2, can be derived from the first one experimentally by opening the container of calcite and water and passing air or other gas with fixed CO_2 content through the system continuously. Equations (3.1) through (3.5) apply for this case.

However, for this case equation (3.8) is not valid, because carbonate species are derived from the external CO_2 source as well as from the

[4] Experiments by R. M. Garrels and R. Siever.

calcite. Equation (3.9), expressing electrical neutrality, does hold, inasmuch as the ionic species present are still only Ca^{++}, H^+, CO_3^{--}, HCO_3^-, and OH^-. Then

$$2m_{Ca^{++}} + m_{H^+} = 2m_{CO_3^-} + m_{HCO_3^-} + m_{OH^-}. \tag{3.9}$$

As a substitute for equation (3.8), the conditions specified—constant pressure of CO_2—can be designated.

$$P_{CO_2} = K. \tag{3.16}$$

From equation (3.5)

$$[H_2CO_3] = 10^{-1.5} P_{CO_2}. \tag{3.17}$$

Then, at a specified P_{CO_2}, equations (3.1) to (3.4), (3.9), and (3.17) describe the system. As a specific example, let us solve for equilibrium with the atmosphere in which $P_{CO_2} \simeq 10^{-3.5}$. From equation (3.17)

$$[H_2CO_3] = 10^{-1.5} \times 10^{-3.5} = 10^{-5.0}. \tag{3.18}$$

Again substituting so as to express equation (3.9) in terms of $[H^+]$, and assuming as a first approximation that $\gamma_i \simeq 1$ and $a_i \simeq m_i$, equation (3.2) yields

$$[H^+][HCO_3^-] = 10^{-6.4} \times 10^{-5.0} = 10^{-11.4}$$

$$[HCO_3^-] = \frac{10^{-11.4}}{[H^+]}. \tag{3.19}$$

From (3.3)

$$[CO_3^{--}] = \frac{10^{-10.3}[HCO_3^-]}{[H^+]} = \frac{10^{-21.7}}{[H^+]^2}. \tag{3.20}$$

From (3.4a)

$$[OH^-] = \frac{10^{-14.0}}{[H^+]}. \tag{3.21}$$

From (3.1a)

$$[Ca^{++}] = \frac{10^{-8.3}}{[CO_3^{--}]} = 10^{13.4}[H^+]^2. \tag{3.22}$$

Substituting these values [equations (3.19) to (3.22) into equation (3.9)]

$$2 \times 10^{13.4}[H^+]^2 + [H^+] = 2 \times \frac{10^{-21.7}}{[H^+]^2} + \frac{10^{-11.4}}{[H^+]} + \frac{10^{-14.0}}{[H^+]}.$$

Multiplying through by $[H^+]^2$ and putting all numbers in powers of 10,

$$10^{13.7}[H^+]^4 + [H^+]^3 = 10^{-21.4} + 10^{-11.4}[H^+] + 10^{-14}[H^+].$$

Collecting terms

$$10^{13.7}[H^+]^4 + [H^+]^3 - 10^{-11.4}[H^+] = 10^{-21.4}. \tag{3.23}$$

CARBONATE EQUILIBRIA

Solving by trial and error

$$[H^+] = 10^{-8.4}.$$

Then $[Ca^{++}] = 10^{-3.4}$; $[CO_3^{--}] = 10^{-4.9}$; $[HCO_3^-] = 10^{-3.0}$; $[OH^-] = 10^{-5.6}$; $[H_2CO_3] = 10^{-5.0}$.

Therefore, the pH of a system containing $CaCO_3$ in water in equilibrium with the atmosphere is 8.4. The ionic strength is still so low that correction for the difference of molality and activity is hardly worthwhile. The molality of Ca^{++} is about $10^{-3.4}$, which is 3.2 times as great as in Case 1 in the absence of the atmosphere. Note the marked lowering of the pH because of the influence of atmospheric CO_2, showing that large errors in pH may result from permitting water samples out of contact with the atmosphere in their native state to come in contact with it before pH measurement. Experiments by R. M. Garrels and R. Siever again show excellent correspondence between the calculated pH and observed pH for water in equilibrium with calcite and the atmosphere.

Case 3. Fixed Total Dissolved Carbonate Species, pH Arbitrarily Selected

Case 3 is probably that most commonly encountered by geologists and geochemists. Chemical analysis of a natural water is available, so that pH, total calcium, and total carbonate species, as well as ionic strength, are known. Is the water in equilibrium with calcium carbonate, or is it oversaturated or undersaturated? The problem can be solved approximately, but in the specific instance of sea water, it has led to a long-standing and sometimes heated controversy. The sea water problem is treated in Chapter 4.

The procedure in Case 3 is to set up the appropriate equations in terms of molalities and activity coefficients:

$$m_{Ca^{++}} m_{CO_3^{--}} = \frac{10^{-8.3}}{\gamma_{Ca^{++}} \gamma_{CO_3^{--}}} \quad (3.1a)$$

$$\frac{a_{H^+} m_{HCO_3^-}}{m_{H_2CO_3}} = \frac{10^{-6.4}}{\gamma_{HCO_3^-}} \quad (3.2a)[5]$$

$$\frac{a_{H^+} m_{CO_3^{--}}}{m_{HCO_3^-}} = \frac{10^{-10.3} \gamma_{HCO_3^-}}{\gamma_{CO_3^{--}}} \quad (3.3)$$

$$a_{H^+} m_{OH^-} = \frac{10^{-14.0}}{\gamma_{OH^-}}. \quad (3.4a)$$

[5] Because pH is given, a_{H^+} is known, and m_{H^+} is almost never required in defining equations. Also $\gamma_{H_2CO_3}$ is so close to unity in natural solutions that it does not have to be considered, except for scrupulous work.

Total carbonate is given by analysis

$$m_{H_2CO_3} + m_{HCO_3^-} + m_{CO_3^{--}} = k_{analysis} \qquad (3.24)$$

and pH is also given in the analysis

$$a_{H^+} = k_{H^+}. \qquad (3.25)$$

Then the system is defined by equations (3.1) to (3.4), and (3.24) and (3.25), plus knowledge of ionic strength from the analysis, which permits calculation of approximate values of all the activity coefficients.

As an example of a specific problem of this type, let us select an analysis of a natural water at random, and attempt to determine whether or not it is in equilibrium with calcite. The analysis selected (Table 3.1) is from water from Lake Earl, California.[6]

TABLE 3.1. Analysis of Water from Lake Earl (ppm)

pH	Ca	Mg	Na	K	HCO$_3$	CO$_3$	SO$_4$	Cl	F	NO$_3$	B
7.5	79	261	2110	4.0	84	0	508	3790	0.1	1.2	1.1

Recalculation of these values, ignoring F, NO$_3$, and B, shows that the ionic strength is 0.140; $m_{Ca^{++}}$ is 0.002 ($10^{-2.7}$); and $m_{HCO_3^-}$ is 0.0014 ($10^{-2.85}$). In analyses of the type made neither CO_3^{--} nor H_2CO_3 can be determined; as will be seen the calculated values are below analytical detection. Activity coefficients for the individual ions[7] are $\gamma_{Ca^{++}} = 0.35$; $\gamma_{HCO_3^-} = 0.77$; $\gamma_{CO_3^{--}} = 0.35$.

Substituting a_{H^+}, $m_{HCO_3^-}$, $\gamma_{HCO_3^-}$, and $\gamma_{CO_3^{--}}$ in equation (3.3)

$$\frac{10^{-7.5} m_{CO_3^{--}}}{10^{-2.85}} = \frac{10^{-10.3} 10^{-0.12}}{10^{-0.46}}$$

$$m_{CO_3^{--}} = 10^{-5.3}.$$

Using this value and $\gamma_{Ca^{++}}$ and $\gamma_{CO_3^{--}}$ in equation (3.1a)

$$m_{Ca^{++}} 10^{-5.3} = \frac{10^{-8.3}}{10^{-0.46} 10^{-0.46}}$$

$$m_{Ca^{++}} = 10^{-2.1}.$$

Thus, calculated calcium is $10^{-2.1}$, and analytical calcium is $10^{-2.7}$, so the water is somewhat undersaturated with calcium carbonate, according to these calculations.

[6] William Back, Geology and ground water features of the Smith River Plain, Del Norte County, California: *U.S. Geol. Surv. Water-Supply Paper* 1254 (1957), p. 70.

[7] The value of $\gamma_{Ca^{++}}$ is taken from Figure 2.15; values of $\gamma_{HCO_3^-}$ and $\gamma_{CO_3^{--}}$ are from curves A and D, respectively, of Figure 4.5.

CARBONATE EQUILIBRIA

The calculations can be continued to show that $m_{HCO_3^-}$ given by analysis is much higher than that expected at equilibrium with the atmosphere at the listed pH of 7.5. The calculated value of $m_{HCO_3^-}$ is $10^{-3.8}$, as opposed to the analyzed value of $10^{-2.9}$. If the bicarbonate present is a measure of the average conditions, the pH of the water, if in equilibrium with the atmosphere, should be higher than 8. If so, the water would be approximately saturated with calcium carbonate at this higher pH. Then we might speculate that the pH was measured at a low point in the cycle of diurnal variation.

Case 4. System in Equilibrium With Calcite at a Given P_{CO_2} and at an Arbitrary pH

The greatest use of Case 4 is probably in calculation to determine the activity and molality of calcium ion in equilibrium with calcite under hypothetical conditions. For example, what are the activity and molality of calcium ion in a solution which has an ionic strength of 0.10, and has reached a pH of 10 because of the presence of volcanic glass, yet is open to the atmosphere and contains solid calcium carbonate?

Equations (3.1a), (3.2a), (3.3), and (3.4a) are applicable, and the defining equations are based on the constancy of P_{CO_2} and knowledge of pH. These necessary relations are

$$m_{Ca^{++}}\gamma_{Ca^{++}}m_{CO_3^{--}}\gamma_{CO_3^{--}} = 10^{-8.3} \quad (3.1a)$$

$$\frac{a_{H^+} m_{HCO_3^-}\gamma_{HCO_3^-}}{m_{H_2CO_3}} = 10^{-6.4} \quad (3.2a)$$

$$\frac{a_{H^+} m_{CO_3^{--}}\gamma_{CO_3^{--}}}{m_{HCO_3^-}\gamma_{HCO_3^-}} = 10^{-10.3} \quad (3.3)$$

$$a_{H^+} m_{OH^-}\gamma_{OH^-} = 10^{-14.0} \quad (3.4a)$$

$$a_{H^+} = k_{H^+} \quad (3.25)$$

$$m_{H_2CO_3} = 10^{-1.5} P_{CO_2}; \quad \gamma_{H_2CO_3} = 1. \quad (3.17a)$$

First, activity coefficients for the individual ions must be determined. For an ionic strength of 0.1, either the Debye-Hückel method or the mean salt method can be used (Chapter 2).

Using the Debye-Hückel equation (2.76) with appropriate values for A and B and for $å_i$ from Tables 2.6 and 2.7, the following values are obtained: $\gamma_{Ca^{++}} = 0.40$; $\gamma_{CO_3^{--}} = 0.36$; $\gamma_{HCO_3^-} = 0.78$; $\gamma_{OH^-} = 0.76$. As before, $\gamma_{H_2CO_3}$ can be considered to be unity, without serious error. Substitution of these γ values in the following equations and the conditions that pH = 10 and that the system is in equilibrium with the atmosphere ($P_{CO_2} = 10^{-3.5}$) in equations (3.25) and (3.17a), respectively, then provide relations all in terms of molalities, with the exception of a_{H^+}, which is known directly by measurement of the solution.

$$m_{Ca^{++}} m_{CO_3^{--}} = \frac{10^{-8.3}}{\gamma_{Ca^{++}} \gamma_{CO_3^{--}}} = 10^{-7.5} \quad (3.1a)$$

$$\frac{a_{H^+} m_{HCO_3^-}}{m_{H_2CO_3}} = \frac{10^{-6.4}}{\gamma_{HCO_3^-}} = 10^{-6.3} \quad (3.2a)$$

$$\frac{a_{H^+} m_{CO_3^{--}}}{m_{HCO_3^-}} = \frac{10^{-10.3} \gamma_{HCO_3^-}}{\gamma_{CO_3^{--}}} = 10^{-10.0} \quad (3.3)$$

$$a_{H^+} m_{OH^-} = \frac{10^{-14.0}}{\gamma_{OH^-}} = 10^{-13.9} \quad (3.4a)$$

$$a_{H^+} = 10^{-10.0} \quad (3.25)$$

$$m_{H_2CO_3} = 10^{-1.5} P_{CO_2} = 10^{-5.0}. \quad (3.17a)$$

The solution is then easy. Putting $m_{H_2CO_3} = 10^{-5.0}$ and $a_{H^+} = 10^{-10}$ from equations (3.17a) and (3.25) into (3.2a)

$$\frac{10^{-10} m_{HCO_3^-}}{10^{-5}} = 10^{-6.3}; \quad m_{HCO_3^-} = 10^{-1.3}.$$

Substituting this value into (3.3)

$$\frac{10^{-10} m_{CO_3^{--}}}{10^{-1.3}} = 10^{-10.0}; \quad m_{CO_3^{--}} = 10^{-1.3}.$$

Then, from (3.1a),

$$10^{-1.3} m_{Ca^{++}} = 10^{-7.5}; \quad m_{Ca^{++}} = 10^{-6.2}.$$

Finally,

$$\gamma_{Ca^{++}} m_{Ca^{++}} = a_{Ca^{++}}$$

$$10^{-0.4} 10^{-6.2} = 10^{-6.6} = a_{Ca^{++}}.$$

In the particular ground water considered, the molality of calcium ion is thus calculated to be vanishingly low—a few hundredths of a part per million. The order of magnitude is certainly trustworthy for $m_{Ca^{++}}$. If actual analysis of the water yields high values for dissolved calcium, additional ionic or molecular species containing calcium must be present.

Case 5. Equilibrium in a System of Water Originally Open to Atmospheric CO_2, Then Closed Before Addition of $CaCO_3$

Case 5 is actually a combination of Cases 1 and 2, but it has peculiar geologic interest because it represents the situation where rainwater,

CARBONATE EQUILIBRIA

which is roughly equivalent to pure water in equilibrium with atmospheric CO_2, descends through a nonreactive soil or rock layer and then comes in contact with calcite-bearing material. How much carbonate is leached per unit of descending solution? The question is a fundamental one in the attempt to determine the leaching action of rainwater quantitatively.

The basic equilibria are as before, but with a slight difference in the defining equations. Considering the rainwater first, only the relations for dissolved carbonate species, plus knowledge of atmospheric P_{CO_2} and an electric neutrality equation are necessary:

$$\frac{[H^+][HCO_3^-]}{[H_2CO_3]} = 10^{-6.4} \tag{3.2}$$

$$\frac{[H^+][CO_3^{--}]}{[HCO_3^-]} = 10^{-10.3} \tag{3.3}$$

$$[H_2CO_3] = 10^{-1.5} P_{CO_2} \tag{3.17}$$

$$P_{CO_2} = 10^{-3.5} \tag{3.26}$$

$$[H^+] = [HCO_3^-] + 2[CO_3^{--}] \tag{3.27}$$

(ignoring differences between m_i and a_i in (3.27) and subsequent electroneutrality equations).

From (3.26) and (3.17), $[H_2CO_3] = 10^{-5.0}$. Also, inasmuch as the reaction of CO_2 with water alone results in production of H^+, the concentration of carbonate ion must be negligible ($K_{HCO_3^-} = 10^{-10.3}$), so from equations (3.27) and (3.2)

$$[H^+] = [HCO_3^-] \quad \text{(3.27, modified)}$$

$$\frac{[H^+]^2}{[H_2CO_3]} = 10^{-6.4} \quad \text{(substitution in 3.2)}$$

$$[H^+]^2 = 10^{-6.4} 10^{-5.0} = 10^{-11.4}$$

$$[H^+] = 10^{-5.7}$$

$$[HCO_3^-] = 10^{-5.7}. \quad \text{(3.27, modified)}$$

Then the total dissolved carbonate in rainwater is $[H_2CO_3] + [HCO_3^-] = 10^{-5.0} + 10^{-5.7} = 10^{-4.8}$. Again the difference between activities and molalities can be ignored, inasmuch as the ionic strength is very low.

When the rainwater reaches the calcite and reacts with it, the original dissolved carbonate species are augmented by carbonate from the calcite.

Every carbonate or bicarbonate ion or H_2CO_3 molecule from the calcite also produces a calcium ion, so one defining equation is

$$[Ca^{++}] + 10^{-4.8} = [H_2CO_3] + [HCO_3^-] + [CO_3^{--}]. \quad (3.28)$$

From electrical neutrality relations

$$2[Ca^{++}] + [H^+] = [HCO_3^-] + [OH^+] + 2[CO_3^{--}].$$

Then we can proceed, using the other relations:

$$[Ca^{++}][CO_3^{--}] = 10^{-8.3} \quad (3.1a)$$

$$\frac{[H^+][HCO_3^-]}{[H_2CO_3]} = 10^{-6.4} \quad (3.2)$$

$$\frac{[H^+][CO_3^{--}]}{[HCO_3^-]} = 10^{-10.3} \quad (3.3)$$

$$[H^+][OH^-] = 10^{-14.0}. \quad (3.4a)$$

The procedure used in solving for the activities of the ions follows that used in Case 1, and yields

$$[Ca^{++}] = 10^{-3.85}$$
$$[H^+] = 10^{-9.9}$$
$$[H_2CO_3] = 10^{-7.5}$$
$$[HCO_3^-] = 10^{-4.0}$$
$$[CO_3^{--}] = 10^{-4.4}.$$

It is of interest that these values deviate only slightly (hardly more than the error of calculation) from those of Case 1, in which the reaction is entirely between $CaCO_3$ and H_2O. In fact, the increase in solubility of calcite is only about 10 percent over that in pure water. The reason for this emerges when we note that the total dissolved carbonate in the rainwater is only $10^{-4.8}$ moles per liter, so that its reacting value is low. On the other hand, if there are rootlets in the soil through which the rainwater passes, or other sources of CO_2, there will be a corresponding increase in calcite solubility and a lowering of the equilibrium pH. The very high value of pH achieved by the reaction of rainwater on calcite has perhaps not been appreciated in consideration of reactions in the weathering zone.

In summary, it can be said that the role of CO_2 in rainwater probably has been overrated, whereas the effects of hydrolysis and of CO_2 in the soil atmosphere have been underrated.

THE EFFECT OF TEMPERATURE ON CARBONATE EQUILIBRIA

Values of the equilibrium constants $K_{H_2CO_3}$, $K_{HCO_3^-}$, and K_{CaCO_3} for various temperatures are given in Table 3.2. Calculations of carbonate

TABLE 3.2. Carbonate Equilibria

Temperature, °C	$pK_1{}^a$	$pK_2{}^b$	$pK_s{}^c$	$pK_{CO_2}{}^d$
0	6.58	10.62	8.02	1.12 (0.2°)
5	6.52	10.56	8.09	
10	6.47	10.49	8.15	
15	6.42	10.43	8.22	
20	6.38	10.38	8.28	
25	6.35	10.33	8.34	1.47
30	6.33	10.29	8.40	
40	6.30	10.22	8.52	1.64
50	6.29	10.17	8.63	
80	(6.32)	(10.12)	8.98	

$pK_1 = -\log K_{H_2CO_3}$; $pK_2 = -\log K_{HCO_3}$; $pK_s = -\log K_{CaCO_3}$; $pK_{CO_2} = -\log K_{CO_2}$. Values of original authors rounded to two decimal places in present table.

[a] H. S. Harned and F. T. Bonner, *J. Am. Chem. Soc.*, 67, 1026 (1945).
[b] H. S. Harned and S. R. Scholes, *J. Am. Chem. Soc.*, 63, 1706 (1941).
[c] T. E. Larson and A. M. Buswell, *J. Am. Water Works Assoc.*, 34, 1667 (1942).
[d] Calculated from solubility data of A. A. Markam and K. A. Kobe, *J. Am. Chem. Soc.*, 63, 449 (1941).

equilibria involving systems at temperatures different from 25 °C, should, of course, be made using the values of the equilibrium constants appropriate to the measured temperature. Solubility changes with changes in total pressure become important at pressures of more than 25 atmospheres; this effect will be considered in Chapter 9.

THE EFFECT OF SOLID SOLUTION ON CARBONATE EQUILIBRIA

In nature the carbonate phase encountered is rarely a pure compound. There is substitution of a variety of elements in the cation positions. How can this substitution be taken into account when making calculations involving mineral carbonates? The basis for handling the problem has been discussed in Chapter 2; here the specific example of strontium substitution in aragonite will be used as an illustrative example. The only basic change in the calculations comes from a modification of equation (3.1). For the pure compound, the activity of the solid is unity, and the term $[CaCO_3] = 1$ is carried implicitly throughout the calculations. But if there is cation substitution, the activity of the solid is changed, and the correct activity of the solid kept in equation (3.1).

A strontian aragonite is a two-component system, and accordingly, in order to describe the aqueous phase-solid relations, we need two equations corresponding to equation (3.1); these are

$$\frac{[Ca^{++}][CO_3^{--}]}{[CaCO_3]_{arag}} = K_{aragonite} \qquad (3.29)$$

$$\frac{[Sr^{++}][CO_3^{--}]}{[SrCO_3]_{arag}} = K_{strontianite}, \qquad (3.30)$$

where $[CaCO_3]_{arag}$ and $[SrCO_3]_{arag}$ denote the respective activities of $CaCO_3$ and $SrCO_3$ in the impure aragonite. For mass balance, as exemplified in the simple case of a pure carbonate by equation (3.8), we now write

$$m_{Ca^{++}} + m_{Sr^{++}} = m_{CO_3^-} + m_{HCO_3^-} + m_{H_2CO_3}.$$

For electrical neutrality we modify equation (3.9) to

$$2m_{Ca^{++}} + 2m_{Sr^{++}} + m_{H^+} = 2m_{CO_3^-} + m_{HCO_3^-} + m_{OH^-}.$$

Combining equations (3.29) and (3.30), and substituting activity coefficients and concentrations, we get

$$\frac{\gamma_{Ca^{++}} m_{Ca^{++}}}{\gamma_{Sr^{++}} m_{Sr^{++}}} = \frac{\lambda_{CaCO_3} N_{CaCO_3} K_{arag}}{\lambda_{SrCO_3} N_{SrCO_3} K_{stron}}. \qquad (3.31)$$

Thus, for a strontian aragonite of known composition, the ratio of concentration of Ca^{++} and Sr^{++} in the aqueous solution can be calculated, if the several activity coefficients appearing in equation (3.31) can be evaluated. In the kind of dilute aqueous solution under discussion it can be safely assumed that $\gamma_{Ca^{++}} \simeq \gamma_{Sr^{++}}$. Making this assumption, and substituting the numerical values of the K's, for an aragonite of composition $(Ca_{0.95}Sr_{0.05})CO_3$, we get from equation (3.31)

$$m_{Sr^{++}} = \frac{0.05 \times 10^{-9.14} \times \lambda_{SrCO_3} m_{Ca^{++}}}{0.95 \times 10^{-8.16} \times \lambda_{CaCO_3}},$$

$$m_{Sr^{++}} \simeq 0.005 m_{Ca^{++}} \frac{\lambda_{SrCO_3}}{\lambda_{CaCO_3}}. \qquad (3.32)$$

As shown in Chapter 2, the value of the activity coefficient of the major component will usually not depart markedly from the value of the mole fraction of the major component, and in our example we can assume that $\lambda_{CaCO_3} \cong 0.95$. This leaves the quantity λ_{SrCO_3} in equation (3.32) still unknown, but admitting even relatively large deviations from unity for it, it is apparent that the contribution of Sr^{++} to total solubility is small. Not many data are available for the relations between λ and N for cation substitution, but research on these relations is moving rapidly ahead.

SUMMARY

Five situations, corresponding to those most frequently encountered by the geochemist in attempting to deduce the interactions of carbonates in natural solutions, have been discussed and the detailed procedures for the solution of the problems described. In the examples, values of the constants for 25 °C were used, and the calculations were made for calcite, with the recognition that the same procedures apply to any other metal carbonate. The last case considered was a combination of two of the more fundamental situations, and was illustrated to suggest that a great many real problems can be handled adequately by such combinations. Values of equilibrium constants at temperatures other than 25 °C have been listed for convenience. Methods for handling solid solutions are discussed.

SELECTED REFERENCES

Chave, K. E., K. S. Deffeyes, P. K. Weyl, R. M. Garrels, and M. E. Thompson, Observations on the solubility of skeletal carbonates in aqueous solutions: *Science, 137*, 33 (1962).

Garrels, R. M., M. E. Thompson, and R. Siever, Stability of some carbonates at 25 °C and one atmosphere total pressure: *Am. J. Sci., 258*, 402 (1960).

Harned, H. S., and F. T. Bonner, The first ionization constant of carbonic acid in aqueous solutions of sodium chloride: *J. Am. Chem. Soc., 67*, 1026 (1945).

Harned, H. S., and S. R. Scholes, The ionization constant of HCO_3^- from 0–50°: *J. Am. Chem. Soc., 63*, 1706 (1941).

Kern, D. B., The hydration of carbon dioxide: *J. Chem. Ed., 37*, 14 (1960).

Revelle, Roger, and Rhodes Fairbridge, Carbonates and carbon dioxide, in *Treatise on Marine Ecology and Paleoecology*, Vol. 1, *Ecology, Geol. Soc. Am. Memoir 67*, 1957, pp. 239–295.

   Summary discussion of carbonates, including many data on solubility in the marine environment.

Schoeller, H., Geochemie des eaux souterraines: *Revue de l'Institut Français du Pétrole et Annales des Combustibles Liquides*, Paris, 1956, pp. 21–28.

   Development of equations representing solubility of $CaCO_3$ in natural waters. Concise, useful.

Sverdrup, H. U., Martin W. Johnson, and R. H. Fleming, *The Oceans.* New York, Prentice-Hall, 1946, pp. 192–211.
  Detailed discussion of chemistry of CO_2 in sea water. Illustrates the complexities of the natural system and the mechanical difficulties resulting from use of empirical constants for equilibria.

Turner, R. C., A theoretical treatment of the pH of calcareous soils: *Soil Science,* **86**, 32 (1958).

PROBLEMS

3.1. At a pH of 10 and an ionic strength of 0.10, what is the solubility of strontianite (i.e., $m_{Sr^{++}}$) in a natural water in equilibrium with the atmosphere ($P_{CO_2} = 10^{-3.5}$) at 25 °C? *Ans.* $m_{Sr^{++}} = 10^{-7.1}$.

3.2. For a carbonate of composition $(Sr_{0.90}Ca_{0.10})CO_3$ at 25 °C, in a natural water in equilibrium with the atmosphere ($P_{CO_2} = 10^{-3.5}$ atm), and assuming ideal solution between $CaCO_3$ and $SrCO_3$:
  a. What is the value of the ion product $[Ca^{++}][CO_3^{--}]$? *Ans.* $10^{-9.35}$
  b. What is the value of the ion product $[Sr^{++}][CO_3^{--}]$? *Ans.* $10^{-9.20}$
  c. What is the pH? *Ans.* 8.1
  d. What is the value of $[Sr^{++}]$? *Ans.* $10^{-3.75}$
  e. What is the value of $[Ca^{++}]$? *Ans.* $10^{-3.90}$

3.3. What is the solubility of calcite at 50 °C in a solution that has pH of 8, ionic strength of 0.10, and total dissolved CO_2 ($m_{H_2CO_3} + m_{HCO_3^-} + m_{CO_3^{--}}$) of $10^{-3.0}$? (Assume γ values are unchanged by temperature.)
  Ans. $m_{Ca^{++}} = 10^{-2.9}$.

3.4. What is the solubility of witherite in rainwater open to the atmosphere? What is the equilibrium pH value? *Ans.* $m_{Ba^{++}} = 10^{-3.3}$; pH = 8.1.

3.5. A piston core of marine mud (25 °C) is found to be in equilibrium with calcite and to have a pH of 7.5. What is its internal CO_2 pressure? Assume the internal solution initially was average surface sea water.
  Ans. $10^{-2.6}$ atm.

CHAPTER 4

Complex Ions

INTRODUCTION

In this chapter, we return for a closer look at one of the most important questions encountered by the geochemist—the actual nature of the solute in various natural waters. Previously, in Chapter 2, in the section on aqueous solutions of electrolytes, a distinction was made among several classes of electrolytes and various definitions of these were given. The treatment given in Chapter 2 will be followed here; it should be reread at this time.

In extremely dilute aqueous solutions, such as most rivers and lakes, the major dissolved species can be looked upon as individual free ions. A chemical analysis of such a water, reported in terms of Na^+, K^+, Ca^{++}, Mg^{++}, Cl^-, SO_4^{--}, HCO_3^-, etc., provides an entirely satisfactory model of the internal economy of the solution. The concentrations of such species can be used in conjunction with values of individual ion activity coefficients derived from Debye-Hückel theory, or from the mean salt method, to yield activities of the species that are accurate within a few percent.

As the ionic strength of a solution of a single salt approaches and exceeds 0.1, differences in the ion activity coefficients calculated by the Debye-Hückel and the mean salt methods become larger. This effect can be seen in Figure 2.15; note there, for example, the rapid fall-off of the curve for $\gamma_{NO_3^-}$, calculated from experimental values of $\gamma_{\pm KCl}$ and $\gamma_{\pm KNO_3}$. In solutions of mixed electrolytes, such as sea water (ionic strength 0.7), experimentally determined ionic activity coefficients depart even more markedly from those predicted by either Debye-Hückel or mean salt calculations. For example, from Figure 4.5, the value of $\gamma_{HCO_3^-}$ at $I = 0.7$ in a solution of $KHCO_3$ is 0.67; in sea water $\gamma_{HCO_3^-}$ is calculated to be 0.47.[1]

[1] R. M. Garrels and M. E. Thompson, *Am. J. Sci.*, 260, 57 (1962).

ION PAIRS

These effects can be explained on the basis of a deficiency of free ions in solution, due to short-range interactions between closely adjacent ions, usually in pairs. While other approaches are possible, we shall adopt here the one due to Bjerrum, namely, that a pair of oppositely charged ions, the *ion pair*, can be treated as a thermodynamic entity, where the ion pair is in equilibrium with the remaining free ions, and the equilibrium is described by a thermodynamic equilibrium constant. The equilibrium constant expressed as an association constant, or the reciprocal of this, the dissociation constant, is a measure of the proportion of undissociated salt in the solution, over the concentration range of interest. The attractive force involved in an ion pair is coulombic, as contrasted with the essentially covalent bonding involved in a weak electrolyte, such as a weak acid like acetic acid. As was pointed out previously (Chapter 2), it is not always possible to decide whether an associated electrolyte is an ion pair or a complex of another kind; this distinction is usually not of importance to our treatment.

Various experimental methods are employed for the determination of dissociation constants for ion pairs (and other types of complexes). These include conductivity measurements, determination of activities, pH measurements, spectrophotometric studies, kinetic measurements, and other methods. In every case the behavior of the salt is measured, and then its deviation from a chosen "standard" interpreted in terms of ion association in the form of some specific species. One standard commonly employed is KCl, which is assumed to exist in solution wholly as free ions. The Debye-Hückel equation (in various forms) is also used as a standard. The theory and experimental procedures for arriving at values of dissociation constants for ion pairs are discussed in more detail by Davies.[2]

When one considers all the possible ion pairs that might occur in a natural water containing many dissolved elements, the complications seem at first to be overwhelming. But if we restrict our attention to the elements that make up 99 percent or so of the dissolved material of streams, lakes, underground waters, and the oceans, we find that only ten elements need be considered: H, Na, K, Mg, Ca, Si, Cl, O, S, and C. Furthermore, we can consider that these occur in solution as the well-known ions, radicals, or molecules H^+, Na^+, K^+, Ca^{++}, Mg^{++}, H_2CO_3, HCO_3^-, CO_3^{--}, Cl^-, SO_4^{--}, H_4SiO_4, $H_3SiO_4^-$, or as combinations of no more than two of these. Thus, we might have to appeal to the formation

---

[2] Cecil W. Davies, Incomplete dissociation in aqueous salt solutions, in *The Structure of Electrolytic Solutions*, Walter J. Hamer (ed.). New York, Wiley, 1959, pp. 19–34; idem, *Ion Association*. Washington, D.C., Butterworth, 1962.

of species like $NaCO_3^-$, $CaCO_3^\circ$, $MgHCO_3^+$, or $MgSO_4^\circ$ in solution to explain the properties of a given natural water. As far as we know, complexing of the major dissolved species of natural waters above the level of the ion pair [e.g., $Na(CO_3)_2^{3-}$, $Na_2CO_3^\circ$] is unimportant at low temperatures except in dense brines.

The problem of obtaining activities of minor constituents in natural waters is essentially impossible to attack at present, at least by calculations based upon water analyses; we do not have sufficient information on the equilibrium constants for the many associations that could occur.

The free ions in equilibrium with the ion pairs in a solution of a given ionic strength will have ion activity coefficients appropriate to that particular ionic strength. Presumably, if all the short-range interactions leading to ion pairs and other complexes (including solvent complexes) could be taken into account, only long-range forces would remain to be considered to determine the values of the activity coefficients; hence, these coefficients could be calculated from Debye-Hückel theory. However, as was pointed out, all possible interactions for various solutions of interest here are far from being known. It is found, as might be expected, that in relatively dilute, simple solutions, after known complexing has been taken into account, the Debye-Hückel and mean salt methods lead to about the same values of ion activity coefficients. In more concentrated solutions, the mean salt method is generally used to determine the values of the activity coefficients of the free ions. Some charged ion pairs can be assigned activity coefficients by analogy with more readily measurable counterpart species. This subject is considered in greater detail when the specific case of sea water as a chemical solution is taken up, in a later section.

Dissociation Constants for Ion Pairs

Most of the complexing of major dissolved species, and important aspects of complexing of minor species, can be related to the formation of ion pairs in solution. Dissociation constants for these pairs are available from a number of sources, the most comprehensive of which is the book by Bjerrum, Schwarzenbach, and Sillén[3] on stability constants of inorganic ligands, and the article and book by Davies.[4] Table 4.1 lists dissociation constants for a variety of ion pairs as well as for some weak electrolytes such as HCO_3^-.

A useful predictive device for estimating dissociation constants for dissolved species as yet not investigated in detail can be derived from the

[3] Jannik Bjerrum, Gerold Schwarzenbach, and Lars Gunnar Sillén, *Stability Constants. Part II: Inorganic Ligands.* London, The Chemical Society, 1958.
[4] Cecil W. Davies, *op. cit.*

TABLE 4.1. Dissociation Constants for Some Dissolved Species in Aqueous Solutions at 25 °C

Cations	Anions				
	OH^-	HCO_3^-	CO_3^{--}	SO_4^{--}	Cl^-
	Dissociation constants as negative logarithms				
H^+	14.0	6.4	10.33	2	—[a]
K^+	—	—	—	0.96	—
Na^+	−0.7	−0.25	1.27	0.72	—
Ca^{++}	1.30	1.26	3.2	2.31	—
Mg^{++}	2.58	1.16	3.4	2.36	—

[a] Denotes no measurable association.

EXAMPLE: $MgOH^+_{aq} = Mg^{++}_{aq} + OH^-_{aq}$

$$\frac{[Mg^{++}][OH^-]}{[MgOH^+]} = 10^{-2.58}$$

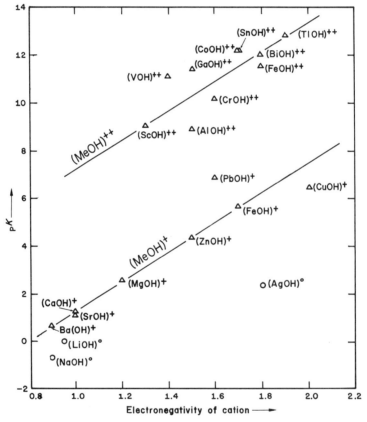

FIG. 4.1. Dissociation constants (25 °C) of dissolved species of hydroxides plotted as pK (−log K) values vs. the electronegativities of the cations. [K values calculated from ΔF_f° data for the species involved; electronegativity values from W. Grody and W. J. O. Thomas, Electronegativities of the elements: *J. Chem. Phys.*, 24, 439 (1956).]

work of Clifford.[5] He showed that when $pK_{sp} = -\log$ (activity product) is plotted against the difference in electronegativity of the anion and cation of the compounds of a given type—for example, the series of monovalent cation sulfides Me_2S—a straight line results. Clifford has applied this type of plot to sulfides, tellurides, selenides, hydroxides, and halides. Following Clifford's lead, we plotted dissociation constants for various metal-hydroxyl ion pairs versus the electronegativity of the cation. The results are shown in Figure 4.1. These results can be compared with relations for the activity (solubility) products of the solid hydroxides shown in Figure 4.2.

The electronegativity of an atom or group of atoms (with appropriate charge) may be defined as its relative electron-withdrawing power. The more dissimilar one kind of atom is from another, the greater the

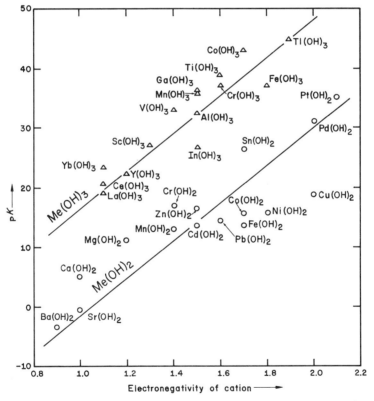

FIG. 4.2. Ion products (activity products), at 25 °C, of solid hydroxides plotted as pK ($-\log K$) values vs. the electronegativities of the cations. [K values obtained by same methods as those described under Figure 4.1.]

[5] A. F. Clifford, J. Am. Chem. Soc., 79, 5404 (1957); J. Phys. Chem., 63, 1227 (1959).

electronegativity difference between the two atoms. For example, F^- is assigned an electronegativity of 4, and Na^+ that of 1, so that the difference is 3.[6] This means that the bonding between Na^+ and F^- is extremely polar. At the other extreme, two atoms of the same kind have zero difference in electronegativity, with corresponding lack of polarity in their chemical bonding. For example, in the chlorine molecule, Cl_2, the two chlorine atoms are bonded essentially completely by a single covalent bond. The greater the difference in electronegativity between interacting ions, the greater the contribution of coulombic attraction to the chemical bonding in the resulting complex. In Figure 4.1, the bonding involved in the solution complexes becomes more polar, i.e., coulombic, as one goes from right to left, and conversely more covalent, reading from left to right. A detailed discussion of the fundamental principles of electronegativity is given by Pauling.[7]

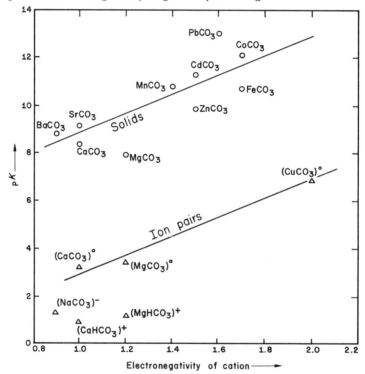

FIG. 4.3. Ion products (activity products), at 25 °C, of solid carbonates, and dissociation constants of dissolved ion pairs, plotted as pK ($-\log K$) values vs. the electronegativities of the cations. [K values obtained by same methods as those described under Figure 4.1.]

[6] A. F. Clifford, *op. cit.*, (1959).

[7] Linus Pauling, *The Nature of the Chemical Bond*, 3rd ed. Ithaca, New York, Cornell, 1960, *cf.* pp. 88–107.

In Figure 4.1 it is seen that there is a good deal of scatter of the points for complexes involving trivalent ions such as $Fe(OH)^{++}$, but this is to be expected, since the $K_{Me(OH)^{++}}$ values depend on the $K_{Me(OH)_3}$ values, which are difficult to obtain and themselves scatter, as shown in Figure 4.2.

Figure 4.3 is a plot of pK values for solid carbonates, as well as dissociation constants for dissolved species, against the electronegativity of the cation. For the solids, a general trend of increase in pK is noted for increase in electronegativity of the cation. However, it is clear that a cation size factor is also involved, because carbonates with smaller cations (Mg, Zn, Fe) lie below the line of best fit. Data for the dissociation constants of the divalent metal carbonate ion pairs in solution are sparse, but the line shown is drawn for the three points available.

The two points[8,9] for $Ca(HCO_3)^+$ and $Mg(HCO_3)^+$ are not at the moment susceptible to interpretation, although it should be noted in passing that these two values are open to considerable question, because the probable error in determining these constants is large.

The dissociation constants for dissolved sulfate species, as well as the solubility products of the solids, follow a pattern entirely different from that of the hydroxides and carbonates. A graph of pK versus electronegativity of the cation shows no regular relation. Instead, a striking plot is obtained by plotting cation size versus pK of the solid, or versus the dissociation constant of the dissolved species. Figure 4.4 is such a plot, and it is seen that for solids, pK increases regularly with cation size. On the other hand, the pK values for ion pairs are independent of cation size, and in fact all plotted divalent $MeSO_4^\circ$ ion pairs in solution have nearly the same pK ($= 2.3$). Similarly, all monovalent $MeSO_4^-$ pairs also have about the same pK ($= 0.8$). The implication is strong that the ion pair in solution is an association between a hydrated metal ion and the sulfate ion, since the hydrated radii are all similar and the bond strengths between such hydrated cations and the sulfate ion might well be similar in magnitude.

None of the major cations of natural waters seems to interact significantly with chloride ion to form ion pairs at earth surface temperatures. This relation is helpful to the geochemist, because it means that synthetic solutions containing the individual metal chlorides can be used as reference solutions in obtaining activity coefficients of the cations, even at high ionic strength. However, although this lack of complexing is applicable to the alkali metals and the alkaline earths, it does not apply

---

[8] I. Greenwald, The dissociation of calcium and magnesium carbonates: *J. Biol. Chem.*, *141*, 789 (1941).

[9] P. B. Hostetler, Complexing of magnesium with bicarbonate: *J. Phys. Chem.*, *67*, 720 (1963).

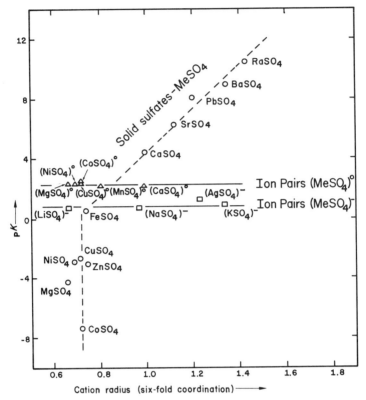

FIG. 4.4. Ion products (activity products), at 25 °C, of crystalline sulfates, and dissociation constants of dissolved ion pairs, plotted as pK (−log K) values vs. the cation radii (for six-fold coordination in the crystal). [K values calculated from ΔF_f° data; ionic radii from L. H. Ahrens, *Geochim. et Cosmochim. Acta*, 2, 155 (1952).]

to many other elements. Lead, zinc, iron, manganese, etc., all can be expected to interact significantly with chloride ion as well as with other halides. Equilibrium constants for the various reactions involving complexes of halides with metals are available in the book by Bjerrum, Schwarzenbach, and Sillén, *op. cit.*

A CHEMICAL MODEL OF SEA WATER

The following discussion on sea water is adapted from the paper on a chemical model of sea water by Garrels and Thompson.[10]

Interaction of major dissolved species in natural waters less concentrated than sea water is apparently relatively slight. A typical stream water containing a few hundred parts per million dissolved solids has an

[10] R. M. Garrels and M. E. Thompson, *op. cit.*

ionic strength of the order of 0.01, and the activities of the individual ions are less than 0.01. For example, if a given stream contains 40 ppm Ca^{++} and 61 ppm HCO_3^-, the molalities of the ions are both 0.001, and their activities somewhat less. The dissociation constant of $CaHCO_3^+$, from Table 4.1, is $10^{-1.26}$. As a first approximation

$$\frac{[Ca^{++}][HCO_3^-]}{[CaHCO_3^+]} = 10^{-1.26};$$

$$\frac{m_{Ca^{++}} m_{HCO_3^-}}{m_{CaHCO_3^+}} \cong 10^{-1.26};$$

$$\frac{10^{-3} \times 10^{-3}}{m_{CaHCO_3^+}} \cong 10^{-1.26};$$

$$m_{CaHCO_3^+} \cong 10^{-4.74}.$$

Use of the activities of Ca^{++} and HCO_3^- will yield a value for the activity of $CaHCO_3^+$ that is even smaller; at any rate, at most a few percent of the calcium and bicarbonate ions would be bound as $CaHCO_3^+$. Similar calculations for other possible interactions between major species yield similar results.

Sea water, which has an ionic strength of 0.7, is sufficiently concentrated to have interactions between ions that markedly modify the thermodynamic properties that would be deduced if complexes were ignored. Table 4.2 gives the composition of average surface sea water at 25 °C. The species listed make up 99.7+ percent of the total dissolved solids.

The table of dissociation constants (Table 4.1) shows that no complexes

TABLE 4.2. Composition of Average Surface Sea Water with 19‰ (parts per thousand) Chlorinity at 25 °C (pH = 8.15)

Ion	Molality
Na^+	0.48
Mg^{++}	0.054
Ca^{++}	0.010
K^+	0.010
Cl^-	0.56
SO_4^{--}	0.028
HCO_3^-	0.0024
CO_3^{--}	0.00027

SOURCE: R. M. Garrels and M. E. Thompson, op. cit.

need be considered between cations and chloride ion, or between K^+ and HCO_3^- or CO_3^{--}. But K^+ interacts with sulfate; Na^+, Ca^{++}, and Mg^{++} with each of the anions SO_4^{--}, HCO_3^-, and CO_3^{--} (the OH^- interactions can be neglected because of the low concentration of OH^-).

Tentatively, then, the important dissolved species in sea water may include K^+, KSO_4^-, Na^+, $NaCO_3^-$, $NaHCO_3^\circ$, $NaSO_4^-$, Ca^{++}, $CaHCO_3^+$, $CaCO_3^\circ$, $CaSO_4^\circ$, Mg^{++}, $MgHCO_3^+$, $MgCO_3^\circ$, $MgSO_4^\circ$, HCO_3^-, CO_3^{--}, SO_4^{--}, and Cl^-—a total of 18 species.

To determine the distribution of these species, 18 independent equations are required. Two kinds of information are available: mass balance relations, e.g.,

$$m_{K^+} + m_{KSO_4^-} = m_{K^+ \text{ total}},$$

and dissociation constant equations, e.g.,

$$\frac{[K^+][SO_4^{--}]}{[KSO_4^-]} = K_{KSO_4^-}.$$

To be able to use these two kinds of information simultaneously, activity coefficients are required to convert molalities to activities, and vice versa. We now assume that we know all the kinds of individual species in sea water. We shall need to have values for the activity coefficients for neutral species such as $CaSO_4^\circ$, $MgSO_4^\circ$, etc., for the various free (uncomplexed) ions Na^+, K^+, Cl^-, CO_3^{--}, etc., and for the ion pairs such as $NaCO_3^-$, KSO_4^-.

Activity Coefficients for Various Species in Sea Water

The values of the individual activity coefficients to be assigned to the various species are those appropriate to a solution of ionic strength of sea water, taken as 0.7.[11]

UNCHARGED SPECIES. Information on the activity coefficients of neutral ion pairs such as $CaSO_4^\circ$, $MgSO_4^\circ$, $CaCO_3^\circ$, $MgCO_3^\circ$, and $NaHCO_3^\circ$ is not available. It does not seem likely that these species will behave in aqueous electrolyte solution in the way dissolved gases generally do, for example, so that no analogy readily presents itself. At present, it seems best simply to assume activity coefficients of unity for these species. In view of the various uncertainties entering the calculations in which these activity coefficients are used, this assumption probably does not greatly affect the final results.

CHARGED SPECIES. Individual ion activity coefficients for K^+ and Cl^- are obtained from the mean activity coefficient of KCl, assuming that $\gamma_{K^+} = \gamma_{Cl^-} = \gamma_{\pm KCl}$. Determination of K^+ activity with a glass electrode, by the methods described in Chapter 8, yields a value in good

---

[11] Strictly speaking, the ionic strength can be calculated only by successive approximations, after the complexing is taken into account. Because the total amount of complexing turns out to be small and there is sufficient uncertainty in the values of the various constants used, any change calculated in the value of I would not be significant.

agreement with that obtained from $\gamma_{\pm\text{KCl}}$. Similarly, values for γ_{Na^+} are obtained from $\gamma_{\pm\text{NaCl}}$ and $\gamma_{\pm\text{KCl}}$. Again, measurement with a sodium-sensitive glass electrode checks the mean salt value within 2 percent.

Activity coefficients for HCO_3^- and CO_3^{2-} are taken from the data plotted in Figure 4.5 for $KHCO_3$ and K_2CO_3 solutions, respectively. The activity coefficients of $NaCO_3^-$, $MgHCO_3^+$, $CaHCO_3^+$, KSO_4^-, and $NaSO_4^-$ are considered to be the same as that of HCO_3^-. Similar values would be expected for all singly charged species of about the same size.

Values for $\gamma_{\text{Ca}^{++}}$ and $\gamma_{\text{Mg}^{++}}$ are obtained by using $\gamma_\pm$ values for the respective chlorides, in conjunction with $\gamma_{\pm\text{KCl}}$. There is apparently no complexing in $CaCl_2$ or $MgCl_2$ solutions at 25 °C.

The activity coefficient of SO_4^{2-} presents a special problem. All salts for which mean activity coefficient data are available exhibit significant complexing at ionic strengths comparable to that of sea water. However, by using the preceding values for γ_{K^+} and $\gamma_{\text{KSO}_4^-}$, the dissociation constant for KSO_4^-, and $\gamma_\pm$ values for K_2SO_4, it is possible to calculate $\gamma_{SO_4^{2-}}$ as a function of the ionic strength.

Table 4.3 summarizes the values of the individual ion activity coefficients and the method used to obtain each.

TABLE 4.3. Activity Coefficients of Individual Species in Sea Water.[a]
(Ionic Strength, 0.7; Chlorinity, 19‰, 25 °C)

Dissolved Species	Activity Coefficient	Method Used
$NaHCO_3^0$	1.13	Analogy with H_2CO_3
$MgCO_3^0$	1.13	Same
$CaCO_3^0$	1.13	Same
$MgSO_4^0$	1.13	Same
$CaSO_4^0$	1.13	Same
HCO_3^-	0.68	Figure 4.5, A
$NaCO_3^-$	0.68	Analogy with HCO_3^-
$NaSO_4^-$	0.68	Same
KSO_4^-	0.68	Same
$MgHCO_3^+$	0.68	Same
$CaHCO_3^+$	0.68	Same
Na^+	0.76	Meas. glass electrode
K^+	0.64	$\gamma_{K^+} = \gamma_{\pm\text{KCl}}$
Mg^{++}	0.36	$\gamma_{Mg^{++}} = (\gamma_{\pm\text{MgCl}_2}^3)/(\gamma_{\pm\text{KCl}}^2)$
Ca^{++}	0.28	$\gamma_{Ca^{++}} = (\gamma_{\pm\text{CaCl}_2}^3)/(\gamma_{\pm\text{KCl}}^2)$
Cl^-	0.64	$\gamma_{Cl^-} = \gamma_{\pm\text{KCl}}$
CO_3^{2-}	0.20	Figure 4.5, D
SO_4^{2-}	0.12	See text

[a] The values of the activity coefficients listed in Table 4.3 are taken from the paper of Garrels and Thompson, op. cit. However, the values used for both the neutral and charged ion-pair species, as well as the individual ions, are somewhat controversial, because it is not possible to put accurate limits of error on these numbers. Perhaps the best method of indicating the degree of uncertainty is to cite alternate values representing the extreme differences that have been suggested for some of these values. Garrels and Thompson used 1.13 for uncharged species; it has been suggested that the correct value may be 1.0 or a few percent less. They used 0.36 for $\gamma_{Mg^{++}}$, a value derived from a given published set of data for $\gamma_{\pm\text{MgCl}_2}$; another set of published data yields $\gamma_{Mg^{++}} = 0.29$. The value of γ_{Na^+} measured with a glass electrode was 0.76; the value calculated from published data on $\gamma_{\pm\text{NaCl}}$ is 0.71. These differences are probably representative of the maximum errors to be expected.

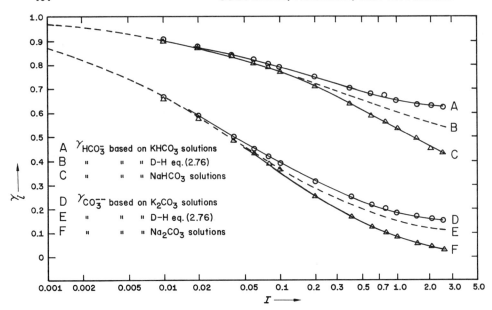

FIG. 4.5. Activity coefficients for bicarbonate and carbonate ions. [Curves A, C, D, F plotted from the experimental data of A. C. Walker, U. B. Bray, and John Johnston, *J. Am. Chem. Soc.*, 49, 1255 (1927). B and E are the Debye-Hückel plots of equation (2.76).]

CALCULATION OF DISTRIBUTION OF SPECIES IN SEA WATER

To determine the molalities of the various species assumed to occur in sea water, a mass balance equation for one of the analyzed species is written and then used in conjunction with the various dissociation constants involving that species. The analytical value given for SO_4^{--}, which can be designated $m_{SO_4^{--}\ total}$, can be equated to the sum of the various species containing sulfate, as follows:

$$m_{SO_4^{--}\ total} = m_{NaSO_4^-} + m_{KSO_4^-} + m_{CaSO_4^0} + m_{MgSO_4^0} + m_{SO_4^{--}\ free}. \tag{4.1}$$

Simultaneously, the equilibrium dissociation relations for the various species must hold

$$\frac{\gamma_{Na^+} m_{Na^+} \gamma_{SO_4^{--}} m_{SO_4^{--}}}{\gamma_{NaSO_4^-} m_{NaSO_4^-}} = K_{NaSO_4^-} \tag{4.2}$$

$$\frac{\gamma_{K^+} m_{K^+} \gamma_{SO_4^{--}} m_{SO_4^{--}}}{\gamma_{KSO_4^-} m_{KSO_4^-}} = K_{KSO_4^-} \tag{4.3}$$

$$\frac{\gamma_{Ca^{++}} m_{Ca^{++}} \gamma_{SO_4^{--}} m_{SO_4^{--}}}{\gamma_{CaSO_4^0} m_{CaSO_4^0}} = K_{CaSO_4^0} \tag{4.4}$$

$$\frac{\gamma_{Mg^{++}} m_{Mg^{++}} \gamma_{SO_4^{--}} m_{SO_4^{--}}}{\gamma_{MgSO_4^0} m_{MgSO_4^0}} = K_{MgSO_4^0}. \tag{4.5}$$

Thus, five equations are available to describe the various sulfate species. However, even though $m_{SO_4^{--}\text{ total}}$, all the K values, and all the γ values are known, there are still nine unknowns (the molalities of SO_4^{--}-bearing species, plus the molalities of the ions Na^+, K^+, Ca^{++}, and Mg^{++}, since these latter may enter into other complexes).

One way to solve the problem would be to write a mass balance equation for CO_3^{--} bearing species, as well as their dissociation constants, and for the HCO_3^- bearing species, plus their dissociation constants. Then the resultant equations could be solved simultaneously with a digital computer, as has been demonstrated by Helgeson.[12]

A fairly rapid manual method of solving the problem is to make the approximation that the cations Na^+, K^+, Ca^{++}, and Mg^{++} are not significantly complexed; that is, $m_{Na^+\text{ total}} = m_{Na^+\text{ free}}$, etc.

The assumption is not unreasonable because the cations in sea water are balanced chiefly by Cl^-, and the total molality of cations far exceeds that of the complexing ions SO_4^{--}, CO_3^{--}, and HCO_3^-. Then the number of unknowns in equations (4.1) to (4.5) is reduced by 4, and the equations can be solved for the molalities of the various sulfate species. Solution of the equations shows that the initial assumption is justified, and only small portions of Ca^{++} and Mg^{++}, and entirely insignificant portions of Na^+ and K^+, are tied up as complexes.

By similar assumptions and procedures, the molalities of the HCO_3^- and the CO_3^{--} bearing complexes can be calculated. Finally, the entire problem is solved a second time, adjusting the molalities of Ca^{++} and Mg^{++} for the amounts complexed by SO_4^{--}, HCO_3^-, and CO_3^{--}. One such recalculation is sufficient to provide results within the accuracy of the various other numerical values used in the calculations. (If the assumption that the cations are not significantly complexed had been seriously in error, several recalculations might have been required to achieve satisfactory roots of all the equations).

The results of the calculations are shown in Table 4.4.

TABLE 4.4. Distribution of Major Dissolved Species in Sea Water (19‰ Chlorinity, 25 °C, pH 8.15)

Ion	Molality (Total)	Free Ion (Percent)	Me-SO$_4$ Pair (Percent)	Me-HCO$_3$ Pair (Percent)	Me-CO$_3$ Pair (Percent)
Na$^+$	0.48	99	1+	—	—
K$^+$	0.010	99	1	—	—
Mg^{++}	0.054	87	11	1	0.3
Ca^{++}	0.010	91	8	1	0.2

[12] Harold C. Helgeson, *Complexing and Hydrothermal Ore Deposition.* New York, Pergamon, 1964, pp. 38–59.

Ion	Molality (Total)	Free Ion (Percent)	Ca-Anion Pair (Percent)	Mg-Anion Pair (Percent)	Na-Anion Pair (Percent)	K-Anion Pair (Percent)
SO_4^{--}	0.028	54	3	22	21	0.5
HCO_3^-	0.0024	69	4	19	8	—
CO_3^{--}	0.00027	9	7	67	17	—
Cl^-	0.56	100	—	—	—	—

DISCUSSION OF THE CHEMICAL MODEL OF SEA WATER

It is inevitable that we intuitively use the form in which chemical analyses are presented to us in visualizing the properties of the material analyzed. This is true even though we are quite aware that the information is misleading. Analyses of igneous rocks are reported in terms of oxides; there is little doubt that thinking concerning the nature of magmas, which are highly ionized, has been hampered by the difficulty of mentally transposing from oxides to ions. In the case of the sea water model, analytical results are much closer to representing the nature of the medium, but the modifications shown in Table 4.4 are helpful in understanding numerous phenomena of sea water chemistry.

The model presented does not markedly change the picture of the state of the major cations. They apparently are present in solution chiefly as individual free species. There is a small amount of ion pairing with SO_4^{--}, especially by Ca^{++} and Mg^{++}. But any increase in the SO_4^{--} of sea water (for example, by evaporation) will increase the fraction of these cations present as $CaSO_4^{\circ}$ or $MgSO_4^{\circ}$. Note also that a nearly constant maximum value of the activity of these ion pairs is reached when the solution equilibrates with solid sulfates.

If gypsum begins to precipitate at equilibrium from concentrated sea water, we can write

$$CaSO_4 \cdot 2H_2O_c = CaSO_4^{\circ}{}_{aq} + 2H_2O_l \qquad (4.6)$$

$$[CaSO_4^{\circ}][H_2O]^2 = K_{gypsum}$$

$$CaSO_4^{\circ}{}_{aq} = Ca^{++}_{aq} + SO_4^{--}{}_{aq}$$

$$\frac{[Ca^{++}][SO_4^{--}]}{[CaSO_4^{\circ}]} = K_{CaSO_4^{\circ}}. \qquad (4.7)$$

For a fixed activity of water, the activity of $CaSO_4^{\circ}$ also is fixed.

The major change in our thinking that results from the chemical model of sea water presented here is in terms of the anions SO_4^{--}, HCO_3^-, and CO_3^{--}. More than 30 percent of each of the first two is tied up as ion pairs with cations, whereas 90 percent of the total CO_3^{--} is complexed. Recognition of the degree and nature of the CO_3^{--} complexing

COMPLEX IONS

is helpful in attempting to unravel the mysteries of deposition of the various carbonate minerals that form such an important part of the geologic record. There has been a tendency to consider only the relations between CO_3^{--} and H^+ in explaining precipitation of $CaCO_3$; it is now evident that the Mg^{++} and Na^+ content of sea water are nearly as important as H^+ in controlling $CaCO_3$ deposition. For example, an organism that excludes Mg^{++} from its cell fluids would tend to cause precipitation of $CaCO_3$, because of the loss of the complexing power of Mg^{++} for CO_3^{--}. It is true that there tends to be a larger short-term variation in H^+ in ocean waters than Mg^{++} or Na^+, but over geologic time, drastic changes in $CaCO_3$ solubility could have resulted from changes in Mg^{++} and Na^+. This point is illustrated by Figure 4.6, which shows the solubility of calcite at constant CO_2 pressure and constant temperature in NaCl solutions and in NaCl + $MgCl_2$ solutions.

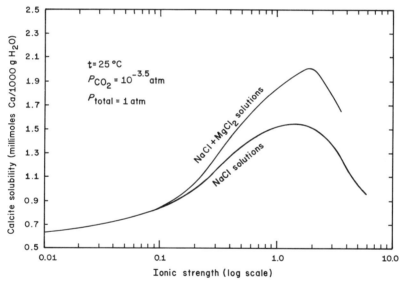

FIG. 4.6. Calculated solubility of calcite as a function of ionic strength, in NaCl solutions, and in solutions of NaCl + $MgCl_2$ that are in the approximate ratio that they have in sea water. [Adapted from R. M. Garrels, M. E. Thompson, and R. Siever, Control of carbonate solubility by carbonate complexes: Am. J. Sci., 259, 43 (1961).]

The distribution of species for normal sea water at 25 °C and 1 atmosphere total pressure will be modified by changes in temperature, pressure, and composition of the water. The effects of pressure and temperature are assessed in a later chapter. However, the qualitative picture holds fairly well over the range of variation in temperature and pressure to which ocean waters are subjected.

COMPLEX IONS IN BRINE

Natural aqueous media more concentrated than sea water are exceedingly difficult to handle in terms of calculations based on ionic equilibria. Whereas the major species in sea water can be considered to be free ions or ion pairs, the amount of interaction between dissolved species in dense natural brine is greater, and species more complex than ion pairs are formed.

There are a few natural waters that are sufficiently simple in composition so that an approximation of the species present can be made. Table 4.5a shows a partial chemical analysis of a brine from Deep Springs Lake, California.[13] The distribution of species, calculated by methods similar to those used for sea water, is shown in Table 4.5b, and the degree of complexing of various species in Table 4.5c.

In this calculation it was necessary to add an extra step. As can be seen from Table 4.5b, there are essentially no divalent anions remaining in the brine; they are all used up by interaction with cations to form uncharged or singly charged species. Therefore, the ionic strength of the brine must be adjusted from the initial value calculated from the chemical analysis. For sea water the change in ionic strength because of ion pairing is insignificant, but for the brine, the initial ionic strength computed from the chemical analysis is about twice the value finally determined after compensating for the reduction in charge caused by interactions of ions.

The great amount of interaction of dissolved species in concentrated salt solutions should warn us that brines differ so markedly in their chemical behavior from dilute surface or ground waters that we are probably safer to think of them as differing in kind rather than simply in degree.

Attempts to calculate the solubilities of minerals in brines are at present nearly hopeless, because of the contributions to solubility from species for which dissociation constants are not known. Also, even if all interactions were known, the number of simultaneous equations to be solved is so large that the results obtained, perhaps after many hours of calculation, must suffer from the accumulation of errors from various dissociation constants, from individual ion activity coefficients, from readjustments of ionic strength, and similar sources.

At present it appears that average open ocean water, with an ionic strength of 0.7, is about the upper limit of concentration of a natural water that is worth while attempting to handle in detail on the basis of calculations from theory and chemical analyses.

A much more promising technique of treating problems involving the thermodynamic properties of natural aqueous solutions is to use a

---

[13] Analysis courtesy Blair Jones, Water Resources Division, U.S. Geological Survey.

COMPLEX IONS 109

chemical analysis of the water in conjunction with direct measurements of individual ionic activities by means of ion-sensitive electrodes (Chapter 8). In this way it is possible to bypass the complexities of calculating

TABLE 4.5a. Partial Analysis[a] of Brine from Deep Springs Lake, California
(Specific gravity 1.272, pH = 10.1, Field No. 39D)

	Parts per Million	Molality[b]
Na^+	101,000	6.48
K^+	15,700	0.59
CO_3^{--}	63,300	1.56
HCO_3^-	12,500	0.30
SO_4^{--}	40,500	0.62
Cl^-	58,100	2.42
Total	291,100	11.97

[a] Total dissolved solids 322,000.
[b] Ionic strength based on species given in analysis is 9.26.

TABLE 4.5b. Calculated Distribution of Dissolved Species in Brine 39D from Deep Springs Lake, California

	Molality[a]	Activity Coefficient	Activity
Na^+	3.50	1.17	4.10
$NaCO_3^-$	0.69	0.60	0.41
$NaSO_4^-$	0.37	0.60	0.22
$Na_2CO_3^\circ$	0.84	1.88	1.58
$NaHCO_3^\circ$	0.20	1.88	0.38
HCO_3^-	0.10	0.60	0.06
CO_3^{--}	0.20	0.11	0.02
K^+	0.50	0.59	0.30
KSO_4^-	0.09	0.60	0.054
SO_4^{--}	0.15	0.11	0.016
Cl^-	2.42	0.59	1.43
Total	9.06		

[a] Ionic strength 4.5.

TABLE 4.5c. Calculated Degree of Complexing in Deep Springs Brine 39D

Ion	Total Molality/Molality Free Ion	Percent Complexed
Na^+	6.48/3.50	46
K^+	0.59/0.50	15
CO_3^{--}	1.56/0.20	87
HCO_3^-	0.30/0.10	67
SO_4^{--}	0.62/0.15	76
Cl^-	2.42/2.42	0

ionic activities by use of dissociation constants, individual ion activity coefficients, and ionic strength.

COMPLEXES INVOLVING MINOR SPECIES IN NATURAL WATERS

No comprehensive effort has yet been made to assemble the data of complex ions involving the 90-odd minor elements present in natural waters. The following discussion is therefore patchy and incomplete, and hardly does more than suggest some of the relationships yet to be assembled, investigated, and systematized.

Trivalent Cations

Trivalent ions, such as Fe^{3+}, V^{3+}, Al^{3+}, Cr^{3+}, and Mn^{3+}, are ordinarily reported in extremely low concentrations in natural waters, except for unusual waters with high acidity. The ions are small and highly charged, so that they not only tend to interact strongly with almost all anions to form soluble complexes, but also tend to form insoluble hydroxides over much of the pH range of natural waters. Figures 4.1 and 4.2 demonstrate well the high pK values of the solid hydroxides. Consequently, even though these trivalent ions are strong complex formers, their total solubility is still held low by the difficultly soluble oxides and hydroxides. In general, it can be said that a report of a measurable concentration of a trivalent ion represents some complex or group of complexes in solution. Figure 4.1 shows, from the large pK values of $(MeOH)^{++}$ ions, that powerful interaction of the trivalent ions with OH^- to form dissolved species occurs. The trivalent ions interact as well with the halides, with sulfate, and probably with bicarbonate and carbonate, although these latter interactions are obscured by the extreme insolubility of other compounds, which make the equilibria difficult to investigate.

On the acid side, then, free trivalent ions tend to be found only under unusual conditions of high acidity and low ionic strength. With gradual neutralization, trivalent ions tend to form OH^- complexes and to precipitate as insoluble hydroxides. At high pH values there is some tendency to form anionic complexes with OH^-, such as FeO_2^- [$Fe(OH)_4^-$ less $2H_2O$], but the pH values required to provide measurable solubility of trivalent ions as anions is commonly out of the geologic range. A notable exception of geological importance is the AlO_2^- ion, which may permit substantial solubility of aluminum in solution of unusual but not extremely high pH.

In summary, analytically determined trivalent ions at low pH probably are present in solution as cationic complexes; whereas at high pH they occur chiefly as anionic hydroxyl complexes.

Divalent Cations

The mere fact that a heading "Divalent Cations" can be used for this section, without subheading into alkaline earths, transition metals, etc., is indicative of the paucity of coverage available.

The behavior of Ca^{++} and Mg^{++} already has been discussed in part in terms of their roles as major cations in natural waters. As can be seen from the trend on Figure 4.1, there is a tendency for greater association of the divalent ions of higher electronegativity, reflecting increasing covalent character in the bonding. In general, then, OH^-, S^{--}, Se^{--}, and halide (except F^-) complexes will tend to be weak for alkaline earth metals, but to strengthen in the transition and post-transition metals.

Of great interest to economic geologists is the relation that the larger the pK value for the activity product of a covalently bonded compound such as a sulfide, the larger the pK value for the dissociation constant of its analogous dissolved complexes. This means that ionic association in solution tends to cause a greater relative solubility increase for a slightly soluble sulfide than for a moderately soluble one. Increase in ionic strength therefore has a tendency to increase the solubility of compounds such as sulfides, and it also tends to equalize their solubilities. Students of ore deposits have long recognized that the concentrations of various metals in certain ore fluids must be approximately equal, even though their solubilities, if described solely in terms of free metal ions, differ by several orders of magnitude.

The general sequence of increasing association of divalent ions in the mixed electrolyte solutions of nature is probably Ba < Sr < Ca (see Figures 4.1 to 4.3). The behavior of sulfate complexes, as pointed out previously, is not in accord with this generalization. High sulfate content of a solution tends to increase the solubility of all metal sulfates proportionately at 25 °C, inasmuch as Me^{++} sulfate dissociation constants all range around $10^{-2.3}$.

Tetravalent and Higher Valence Cations

Cations with charges of 4+ or higher show a uniform general behavior, but one that differs in details of great geological importance. Small, highly charged ions have such an affinity for oxygen that they convert into stable anionic complexes: silicates, permanganates, vanadates, vanadites, chromates, molybdates, tungstates, etc.

The behavior of quinquevalent vanadium illustrates well the general principles of metal-anionic complex behavior. In extremely acid solutions (pH < 2), V^{5+} exists in solution as a cationic complex VO_2^+, and in extremely alkaline solution (pH $\cong$ 14?) as orthovanadate ion VO_4^{3-}. If an alkaline solution is acidified slowly, a series of transformations take place by reactions involving H^+ and H_2O,

$$2VO_4^{3-} + 2H^+ = V_2O_7^{4-} + H_2O \quad (pH \cong 13)$$
orthovanadate (colorless) pyrovanadate (colorless)

$$2V_2O_7^{4-} + 4H^+ = V_4O_{12}^{4-} + 2H_2O \quad (pH \cong 10)$$
metavanadate (colorless)

$$5V_4O_{12}^{4-} + 8H^+ = 2V_{10}O_{28}^{6-} + 4H_2O \quad (pH \cong 6.5)$$
polyvanadate (orange)

$$V_{10}O_{28}^{6-} + H^+ = HV_{10}O_{28}^{5-} \quad (pH \cong 6)$$
hydrogen-polyvanadate (orange)

$$HV_{10}O_{28}^{5-} + H^+ = H_2V_{10}O_{28}^{4-} \quad (pH \cong 4)$$
dihydrogen polyvanadate (deep orange)

$$H_2V_{10}O_{28}^{4-} + 4H^+ + 2H_2O = 5V_2O_5 \cdot H_2O \quad (pH \cong 2.5)$$
vanadium pentoxide hydrate (solid, dark brown)

$$V_2O_5 \cdot H_2O + 2H^+ = 2VO_2^+ + 2H_2O \quad (pH \cong 1.5).$$
pervanadyl (pale yellow)

The sequence, from alkaline to acid, is from a simple ion (VO_4^{3-}) through a series of progressively more highly polymerized ions, to the formation of an insoluble oxide, and finally to re-solution as an oxygenated cation VO_2^+. The reactions to form polymers are slow, and the reverse reactions even slower. In fact, it is essentially impossible to achieve equilibrium between the vanadate species in the laboratory by starting in acid solution and neutralizing.

In general, the simple alkali metal salts of the vanadates are fairly soluble; those of the divalent ions much less so. However, except for the development of some compounds with mixed cations, such as carnotite, the potassium uranyl vanadate, the vanadates are rarely persistent minerals.

The other "ates" such as molybdates, tungstates, and silicates, follow a similar pattern in solution, differing chiefly in the pH values required to produce simple ortho-ions, the pH at which an insoluble oxide or hydroxide is formed, and the pH range over which the oxide or hydroxide is stable. Molybdenum forms polymolybdates under neutral conditions, and a group of complex hydroxides under more acid conditions. Sufficiently low pH to redissolve these oxides as cations is not achieved

COMPLEX IONS

in nature. The chemistry of silica in natural solutions at low temperatures is simplified by the fact that the polyions form only in highly alkaline solutions (pH ≥ 9), and only the simple neutral monomer H_4SiO_4 occurs in most natural waters; pH values low enough to produce a cation like $HSiO_2^+$ do not exist.

Monovalent Cations

At 25 °C, monovalent cations tend to complex only slightly, although the effect may become important in dense brines. Potassium ion, except in sulfate solutions, is remarkably "noble"; it shows essentially no association with halides, bicarbonate, carbonate, bisulfide, or sulfide. Sodium ion, on the other hand, is more interactive: under favorable conditions it may complex significantly by forming ion pairs with OH^-, HCO_3^-, and CO_3^{2-}, as well as with SO_4^{2-}. Also, it can be shown that in strong Na_2CO_3 solutions, for example in some soda lakes, there is a significant concentration of undissociated $Na_2CO_3^\circ$.[14] Water saturated with Na_2CO_3 (total molality about 2.5) contains approximately 30 percent of its dissolved salt as $Na_2CO_3^\circ$, and 30 percent as $NaCO_3^-$. The behavior to be expected from other monovalent species is best shown by the incomplete data and suggested trends illustrated in Figures 4.1 through 4.4.

Summary

Many of the preceding points are demonstrated by Figure 4.7, which is a plot of the logarithm of the ratio of the amount of an element in sea water to the total amount that has been delivered to the ocean, from the weathering of rocks, versus the ratio of the charge of the cation of the element to its radius (ionic potential). The ordinate is one kind of a measure of solubility; the abscissa is a rough measure of the tendency of the cation to bond with anions. The diagram shows the general trends that have been discussed, as well as the many specific differences from the trends. Note the relatively high solubility of compounds of the typically monovalent cations, such as the alkali metals; the relatively high solubility of the alkaline earths; the lower solubilities of the smaller, doubly charged cations, which form a variety of slightly soluble species with a variety of anions; the vanishingly low solubility of those species with a trivalent ion easily formed under the range of natural conditions, such as iron, aluminum, and cobalt. Such trivalent species, despite their tendency to form soluble complexes, are prevented from dissolving because they precipitate out as oxides or hydroxides. Finally, for those elements that form small high valence cations, which react with oxygen to develop

[14] Unpublished experiments of R. M. Garrels and M. E. Thompson, Harvard University.

TABLE 4.6. Effect of Complexes on Minor Elements in Natural Waters

Complex	Dissociation Constant	Solution	Ratio Complex/Simple Cation	Solution	Ratio Complex/Simple Cation
$Cu(CO_3)_{2\ aq}^{--}$	$\dfrac{[Cu^{++}][CO_3^{--}]^2}{[Cu(CO_3)_2^{--}]} = 10^{-10}$	$[CO_3^{--}] = 10^{-5}$	10^0	$[CO_3^{--}] = 10^{-1}$	10^8
$CuCO_{3\ aq}^{\circ}$	$\dfrac{[Cu^{++}][CO_3^{--}]}{[CuCO_3^{\circ}]} = 10^{-6}$	$[CO_3^{--}] = 10^{-5}$	10^1	$[CO_3^{--}] = 10^{-1}$	10^5
$UO_2(CO_3)_{3\ aq}^{4-}$	$\dfrac{[UO_2^{++}][CO_3^{--}]^3}{[(UO_2^{++})(CO_3^{--})_3^{4-}]} = 10^{-24}$	$[CO_3^{--}] = 10^{-10}$	10^{-6}	$[CO_3^{--}] = 10^{-6}$	10^6
$PbCl_{aq}^{+}$	$\dfrac{[Pb^{++}][Cl^{-}]}{[PbCl^{+}]} = 10^{-1.5}$	$[Cl^{-}] = 10^{-3}$	$10^{-1.5}$	$[Cl^{-}] = 10^{0}$	$10^{1.5}$
$CdI_{4\ aq}^{--}$	$\dfrac{[Cd^{++}][I^{-}]^4}{[CdI_4^{--}]} = 10^{-6.3}$	$[I^{-}] = 10^{-2}$	$10^{-1.7}$	$[I^{-}] = 10^{0}$	$10^{6.3}$
$CuS_{5\ aq}^{-}$	$\dfrac{[Cu^{+}][S_{5\ aq}^{--}]}{[CuS_5^{-}]} = 10^{-21}$	$[S_5^{--}] = 10^{-21}$	10^0	$[S_5^{--}] = 10^{-10}$	10^{11}

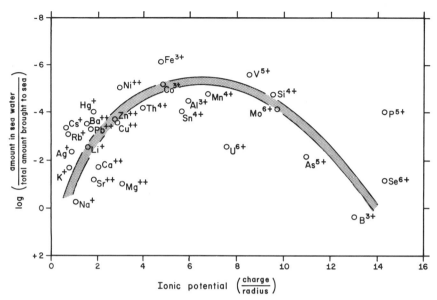

FIG. 4.7. Plot showing the relation between the ionic potential of a cationic species and the logarithm of the ratio of the total amount of the species in sea water to the total amount calculated to have been released to the oceans by the weathering of rocks. [Ionic radii from L. H. Ahrens, *Geochim. et Cosmochim. Acta*, 2, 155 (1952); element abundance data from K. Rankama and Th. G. Sahama, *Geochemistry*. Chicago, Ill., The University of Chicago Press, 1950, p. 295.]

metal-oxygen anions, solubility again tends to be enhanced. It can be considered, in the broadest terms, that the insoluble oxides and hydroxides of trivalent cations and some tetravalent cations have their minimum solubility at pH values commonly encountered, whereas some tetravalent and higher-valent cations have their solubility minima under strongly acid conditions, but tend to dissolve as oxy-polyions under neutral or alkaline conditions.

Table 4.6 is given in an attempt to show a few of the interactions that can take place between less abundant cations and the common anions of natural waters. The table shows the tremendous effect of a change in the complexing anion upon the activity of free metal ion, especially when the complex involves more than one anion.

EFFECT OF TEMPERATURE ON FORMATION OF COMPLEXES IN AQUEOUS SOLUTION

A detailed study of the formation of complex species in aqueous solutions at elevated temperatures has been made by Helgeson.[15] The following discussion is largely a résumé of a part of his work.

[15] Harold C. Helgeson, *op. cit.*

Data on dissociation constants of complex ions at elevated temperatures are still too fragmentary to permit effective generalizations concerning stability changes with temperature. However, a hypothetical model of the effect of temperature and pressure on aqueous salt solutions permits useful conclusions concerning the kinds of conditions under which the formation of complexes will be inhibited, and those that will promote ionic association.

If a solution is heated at nearly constant density, the kinetic energy of the water molecules is increased, but because the relative ion-water molecule separation is unchanged, the degree of hydration of dissolved ionic species tends to be similar to that at room temperature. Therefore, one would expect the degree of ion-ion association in a given solution to be relatively unchanged with temperature by any change in ion hydration, if density is held constant. If, on the other hand, a solution is heated so that it expands significantly, the dipole water molecules no longer are packed so tightly around dissolved ions, and their polar effects, which are the underlying cause of the high degree of ionization of salts in water, are markedly diminished. Thus, the tendency for ions like Na^+ and Cl^- in such a solution to unite to form $NaCl°$ would be enhanced.

Figure 4.8 shows pressure-temperature-density relations for water. Also plotted are lines showing equal values for the dissociation constant of KCl and HCl. The iso-K lines are nearly parallel to the isodensity contours. Similar behavior has been observed by Franck[16] for LiCl, NaCl, RbCl, CsCl, KOH, and HF.

Figure 4.9 shows three typical geothermal gradients plotted on the pressure-temperature-density plot for water. Gradient A represents the conditions in pore waters subjected to lithostatic pressure of rocks of density 2.5 g cm^{-3}, and with a typical "normal" geothermal gradient of 1 °C per 100 feet. Because the density of the pore waters is nearly constant, the degree of association of electrolytes would not be expected to change drastically. Gradient B is a "normal" hydrostatic gradient of 1 °C per 100 feet. Under these conditions the density of water diminishes somewhat at depth, and complexing should increase over that of gradient A.

Gradient C is an abnormally high geothermal gradient, and is a first approximation to the temperature-pressure conditions that might be encountered between the surface and a magma at shallow depth. Note that this steep gradient crosses more isodensity lines per unit of pressure than either gradient A or gradient B.

It is a reasonable conclusion that under conditions of geosynclinal sedimentation, the geothermal gradient would be somewhere between gradients A and B; the density of pore waters would be high, and the

[16] E. U. Franck, *Angew. Chemie*, 73, 309 (1961).

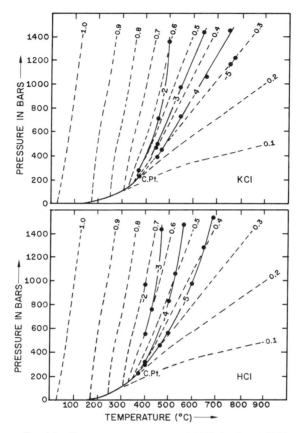

FIG. 4.8. Dissociation constants of KCl (top) and HCl (bottom), superimposed as iso-log values (solid lines) on a plot of temperature-pressure-density relations for water. The dashed lines are iso-density lines for water, with values shown. Note that equal values of log K tend to parallel equal values of the density of water, and that the values of log K become small in the low density region. [From H. C. Helgeson, *Complexing and Hydrothermal Ore Deposition*, New York, Pergamon, 1964, p. 12.]

distribution of dissolved species between free ions and complexes would not be markedly changed. To put it another way, water under geosynclinal conditions, even at depth, is much like water at the earth's surface—it is a highly polar solvent. On the other hand, any increase in the normal thermal gradient, perhaps caused by an intrusive body at shallow depth, diminishes water density markedly. Consequently, the water tends to lose its characteristic polar character, and association of salts occurs. High temperature and low density H_2O solutions are thus

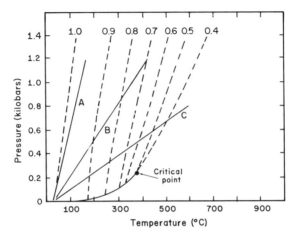

FIG. 4.9. Plot of the temperature-pressure-density relations for water, with three typical depth-temperature relations for earth conditions superimposed. Gradient A represents a lithostatic load and normal thermal gradient; gradient B represents the hydrostatic load and normal thermal gradient; gradient C is for hydrostatic load and abnormal thermal gradient. The dashed lines are isodensity lines for water, with the values shown.

quite unlike the low-temperature dense fluids with which we are most familiar. The relation of a magma to a coexisting aqueous phase is rather like the relation of an aqueous salt solution to a coexisting organic solvent at room temperature. The magma dissolves its solutes ionically; the coexisting water tends to dissolve them molecularly. At 600 °C and a water density of 0.4 g cm^{-3}, HCl, NaCl, and KCl are almost completely associated in aqueous solution ($K_{\text{dissoc}} = 10^{-4}$ to 10^{-6}). Such behavior is in striking contrast to their "complete" dissociation at room temperature.

Although the qualitative picture presented here is useful in picturing the general kinds of changes to be expected with increasing temperature, serious difficulties are encountered in attempts to predict the behavior of a given species. In a straightforward dissociation of a salt like NaCl, there is not much question that a decrease in the polar effects of the water solvent will tend to cause association. But what about a species like $PbCl_4^{--}$? Is it more stable than the constituent individual ions in a strongly polar medium? Is it more stable than $PbCl_3^-$? The answer depends upon the details of the relations of the species to the surrounding water dipoles as contrasted to the relations of the individual ions to the water dipoles.

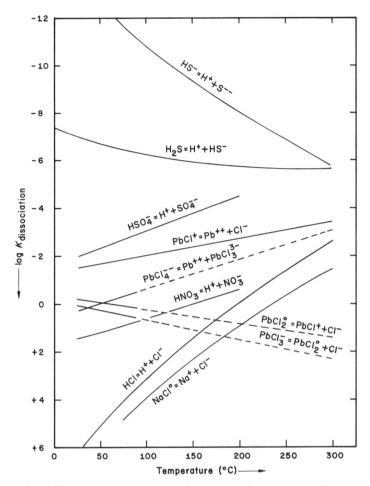

FIG. 4.10. Dissociation constants of various dissolved species as a function of temperature. Dashed lines indicate extrapolations. [Compiled from H. C. Helgeson, *Complexing and Hydrothermal Ore Deposition*, New York, Pergamon, 1964.]

Figure 4.10 shows the temperature dependence of the dissociation constant for a number of species, and serves chiefly to demonstrate the varieties of behavior. However, note that K_{dissoc} values for most ion pairs diminish with increasing temperature.

SUMMARY

Natural waters are multicomponent electrolyte solutions. At present, enough chemical information is available to permit calculation of the interactions that take place among the major dissolved species at earth surface temperatures in media as concentrated as sea water (ionic strength 0.7). A useful model results if it is considered that the "complete" ionization of extremely dilute waters is modified by increasing concentration only to the extent of the formation of new species made up of no more than pairs of those present at high dilution.

Modification of the concept of "complete" ionization of the major components dissolved in natural waters seems to be unnecessary for most purposes at ionic strengths below about 0.05. At higher values, increasing interaction takes place, and in sea water large percentages of carbonate, bicarbonate, and sulfate can be considered to form ion pairs with cations. In dense brines, interaction is profound; complex species of higher order than ion pairs must be taken into account. So much interaction takes place that for some brines the effective ionic strength is probably only about half the ionic strength calculated assuming complete dissociation.

The effect of temperature on the internal structure of aqueous solutions is apparently relatively small if there is little change in density with increasing temperature. In general, a lowering of density with increasing temperature tends to cause association of individual ions; alkali halides and inorganic acids and bases are weak electrolytes at high temperature and low density.

Much more work needs to be done before useful calculations can be performed on the interactions of the minor constituents of natural waters.

SELECTED REFERENCES

A large literature exists on the subject of complex species in aqueous solution. Here are presented only a few titles that provide summaries of data or guides to the literature.

Bjerrum, Jannik, Gerold Schwarzenbach, and Lars Gunnar Sillén, *Stability Constants. Part II: Inorganic Ligands.* London, The Chemical Society, 1958.
    The most comprehensive compilation available. Somewhat difficult to use, and is not selective.

Davies, Cecil W., *Ion Association.* Washington, D.C., Butterworth, 1962.
    Well-presented and complete treatment by one of the most active workers in the field.

Hamer, Walter J., (ed.), *The Structure of Electrolytic Solutions.* New York, Wiley, 1959.
    Excellent guide to the literature, and contains numerous excellent summary articles by specialists.

Helgeson, Harold C., *Complexing and Hydrothermal Ore Deposition.* New York, Pergamon, 1964.

COMPLEX IONS

Summary of the literature and data available on the effect of temperature on the formation of complexes. Also contains discussion of the requirements for solving species-distribution problems by digital computer methods.

Parsons, Roger, *Handbook of Electrochemical Constants*. London, Butterworth, 1959.

Compact and convenient source of a variety of thermodynamic data on solutions of electrolytes, including tables of dissociation constants.

PROBLEMS

4.1. If pK for the dissociation of $MgCO_3^\circ$ is 3.4, what is the ratio of the activity of CO_3^{--} to $MgCO_3^\circ$ in a solution in which the activity of Mg^{++} is $10^{-2.1}$? *Ans.* $10^{-1.3}$.

4.2. A solution containing 0.01 mole of $NaHCO_3$ per 1000 g of H_2O has a pH of 8.4 at 25 °C:

a. What is the ratio of carbonate ion activity to bicarbonate activity?

b. What dissolved species should be considered in the internal equilibria of such a solution?

c. What is the ionic strength of the solution if the $NaHCO_3$ is considered to dissociate completely into Na^+ and HCO_3^- ions, and no others?

d. Assuming complete dissociation as in (c), what are γ_{Na^+} and $\gamma_{HCO_3^-}$ from Debye-Hückel theory?

e. What is the mass balance relation for Na-bearing species in a $0.01m$ solution of $NaHCO_3$, considering the various dissolved species known to occur?

Ans. a. $10^{-1.93}$.
b. Na^+, $NaHCO_3^\circ$, $NaCO_3^-$, HCO_3^-, CO_3^{--}, H^+, H_2CO_3.
c. $I = 0.01$.
d. $\gamma_{Na^+} = 0.90$; $\gamma_{HCO_3^-} = 0.90$.
e. $m_{Na\ total} = m_{Na\ free} + m_{NaHCO_3^\circ} + m_{NaCO_3^-} = 0.01$.

4.3. Calculate the distribution of dissolved species in $0.1m$ $NaHCO_3$ solution at 25 °C, using mass balance relations, dissociation constants, and activity coefficients; pH of the solution is 8.4.

Ans. $m_{Na^+\ free} \cong 0.095$; $m_{NaHCO_3^\circ} \cong 0.003$; $m_{NaCO_3^-} \cong 0.002$; $m_{H_2CO_3} \cong 0.001$; $m_{CO_3^{--}} \cong 0.002$; $m_{HCO_3^-} \cong 0.092$.

CHAPTER 5

Measurement of Eh and pH

INTRODUCTION

Measurement of Eh and pH has become a routine procedure in many branches of chemistry and biology. The method of using Eh and pH as characterizing variables in reactions involving dissolved species permits the presentation of mineral stability diagrams in a most convenient form. Such Eh-pH diagrams for mineral stability relations are the subject of Chapter 7. Eh and pH measurements are also sometimes used in a qualitative or semiquantitative way to study certain more indefinite geochemical systems.

The details of the theory and practice of pH measurement are admirably treated in books by Bates,[1] Gold,[2] and Mattock.[3] A somewhat more specialized treatment, but one of considerable geochemical interest because of its discussion of the electrode measurement of P_{CO_2} and P_{O_2}, is the book edited by Woolmer.[4] Reference electrodes of all kinds are discussed definitively in Ives and Janz.[5] Both Eh and pH theory and measurements are treated in the book by Clark.[6] The book by Glasstone[7] provides a good general background to electrochemical theory.

In the treatment that follows, definitions of Eh and pH will be given, particularly in terms of operational procedures; emphasis will be placed on such practical aspects as the mechanics of measurement and the precautions to be observed in geological applications.

[1] R. G. Bates, *Electrometric pH Determinations*. New York, Wiley, 1954.
[2] Victor Gold, *pH Measurement*. New York, Wiley, 1956.
[3] G. Mattock, *pH Measurement and Titration*. London, Heywood, 1961.
[4] R. F. Woolmer (ed.), *A Symposium on pH and Blood Gas Measurement*. Boston, Little, Brown, 1959.
[5] D. J. G. Ives and G. J. Janz, (eds.), *Reference Electrodes*. New York, Academic, 1961.
[6] W. M. Clark, *Oxidation-Reduction Potentials of Organic Systems*. Baltimore, Williams & Wilkins, 1960.
[7] Samuel Glasstone, *An Introduction to Electrochemistry*. New York, Van Nostrand, 1942. See especially chap. 8.

pH MEASUREMENT

Definition of pH and Operational Procedure for Measurement

The definition of pH that we adopt here is that previously given in Chapter 1, equation (1.15)

$$\mathrm{pH} = -\log[\mathrm{H}^+]. \tag{5.1}$$

This definition, although exact, is a purely formal one, since it involves the activity of a single ion, a quantity that cannot be measured directly. As a result of this, measured pH is defined in terms of an operational procedure that is described in the following discussion.

For all practical purposes, measurement of pH is accomplished today by the use of an EMF cell that consists of a glass electrode dipping into the test solution, together with a reference electrode to complete the circuit. The reference electrode is usually a mercury-mercurous chloride (calomel) electrode in saturated KCl solution, connected to the solution to be measured by a salt bridge of saturated KCl solution. The glass electrode consists of a bulb of special glass containing an acid solution, and an inner electrode of fixed voltage, usually Ag-AgCl, to conduct electrons reversibly into and out of the solution. When the bulb is immersed in a solution, an electric potential is developed between the inner and outer solutions that is proportional to the logarithm of $[\mathrm{H}^+]$ in the external solution. Thus, the reference electrode has a constant potential, while that of the glass electrode will vary with the hydrogen ion activity, so that the overall voltage is a function of the hydrogen ion activity. This cell may be written

$$\text{Glass electrode, sol. } x \mid \mathrm{KCl(sat.)}, \mathrm{Hg_2Cl_2(s)}; \mathrm{Hg(l)}. \tag{5.2}$$

The theory of the glass electrode is discussed in some detail in Chapter 8. From equation (8.66a) of that chapter we see that the half-cell potential of a hydrogen ion sensitive glass electrode is given, for a solution having hydrogen ion activity $[\mathrm{H}^+]_x$, by

$$E'_x = C_\mathrm{H} + \frac{2.303 RT}{\mathscr{F}} \log[\mathrm{H}^+]_x \tag{5.3}$$

i.e.,

$$E'_x = C_\mathrm{H} - \frac{2.303 RT}{\mathscr{F}} \mathrm{pH}. \tag{5.4}$$

The overall potential of cell (5.2) is given by the difference of the two half-cell potentials

$$E_x = E'_x - E_\mathrm{ref} = C_\mathrm{H} - \frac{2.303 RT}{\mathscr{F}} \mathrm{pH} - E_\mathrm{ref}, \tag{5.5}$$

(neglecting "liquid junction potential," which will be discussed later). If, then, a given experimental arrangement, as represented by cell (5.2) is used first to determine the EMF developed by a solution whose pH is to be measured, and then the test solution is replaced by a standard reference solution of assigned pH value, pH_s, and the new EMF measured (at the same temperature), we have

$$E_x = C_H - \frac{2.303RT}{\mathscr{F}} pH_x - E_{ref} \tag{5.6}$$

and

$$E_s = C_H - \frac{2.303RT}{\mathscr{F}} pH_s - E_{ref}. \tag{5.7}$$

Subtracting (5.7) from (5.6), and rearranging

$$pH_x = pH_s - \frac{\mathscr{F}(E_x - E_s)}{2.303RT}. \tag{5.8}$$

Equation (5.8) is the basis for the modern operational definition of pH.

In deriving equation (5.8), one property common to EMF cells of the type illustrated by cell (5.2) was neglected. It will be noted in cell (5.2) that a vertical line is drawn between sol. x and KCl(sat.). This denotes a "liquid junction potential," a potential that arises at the interface of two different solutions, i.e., solutions of different electrolytes, or solutions of different concentrations of the same electrolyte. The use of the saturated KCl bridge helps to minimize this potential. If the unknown and the standard solutions are nearly alike in composition and concentration, then the liquid junction potential will be the same for each and will cancel, leaving equation (5.8) as written.

Standard pH Reference Solutions

Certified samples of buffer substances from which standard reference solutions of reproducible and precisely defined pH value can be prepared are furnished by the U.S. National Bureau of Standards. In turn, most manufacturers of pH equipment supply concentrated solutions of these standard substances, to be diluted for use. The pH values established for seven reference solutions, over the temperature range 0° to 95 °C, are given in Appendix 5.

The pH values of these reference solutions are established from EMF measurements of cells having no significant liquid junction potential. The cells used involve chloride ions, and the accuracy of the value of the pH assigned to the reference solution rests ultimately upon the convention used in assigning a value to γ_{Cl^-} in a solution of low ionic strength, and infinitely dilute with respect to chloride ion. The scale of hydrogen ion activities so established by these standard solutions is called a *conventional activity scale*.

Accuracy of pH Measurements

Measurements of pH can be useful for many purposes without interpretation in terms of hydrogen ion activities. However, for most physical chemical calculations a pH measurement is only as meaningful as the accuracy with which equation (5.1) applies to that measurement. That is, we are not interested in pH measurements per se, but rather in $[H^+]$ values. From equation (5.8) it follows that in order to equate pH_x with $-\log [H]_x$, it is necessary that pH_s be known accurately in terms of $[H]_s$. It is believed that the pH reference solutions of the National Bureau of Standards have theoretical significance in terms of $[H^+]$ to about ± 0.01 pH. In addition, experimental measurement of pH_x in terms of pH_s is subject to the error arising from the difference in the liquid junction potential when the unknown solution is in the cell and when the standard solution is in the cell. The more nearly alike the test solution and standard solution are in composition and concentration, the smaller this error will be.

It is difficult to generalize about the magnitude of the liquid junction potential error likely to be encountered in practice. When the pH of both the test solution and the standard solution lie between 2 and 12, and the test solution is an aqueous solution of ionic strength between 0.01 and 0.1, and free from colloids and suspensions, the error due to the liquid junction potential will be negligible. Probably, with most dilute to moderately concentrated aqueous solutions, in this pH range, the overall error to be feared will range from about ± 0.02 pH to about ± 0.04 pH. However, an estimate of the error can only be made in each specific case. The details of these problems are discussed in the references given at the beginning of this chapter.

It will be obvious from the foregoing discussion that by *accuracy* of a pH measurement, we mean the validity of the measurement in terms of hydrogen ion activity. Accuracy of measurement, so defined, is to be contrasted with *reproducibility* of measurement, or with *discrimination* of measurement. Equipment presently available ranges from pocket-sized, battery-operated units readable to 0.1 pH, to elaborate laboratory setups capable of yielding results with an average deviation of ± 0.0004 pH, upon repeated measurement, under very carefully controlled conditions.

The Glass Electrode

The theory of the glass electrode as a pH-sensitive device is covered in Chapter 8. Here, we discuss a few of the practical aspects. Modern glass electrodes can be obtained in a variety of shapes and sizes. The "general purpose" electrode behaves most efficiently in the pH range 0 to 11. For measurements above pH of about 11, "high pH" electrodes must be used for accurate work. In high pH solutions the hydrogen ion activity is so low, and the activity of alkali metal (or alkaline earth) ions usually so high, that the ordinary pH electrode begins to respond to these latter ions. Correction tables for this effect are available, but use of the special electrodes is preferable. Fortunately, few natural solutions have pH values as high as 11.

Although the change in EMF of various glass electrodes for a given change in pH is the same (at the same temperature), it is rare that two electrodes exhibit the same EMF in a given solution. Therefore, each glass electrode is calibrated individually against standard buffer solutions. It is common practice to calibrate against a single buffer solution, usually at pH 7. However, from the previous discussion on the accuracy of pH measurements it is obvious that the pH of the standardizing solution should be near that of the solution to be measured, if accurate results are to be obtained. Also, from equation (5.7) it is seen that the theoretical slope of the line relating the EMF of a glass electrode (measured against a reference electrode) to the pH is $-(2.303RT)/\mathscr{F}$; i.e.,

$$\frac{dE_s}{d\text{pH}} = -\frac{2.303RT}{\mathscr{F}} \quad \text{(at constant } T\text{)}. \tag{5.9}$$

At 25 °C,[8]

$$\frac{dE_s}{d\text{pH}} = -0.05916 \text{ volt/pH}. \tag{5.10}$$

By measuring the EMF against at least two, and preferably three, different buffers, this slope can be checked. At the same time the overall behavior of the whole instrumental setup can be assessed. This calibration of the slope must be carried out for each new glass electrode, and checks of the slope should be run from time to time during the course of using the electrode for making measurements on unknown solutions. When high accuracy is sought, the standard pH solutions used for calibration should cover only a short range and bracket the pH values to be measured.

The Reference Electrode

Various electrodes can be used in conjunction with the glass electrode in the measurement of pH; the chief requisite is that the reference electrode provide a constant voltage at a given temperature. Currently,

---

[8] See Chapter 1 for the evaluation of $(2.303RT)/\mathscr{F}$; tables of numerical values of this expression as a function of temperature are given in G. Mattock, *op. cit.*, p. 395, and R. G. Bates, *op. cit.*, p. 313.

the *saturated calomel electrode* is in most general use. It consists of mercury in contact with mercurous chloride, in turn in contact with a saturated solution of potassium chloride. The half-cell reaction for this electrode is

$$2Hg_l + 2Cl^-_{aq} = Hg_2Cl_{2\,c} + 2e. \tag{5.11}$$

The half-cell potential for (5.11) is given by

$$Eh = E° + \frac{RT}{2\mathscr{F}} \ln \frac{[Hg_2Cl_2]}{[Hg]^2[Cl^-]^2}. \tag{5.12}$$

The activities of Hg_2Cl_2 and Hg are constant and equal to unity (at 1 atmosphere), and since the KCl solution is in equilibrium with solid KCl, the activity of Cl^- is constant. Hence, (5.12) becomes

$$Eh = E° + \frac{RT}{2\mathscr{F}} \ln \frac{1}{k^2}. \tag{5.13}$$

At 25 °C, the value of Eh for the saturated calomel electrode is 0.2444 volt, positive relative to the standard hydrogen electrode. Accurate values for the potential of this electrode at other temperatures are listed in Ives and Janz.[9] Inspection of these values, shows that for temperatures in the neighborhood of 25 °C, the approximate equation

$$Eh = 0.2444 - 0.00066\,(t - 25) \text{ volt} \tag{5.14}$$

is obeyed, where t is the temperature in degrees C. A knowledge of the numerical value of the calomel electrode is not needed for the method of pH measurement developed here, as expressed in equation (5.8); however, its value will be required for Eh measurements to be discussed subsequently.

In practice, the calomel electrode is connected to the solution to be measured by a variety of devices. Some electrodes are designed to be immersed directly in the solution and have an asbestos fiber, or ceramic plug, sealed into the base of the electrode. The fiber or plug becomes saturated with KCl solution, and the K^+ and Cl^- provide the necessary electrical connection to the solution. In others a hole in the side of the electrode is covered by a ground-glass sleeve fitted to the electrode, so that connection is by a thin film of KCl solution between the sleeve and the electrode. In still others the calomel electrode is maintained externally to the solution, and connection is through a salt bridge. The most commonly used electrodes are designed for immersion in the solution to be measured, and a small amount of KCl inevitably diffuses into the solution from the electrode, so that measurements of solutions containing ingredients sensitive either to K^+ or Cl^- may be affected.

[9] D. J. G. Ives and G. J. Janz, *op. cit.*, p. 161.

If so, the use of a salt bridge containing a nonreacting electrolyte is indicated.

If an electrode is immersed in a solution for hours or days, not only will KCl contamination of the solution become serious, but diffusion of solution ions into the electrode will also occur and may change its EMF. Thus, it is good practice to replace the KCl solution after each 24 hours or so of immersion.

pH-Measurement Precautions

Directions for operation of various types of pH equipment are available with the apparatus, but some remarks about precautions to be observed, especially in making readings in the field, may be helpful.

The glass electrode is a high-resistance electrode, and the instrument used to measure the EMF between the glass and the calomel electrodes, the "pH meter," is a high-impedance electrometer. Because of the sensitivity of the equipment, glass electrode leads are furnished with metallic sheathing to provide electrical shielding against stray electric fields or capacitance effects generated by the operator and which could lead to erratic pH readings. When long glass electrode leads are required in the field, care must be taken to ensure that these are continuously shielded along their entire lengths.

The role of temperature must be kept continuously in mind. The pH meter must be standardized with buffers at the same temperature as that at which the pH of the test solution is measured.[10] The value of the standard reference buffer at the measured temperature is used. Most instruments have a "temperature compensator," which can be adjusted to the temperature of the test solution. This adjustment will have the effect of putting the correct value of the slope of equation (5.7), i.e., $-(2.303RT)/\mathscr{F}$, into the pH meter.

The several electrodes of a pH measuring system, any standard cell used in the meter, and the batteries of portable instruments change their EMF with temperature. The result is that the instrument zero of the usual pH meter is subject to drift, and must be checked and reset, if necessary, before and after each pH reading is made. This is particularly important in outdoor work. In high mountains, for example, early morning readings may be made at an instrument temperature of 10 °C, whereas in the early afternoon, instrument temperatures may be close to 40 °C. This change has a marked effect on battery and amplifier performance; if frequent zero checks are not made, the error in pH may be as much as 0.5 unit. In a case of temperature change of this kind,

---

[10] A few instruments provide a "buffer compensation" control that permits the use of buffers at a temperature different from that of the measurement (over a limited range).

consideration must also be given to any change in temperature of the solutions being measured, and restandardization of the meter must be carried out at a new temperature, if necessary.

In general, electrode response to homogeneous solutions is rapid, but between readings the glass electrode should be rinsed several times with distilled water to remove adsorbed ions. Because equilibrium is reached at a decreasing rate, readings should be made at regular intervals until drift has ceased. A small change in pH with time should never be accepted as an approximation of the final value; in all too many instances the reaction of the electrodes to a new environment is rapid at first, and then changes to a slow, steady drift over a considerable pH range before final equilibrium with the solution takes place.

pH Readings in Natural Media

A major problem of obtaining satisfactory pH readings from earth surface environments is that of introducing electrodes into the environment without changing it significantly. Good results can be obtained from open waters such as streams, lakes, and oceans, which are already in intimate contact with the atmosphere. On the other hand, accurate values for environments not open to the atmosphere are difficult to obtain because of contamination during or after emplacement of the electrodes. Entry of the electrodes commonly introduces gases or permits them to escape.

Little has been done in the development of special electrodes and techniques for sampling natural waters *in situ*. The tendency has been to bring the water to the instrument, which usually permits exchange with the atmosphere. This is reflected by a steady drift in the pH readings, and the operator is in the unenviable position of attempting to guess what the reading would have been without exchange. Also, the practice of bringing the sample to the instrument allows temperature change, which promotes gain or loss of volatiles. Carbon dioxide and hydrogen sulfide are two of the most important volatile constituents, and both tend to be the constituent of the system controlling the pH, so that extraordinary precautions have to be taken in protecting any sample collected and brought to the meter.

The difficulties of sampling homogeneous solutions are considerable, but they are minor compared with the problems related to measuring and interpreting the results of pH determinations on mineral suspensions or pore waters of rocks. For example, the pH of stirred suspensions may be different from that of the supernatant after settling, indicating an effect on the electrodes of particles with adsorbed hydrogen ion. This effect has been studied by Hauser and Reed[11] in their cation-exchange

[11] E. A. Hauser and C. E. Reed, *Soil Sci.*, *59*, 175 (1937).

capacity study of soils. The electrodes "see" only their immediate environment, so that great care must be exercised to be sure that the medium adjacent to the electrodes is that which the operator wishes to measure. There is a tendency to use pH measurements made in muds or similar environments in chemical calculations valid only for homogeneous solutions; the results are highly dubious.

Innumerable determinations of "soil pH" are available in the literature. The usual practice is to suspend a given weight of soil sample in a given weight of water, and then to determine the pH of the slurry. The results are useful only in a highly qualitative sense, inasmuch as they represent the complex contribution of the hydrogen ions in the original water, plus those released by the soil sample, plus effects of soil colloids on the electrodes.[12] If the suspension used is open to the atmosphere, and the sample uses up hydrogen ions from the added water, the original reaction may be to give a high pH, followed by a downward drift as CO_2 is absorbed. A good example of the kinds of changes that can occur in alkaline media is illustrated by the following experiment. When an excess of finely ground calcite is added to deaerated water, the pH of the supernatant solution is about 9.9. On opening the system to the atmosphere, a downward drift sets in, with a final pH at equilibrium of 8.4. Thus, the difference with and without air is 1.5 pH units.

An interesting study has been made by Gorham[13] of pH readings in soils and lake muds, in which he first read the pH by direct electrode insertion; then he squeezed out the soil water and measured the pH of it before aeration; finally he aerated the expressed water and obtained a third reading. His results are given in Table 5.1. The low pH readings obtained by direct insertion were attributed to the suspension effect, and the high readings after aeration to loss of CO_2 during equilibration with the atmosphere. There is perhaps a question as to whether the differences in pH values obtained by direct insertion and those obtained on freshly expressed waters from lake muds may not represent partial aeration. In measurements on marine sediments, Siever et al.[14] found that the pH values obtained by direct insertion of the electrodes into fresh piston cores ranged from 7.4 to 7.8, whereas the waters expressed from these cores had pH values of 8.1 to 8.2. However, when the waters are expressed under nitrogen, they have the same pH values as are obtained

---

[12] The apparent lowering of the pH value in a sediment with respect to its supernatant solution is considered in G. Mattock, op. cit., pp. 172–173. The cause and magnitude of this "suspension effect" has been the subject of considerable discussion; it appears not to be very well understood. However, where the suspended material is well-aggregated, as with many clays, the effect is not encountered; it becomes important when the particles are truly colloidal and highly charged, and when the system is on the acid side.

[13] E. Gorham, Ecology, 41, 563 (1960).

[14] R. Siever, R. M. Garrels, J. Kanwisher, and R. A. Berner, Science, 134, 1071 (1961).

TABLE 5.1. Three Methods of pH Measurement on Natural Soils of the English Lake District

	pH by Glass Electrode		
	Direct insertion	Expressed solution	Aerated solution
Mor humus layers	3.5	3.9	3.9
	3.4	4.1	4.1
	3.4	4.3	4.3
	3.5	4.5	4.5
	3.3	4.6	4.8
Mull humus layers	4.6	5.9	6.5
	5.3	6.0	6.6
	5.4	6.2	7.1
	5.7	6.5	7.3
	5.4	6.6	7.1
Lake muds[a] depth in core (cm)			
20–40	6.5	6.8	8.1
80–100	6.4	6.8	8.5
100–120	6.3	6.6	8.4
160–180	6.3	6.8	8.0
220–240	6.2	6.7	8.0
250–270	6.2	6.8	8.3
290–310	6.3	7.0	8.4
340–360	6.3	6.8	8.2
385–405[b]	6.6	6.6	7.1

[a] pH of water above mud 6.8, rising to 7.6 after aeration.
[b] Mud less organic and somewhat clayey.
SOURCE: From E. Gorham, *Ecology*, 41, 563 (1960).

by direct insertion,[15] indicating that when the squeezing takes place in the atmosphere, the waters are aerated and CO_2 is lost. We tend to conclude, at the present time, that the pH of pore waters measured by direct insertion of electrodes is a satisfactory measure of the pore water environment, except for acid samples, where the suspension effect may become important.

A very useful new instrument, called the "carbonate saturometer," has been developed by Weyl.[16] This consists essentially of a pH electrode, reference electrode, and portable high-impedance millivolt meter. The EMF cell is so arranged that powdered carbonate can be introduced in the vicinity of the pH electrode.

The instrument takes advantage of the fact that the pH of an aqueous solution will change if a carbonate is either dissolved or precipitated.

[15] R. Siever, Harvard University, personal communication.
[16] P. K. Weyl, *Jour. Geol.*, 69, 32 (1961).

Thus, if a powdered carbonate is introduced into a solution undersaturated with respect to that particular carbonate, the system will change toward equilibrium and the hydrogen ion activity will be lowered. Conversely, if the solution is supersaturated, precipitation will take place and H^+ will be released. Weyl has developed the theory of the carbonate saturometer so that it can be calibrated in terms of a particular carbonate for any type of solution. In applying the instrument to artificial sea water, he has shown that in the pH range from 7.5 to 8.3, a sensitivity of 1 millivolt for a change of 1 ppm of calcite is obtained.

Eh MEASUREMENT

Definitions, Conventions, and Procedures

In Chapter 1, the oxidation potential Eh was defined as the potential of a half-cell, referred to the standard hydrogen half-cell, the EMF of the standard hydrogen half-cell being taken as zero at all temperatures, by definition. The relationships among Eh, $E°$, ΔF, $\Delta F°$, the equilibrium constant K, and the reaction quotient Q were also considered in that place.

In order to fix unequivocally the *sign* of any half-cell potential, relative to the hydrogen half-cell, certain conventions must in turn be adopted. These conventions involve the way half-cell reactions are written and the sign used in the relationship between ΔF and $nE\mathscr{F}$. In this book, we write half-cell reactions with the electrons appearing on the right side. Thus, for example, for the formation of Zn^{++}_{aq} from metallic zinc, we write

$$Zn = Zn^{++}_{aq} + 2e. \qquad (5.15)$$

In Chapter 1, we adopted the relationships (with sign as shown)

$$\Delta F = +nE\mathscr{F},$$

and

$$\Delta F° = +nE°\mathscr{F} = -RT \ln K.$$

These conventions are sufficient to fix the sign of any half-cell potential.

Thus, for the half-reaction (5.15), $E° = -0.763$ volt at 25 °C. The sign convention adopted here is the same as the so-called European sign convention. Many American books adopt the same way of writing the half-reactions as shown in (5.15), but set

$$\Delta F° = -nE\mathscr{F} = -RT \ln K.$$

By the American convention the half-reaction (5.15), at 25 °C, has an $E° = +0.763$ volt. The half-cell potentials on the European and American systems have the same numerical values, but opposite signs.

Oxidation potential is measured with an electrode pair consisting of an inert electrode and a reference electrode. The same reference electrode is used for most Eh measurements as for pH measurements—the saturated calomel.

The inert electrode used most is bright platinum, although the gold electrode has a fair number of applications. As in pH measurements, the role of the calomel electrode is to supply a known EMF and to make electric connection with the system to be measured. The inert electrode acts as an electron acceptor or donor to the ions in the measured solution. When connected to the calomel electrode, the platinum electrode can accept electrons from dissolved ionic species, or it can give them up, depending on whether the potential of the half-cell containing the dissolved species is greater or less than that of the calomel reference electrode.

As an example of an oxidation-reduction system whose oxidation potential is to be measured, we consider an aqueous solution containing Fe^{++} and Fe^{3+}, with $[Fe^{++}] = 0.001$, and $[Fe^{3+}] = 0.01$. The appropriate EMF cell may be represented as

$$\text{Pt (inert)} \left| \begin{matrix} Fe^{3+}_{aq} & 0.01 \\ Fe^{++}_{aq} & 0.001 \end{matrix} \right| \left| \begin{matrix} Hg_2Cl_{2\,c} \\ KCl_{aq\,sat} \end{matrix} \right| Hg_l. \qquad (5.16)$$

Following the methods outlined in Chapter 1, we consider the EMF of the cell (5.16) in the following way. The two half-reactions and half-cell potential symbols are

$$2Fe^{++}_{aq} = 2Fe^{3+}_{aq} + 2e, \qquad Eh_{Fe^{++},Fe^{3+}} \qquad (5.17)$$

$$2Hg_l + 2Cl^-_{aq} = Hg_2Cl_{2\,c} + 2e, \qquad Eh_{ref}. \qquad (5.18)$$

Subtracting (5.18) from (5.17), we have

$$2Fe^{++}_{aq} + Hg_2Cl_{2\,c} = 2Fe^{3+}_{aq} + 2Hg_l + 2Cl^-_{aq}, \qquad (5.19)$$

and

$$EMF_{(5.16)} = Eh_{Fe^{++},Fe^{3+}} - Eh_{ref}. \qquad (5.20)$$

Equation (5.19) gives the overall cell reaction, and equation (5.20) the relation between the potential of cell (5.16), the oxidation potential of the Fe^{++}, Fe^{3+} couple, and the potential of the calomel reference electrode. The value of Eh_{ref} is constant at a given temperature and pressure, and for the saturated calomel electrode at 25 °C is 0.2444 volt, as discussed in the section on pH measurement.

By virtue of equation (1.24a), we write, for 25 °C, for the half-reaction (5.17)

$$Eh_{Fe^{++},Fe^{3+}} = E^\circ_{Fe^{++},Fe^{3+}} + \frac{0.0592}{2} \log \frac{[Fe^{3+}]^2}{[Fe^{++}]^2}.$$

Simplifying,

$$Eh_{Fe^{++},Fe^{3+}} = E^\circ_{Fe^{++},Fe^{3+}} + 0.0592 \log \frac{[Fe^{3+}]}{[Fe^{++}]}. \quad (5.21)$$

Substituting numerical values for $E^\circ_{Fe^{++},Fe^{3+}}$ and for the activities of the ions, we have

$$Eh_{Fe^{++},Fe^{3+}} = 0.771 + 0.0592 \log \frac{0.01}{0.001}$$

$$= 0.830 \text{ volt}. \quad (5.22)$$

Substituting this value and the value of the potential of the calomel electrode into equation (5.20), we have

$$EMF_{(5.16)} = 0.830 - 0.244 = 0.586 \text{ volt}.$$

The potential of 0.586 volt is the value that would be obtained upon measurement of cell (5.16) (neglecting liquid junction potential, which is minimized through use of the saturated KCl bridge).

A little consideration will show how the methods outlined in the foregoing may be generalized. The half-reaction for every oxidation-reduction system can be written

$$\text{reduced state} = \text{oxidized state} + ne, \quad (5.23)$$

and the corresponding half-cell potential as

$$Eh = E^\circ + \frac{RT}{n\mathscr{F}} \ln \frac{[\text{oxidized state}]}{[\text{reduced state}]}, \quad (5.24)$$

The reference electrode simply acts as a half-cell of constant potential (at a constant temperature), and the overall cell EMF is given by

$$EMF_{cell} = Eh_{\text{reduced state, oxidized state}} - Eh_{ref}. \quad (5.25)$$

The use of numerical values obtained in Eh measurements to obtain activity ratios of ions, by means of a relation of the type given by equation (5.24), is predicated upon certain assumptions. These assumptions include the following:

1. All species involved in the oxidation-reduction system are in internal equilibrium; that is, the measured EMF truly indicates the tendency of the system to oxidize or reduce. For example, if a solution containing ferric ion is added to one containing vanadous ion, it is assumed that they will interact

$$Fe^{3+} + V^{3+} + H_2O = Fe^{++} + VO^{++} + 2H^+,$$

and that the resulting proportions of ions are equilibrium proportions. In general, this assumption is satisfactory for simple dissolved species, with the notable exceptions, for inorganic ions, of sulfate or bisulfate ions and dissolved oxygen.

2. The platinum or gold electrode functions as an inert electrode. In some solutions the platinum or gold may react and become coated with another substance, in which case readings are useless for calculations involving dissolved species. Notable offenders are solutions containing divalent sulfur, in which platinum electrode behavior may become erratic. Fortunately, the worst offenders in causing reaction of gold or platinum are strong oxidizing agents not encountered in nature.

Eh Electrodes

The platinum electrode functions best if it has a large surface, so that "thimble type" electrodes generally are more satisfactory than "button type" or single platinum wires. As indicated before, the platinum electrode may become coated by reaction with the solution to be measured or by precipitates from the solution. Various cleaning techniques have been employed, but the most satisfactory seems to be mechanical cleaning with fine emery paper. Drastic chemical cleaning may in many instances simply lead to the formation of a new reaction coating.

Various oxidation-reduction systems of experimentally determined potentials may be prepared against which the platinum-calomel electrode pair can be checked. In a description of a study of the oxidation potentials of marine sediments, ZoBell[17] suggests the use of a solution $1/300$ M in potassium ferrocyanide, $1/300$ M in potassium ferricyanide, and $1/10$ M in KCl. He gives the Eh for this system as 0.430 volt at 25 °C. The "formal potential" of a solution, as proposed by Swift,[18] is of utility in this regard. The formal potential of a solution is the experimentally determined potential of a solution containing both the oxidized and reduced species, each at a concentration of 1 formal, together with other substances at given concentrations. Lists of values of formal potentials are given by Swift.

The reference electrode usually used in Eh measurement is the saturated calomel electrode, and the remarks made about it in the discussion on pH measurement are applicable here. In addition, it appears that contamination of the calomel electrode by diffusion of solution into the electrode is even more serious in the case of Eh measurements than it is for pH, and electrodes must be cleaned after every few hours of use. When the calomel electrode is dipped into a solution 5 to 10 degrees above the initial temperature of the electrode, a half-hour or so may be necessary for the electrode to come to equilibrium.

[17] C. E. ZoBell, *Bull. Am. Assoc. Petrol. Geol.*, **30**, 477 (1946).
[18] E. H. Swift, *A System of Chemical Analysis*. New York, Prentice-Hall, 1939, p. 540.

Eh Measurement Precautions

Eh measurement is subject to most of the precautions applicable to pH measurement in terms of instrumentation, except that the measuring circuit has relatively low resistance, and fewer difficulties are experienced in the shielding of electrode leads.

In Eh measurements, most of the difficulties result from contamination of the platinum electrode, and it is a useful experimental device to have two platinum electrodes available so that they can be alternated in the circuit. The poisoning of the platinum electrode is usually an erratic process, so that any such effects show up as differences in potential from the two electrodes. Experience shows that even careful work cannot be duplicated closer than about 5 millivolts, and that differences in replication of measurements on natural media may be of the order of 10 to 20 millivolts.

A convenient method of Eh-pH measurement in the laboratory is by use of multichannel continuous recording equipment. In such permanent setups it is not difficult to achieve good grounding of instruments and solutions, and shielding of leads. Continuous recording permits observation and determination of trends in values over long periods of time.

Eh Readings in Natural Media

The number of Eh readings that have been made in natural media is but a small fraction of the number of pH measurements. The values available from systems in equilibrium with the atmosphere have been disappointing in the sense that they show little range of Eh. The readings are in accord with the "irreversible oxygen potential" as discussed by Merkle.[19] Dissolved oxygen does not exert the potential expected if it is functioning at equilibrium; instead it acts like a much weaker oxidizing agent. Apparently there is a slow oxidation step of low potential in the action of dissolved oxygen on dissolved ionic species. As a result, the effect of the oxygen of the atmosphere is to provide a relatively mild oxidizing effect, with an Eh of about 0.650 to 0.700 volt at pH = 0, and of 0.300 to 0.350 volt at pH = 8. Although this effect is far less than theoretical, admission of air into systems of originally low Eh is followed by rapid reaction to the "irreversible" potential. Also, there seems to be some tendency for direct action of dissolved oxygen on the platinum electrode, so that the potential observed in systems containing dissolved air is to a certain extent independent of the oxidation

[19] F. G. Merkle, Oxidation-reduction processes in soils, in *Chemistry of the Soil*, Firman E. Bear (ed.). (ACA Monograph 126.) New York, Reinhold, 1955, pp. 200–218.

state of the dissolved components. In short, systems exposed to air show potentials (in volts, at 25 °C) according to the approximate relation

$$Eh = 0.70 - 0.06 \text{ pH}. \qquad (5.26)$$

This is in contrast to the expected value at equilibrium of

$$Eh = 1.23 - 0.06 \text{ pH}. \qquad (5.27)$$

It is difficult to measure natural systems out of equilibrium with air because of the problem of introducing electrodes without simultaneously introducing air. The general technique of bringing samples to the instrument is even more unsatisfactory in making Eh measurements than it is for pH measurements because of the rapid and homogenizing effect of oxygen. On the other hand, careful work can yield results of great utility, as shown by the excellent study of environmental relations in aerated and waterlogged soils by Starkey and Wight.[20] They designed a sealed probe that could be driven into waterlogged soils and then opened to permit the soil slurry to come into contact with pH and Eh electrodes. Some of their measurements and correlations with iron corrosion are given in Table 5.2.

TABLE 5.2. Eh-pH Measurements of Waterlogged Soils

Eh	pH	Corrosion Effects on Iron Pipe
0.046	6.5	Crust of FeS, corrosion severe
−0.293[a]	6.7	Crust of FeS, corrosion moderate
0.206	6.5	No corrosion observed
0.370	6.2	Slight corrosion
0.563	7.7	No corrosion
0.544	7.6	Small amount iron oxides, little corrosion

[a] The significance of this low potential is not understood.
SOURCE: Starkey and Wight, op. cit.

Most of the other trustworthy Eh measurements of natural conditions come from studies of deoxygenated bottom environments of lakes and seas. They indicate that the extreme range of reducing environments can be attained, for the lowest potentials are those of waters containing dissolved hydrogen, inasmuch as analyses show the presence of hydrogen gas. Just as for pH, many measurements have been made of the Eh of soils and sediments, in which a given weight of the sample is suspended in a given quantity of water. The significance of such Eh measurements is not at all clear. Most results apparently indicate the fact that the experimenter did not preclude air from suspension. In others in which air is excluded (except for that already in the sample?), the readings

[20] R. L. Starkey and K. M. Wight, *Anaerobic Corrosion of Iron in Soils*. New York, American Gas Association, 1946.

apparently measure the relative proportions of various oxidation-reduction ion pairs that happen to be soluble in water. It certainly is questionable whether the readings bear any relation either to the bulk composition of the sediment or to what might have been expected from a pore water originally in contact with the sediment.

The great unexplored region is the Eh of pore waters in natural contact with rocks. The experimental difficulties of placing electrodes so that they will not permit access of foreign materials to the environment or inhibit natural circulation are great, but solution of the problem will tell us much about oxygen loss from migrating waters and about rates at which pore waters equilibrate with their surroundings.

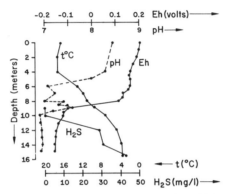

FIG. 5.1. Eh and pH (as well as H_2S concentration and temperature) as a function of depth in a brackish stagnant bay. [After F. Manheim, *In situ* measurements of pH and Eh in natural waters and sediments. Preliminary note on applications to a stagnant environment: *Stockholm Contributions in Geology*, 8, 27 (1961). Reproduced by permission.]

Manheim[21] has successfully studied the Eh and pH, as functions of depth to about 16 meters, in a stagnant bay in the Baltic area. He used an ingenious method, in which he placed a platinum collar as an Eh electrode just above the sensitive bulb of the glass pH electrode, and used the metallic shielding-sheath of the connector lead to the pH electrode as the connecting wire for the platinum electrode. The Eh and pH electrodes were then lowered as a dual electrode at the end of a single shielded cable. The reference electrode was kept at the surface on a float. Temperature corrections were carefully made. Some of the results obtained are shown in Figure 5.1.

[21] F. Manheim, *Stockholm Contributions in Geology*, 8, 27 (1961).

Germanov et al.[22] made measurements of the Eh and pH of ground waters, using care to prevent atmospheric contamination by sampling flowing boreholes, at the surface or underground, with suitable attendant precautions. They found a range of Eh from $+0.550$ to -0.480 volt, and correlated their results with the dissolved gases present in the waters. High (positive) potentials were related to free oxygen, low potentials chiefly to hydrocarbon gases, as might be expected. Oxygenated waters were found, in some cases, as much as 1000 meters below the water table.

MEASUREMENT OF INDIVIDUAL ION ACTIVITIES

An application of EMF measurement that seems to offer considerable potential usefulness for geological purposes is the determination of individual ion activities by use of suitable electrodes. In recent times, increasing emphasis has been placed by scientific workers upon "operational" concepts and procedures, i.e., upon those concepts and procedures that are based on directly measurable quantities. Accordingly, because single ion activities cannot be measured directly and can be evaluated from EMF measurements only after the introduction of one or more assumptions of an extrathermodynamic nature, the tendency has been to neglect the usefulness of EMF cells for deriving even approximate values of single ion activities. Of course, since mean activity coefficients can be measured directly and accurately by a variety of methods, including that of EMF measurement, their use is indicated whenever at all possible. Nevertheless, in many useful physical chemical calculations the employment of single ion activities appears to be unavoidable, and it is desirable to consider means of obtaining even approximate values for these. A good example of the need for a knowledge of single ion activities is furnished by the discussion of ion exchange in Chapter 8.

It is not feasible to consider here in detail the reasons why it is generally considered to be impossible to measure a single ion activity. Theoretical arguments usually revolve around the impossibility of measuring an experimental process by means of which a single charged species is transferred into or out of solution, because of the large electric field that would be created in separating more than a few ions from their oppositely charged counterparts.

It is beyond the scope of this text to attempt to assess the advantages and disadvantages of all possible electrodes that could be used, but it is important to point out that methods are available, or could be developed, to measure the activities of a great many ions in their natural habitat.

[22] A. I. Germanov, G. A. Volkov, A. K. Lisitsin, and V. S. Serebrennikov, *Geokhimiya*, *3*, 322 (1959).

In the following discussion we point out a few examples of these electrodes.

The newly developed cation selectively-sensitive glass electrodes discussed in Chapter 8 offer considerable promise as a very powerful investigative tool to the geochemist. Electrodes selectively sensitive to alkali ions and alkaline earth ions have been studied. Marshall and coworkers[23] have developed electrodes of another type for measurement of alkali metal ion activities in simple solutions. These are essentially concentration cells in which an inert electrode dips into an alkali salt solution, and the difference in activity between the alkali salt solution inside and outside the membrane sets up an EMF that can be calibrated in terms of the activity of the alkali metal salt in the solution. This device is similar in principle to the glass electrode. The equation appropriate to this type of cell is

$$E = \frac{RT}{n\mathscr{F}} \ln \frac{[\text{standard}]}{[\text{unknown}]}. \qquad (5.28)$$

Thus, in measuring [K$^+$], if the standard solution consists of a KCl solution of known, relatively low, ionic strength, where the approximation $\gamma_{K^+} = \gamma_{Cl^-}$ can be expected to be valid, the activity of K$^+$ in the standard solution can be readily calculated from known values of $\gamma_{\pm KCl}$, i.e.,

$$\gamma_{K^+} = \gamma_{\pm KCl}$$

$$a_{K^+} = \gamma_{K^+} \times m_{K^+},$$

and equation (5.28) becomes

$$E = \frac{RT}{\mathscr{F}} \ln \frac{(a_{K^+})_s}{(a_{K^+})_u}. \qquad (5.29)$$

Concentration cells of all kinds have long been investigated; equation (5.28) is the fundamental equation describing the behavior of these. In every case, single ion activities obtained through their use would be based on some reference standard, as shown.

Several satisfactory electrodes are available for measurement of anion activities. Among them is the silver-silver chloride electrode (commonly employed as a reference electrode), which may be used to determine chloride ion activity. The half-cell reaction is

$$Ag_c + Cl^-_{aq} = AgCl_c + e,$$

and the half-cell potential is

$$Eh = E° - \frac{RT}{\mathscr{F}} \ln a_{Cl^-}. \qquad (5.30)$$

---

[23] Cf. C. E. Marshall and W. E. Bergman, *J. Phys. Chem.*, **46**, 52 (1942).

The Ag-AgBr, and Ag-AgI electrodes can be used similarly to measure Br^- and I^- activities, respectively.

The silver-silver sulfide electrode responds to sulfide ions

$$2Ag_c + S^{--}_{aq} = Ag_2S_c + 2e,$$

$$Eh = E° - \frac{RT}{2\mathscr{F}} \ln a_{S^{--}}. \tag{5.31}$$

Oxide, oxygen, and sulfide electrodes, all of considerable potential importance to geochemists, are discussed in detail by Ives.[24]

In the discussion of oxide and oxygen electrodes given by Ives will be found the reasons for one of the major barriers to the use of metallic electrodes for measuring ion activities in natural environments. Most metals form oxide or hydroxide coatings and cease to function as metal-metal ion electrodes. In this circumstance, the electrode becomes responsive to hydrogen ions, and in fact many of the metal-metal oxide pairs can be used as pH electrodes. The other major stumbling block lies in the difficulty of reproducing the physical state of metals, owing largely to the introduction of strain in the fabrication of the metal, in all but the softest metals.

EMF MEASUREMENT OF GAS PRESSURE

Electrodes also are available for measurement of oxygen and carbon dioxide pressures.[25,26] The CO_2 electrode is a dipping type of electrode sealed off by a membrane that is permeable to CO_2 but impermeable to solution ions. The admitted CO_2 is dissolved in a solution of sodium bicarbonate contained within the electrode and changes its pH. This measured pH is a function of P_{CO_2}. The CO_2 electrode has been used by Moore et al.[27] to determine the carbon dioxide content of sea water and deep sea sediments.

The oxygen pressure electrode operates on the principle that the electrolysis current of water between inert electrodes at a constant voltage depends upon the concentration of dissolved oxygen.

EMF MEASUREMENTS AT HIGH TEMPERATURES AND PRESSURES

Some progress is being made in developing electrodes for use in studying aqueous solutions and other systems at high temperatures and

---

[24] D. J. G. Ives, in Ives and Janz, *op. cit.*, chap. 7.
[25] J. W. Severinghaus and A. F. Bradley, *J. Appl. Physiol.*, *13*, 515 (1958).
[26] D. B. Cater and I. A. Silver, in Ives and Janz, *op. cit.*, pp. 497–515.
[27] G. W. Moore, C. E. Roberson, and H. D. Nygren, *U.S. Geol. Survey Prof. Paper 450B*, B-83 (1962).

pressures. In a survey of recent developments in high-temperature EMF measurements, Lietzke[28] has pointed out that with suitable electrodes it should be possible to study at higher temperatures all the solution phenomena that can be determined at room temperatures, including activity coefficients, equilibrium constants, and solubilities of salts. He studied the Ag-AgCl and Hg-Hg_2Cl_2 electrode potentials as a function of temperature, and found that the Ag-AgCl electrode has excellent and nearly theoretical response up to 200 °C, and can be used with empirical corrections up to 260 °C. The electrode has a nearly constant temperature coefficient of approximately +0.3 millivolt per °C. A discussion of the Ag-AgCl electrode as a pH electrode at temperatures up to 250 °C and pressures up to 40 atmospheres is given by Janz.[29] Glass electrodes for measuring pH have been used successfully up to 150 °C and 1000 atmospheres.[30]

Finally, there is a nonaqueous system of electrolytes of considerable importance that must of necessity be studied at high temperatures, namely the system of fused salts. Procedures and modes of thought employed in the investigation of fused salts have many points in common with those used in the study of aqueous electrolyte solutions, but differ considerably in other details. The subject of electrodes in fused salt systems is treated at length by Laity,[31] and need not be further considered here.

SUMMARY

The pH of a solution is measured against standard reference buffer solutions, and under appropriate conditions pH may be equated to $-\log [H^+]$. Eh, the oxidation potential, is the potential of a half-cell referred to the standard hydrogen half-cell, taken to be zero by convention.

Measurement of Eh and pH for geological purposes is set about with many difficulties, and numerous precautions must be taken in order to obtain meaningful results. A number of these difficulties are described. Most values for Eh and pH in the literature are for soil and water systems that either are in intimate contact with the atmosphere or are subjected to the atmosphere after collection, but prior to and during measurement.

[28] M. H. Lietzke, Am. Inst. Chem. Engrs., Nuclear Eng. and Sci. Congress, 1955, Preprint 223.
[29] G. J. Janz, in Ives and Janz, *op. cit.*, p. 226.
[30] M. LePeintre, C. Mahieu, and R. Fournie, Commissariat à L'Energie Atomique de France, Rapport No. 1951, (1961).
[31] R. W. Laity, in Ives and Janz, *op. cit.*, chap. 12.

Techniques for introduction of electrodes into natural environments without contamination is a most pressing need today. The several excellent studies that have been made on free waters and on the pore waters of rocks and sediments indicate how useful such results can be geochemically, if they are not overinterpreted.

A field of great future potential lies in the possible development of electrodes selectively sensitive to individual cations and anions. The continued development of electrodes for high-temperature and high-pressure studies will be of great value for geological studies.

SELECTED REFERENCES

Baas Becking, L. G. M., I. R. Kaplan, and D. Moore, Limits of the natural environment in terms of pH and oxidation-reduction potentials: *Jour. Geol.*, *68*, 243 (1960).

Bates, R. G., *Electrometric pH Determinations*. New York, Wiley, 1954.
  Comprehensive reference for theory and measurement of pH.

Glasstone, Samuel, *An Introduction to Electrochemistry*. New York, Van Nostrand, 1942.
  General reference for electrochemical theory.

Ives, D. J. G., and G. J. Janz (eds.), *Reference Electrodes*. New York, Academic, 1961.
  Definitive treatment of the usual reference electrodes and others.

Mattock, G., *pH Measurement and Titration*. London, Heywood, 1961.
  Excellent overall treatment, but especially useful for the practical aspects.

Starkey, R. L., and K. M. Wight, *Anaerobic Corrosion of Iron in Soils*. New York, American Gas Association, 1946.

ZoBell, C. E., Studies on redox potentials of marine sediments: *Bull. Am. Assoc. Petrol. Geol.*, *30*, 477 (1946).

CHAPTER 6

Partial Pressure Diagrams

At 25 °C and 1 atmosphere total pressure, equilibrium relations for reactions of type 1, solid-solid reactions, are essentially impossible to represent, inasmuch as the mineral relations are functions only of temperature and pressure. It is mere chance if the free energy of such a reaction happens to be zero under these arbitrarily chosen conditions. On the other hand, reactions of type 2, involving a gas phase of variable activity, can be shown as a function of the gas phase activity. At the low temperature and pressure conditions of interest, the activity of the gas phase is equal to its partial pressure measured in atmospheres, for almost all gases behave nearly ideally under these conditions. The relations among fugacity, activity, and pressure have been discussed in some detail in Chapter 2.

STABILITY OF WATER

The most fundamental chemical relation of geological interest is the stability of water. If we assume equilibrium between water and its dissociation products hydrogen and oxygen, we can write

$$2H_2O_l = 2H_{2\,g} + O_{2\,g}. \tag{6.1}$$

The equilibrium constant is

$$\frac{P_{H_2}^2 P_{O_2}}{[H_2O]^2} = K \tag{6.2}$$

for nearly pure water $[H_2O] \cong 1$, and

$$P_{H_2}^2 P_{O_2} = K. \tag{6.3}$$

The free energy relations for equation (6.1) are

$$2\Delta F_{fH_2}^\circ + \Delta F_{fO_2}^\circ - 2\Delta F_{fH_2O}^\circ = \Delta F_r^\circ. \tag{6.4}$$

PARTIAL PRESSURE DIAGRAMS

Substituting numerical values and solving

$$0 + 0 + 113.4 = \Delta F_r^\circ$$

$$\Delta F_r^\circ = 113.4 \text{ kcal.} \tag{6.5}$$

Then, from the relation between standard free energy of reaction and the equilibrium constant, $\Delta F_r^\circ = -1.364 \log K$

$$113.4 = -1.364 \log K$$

$$K = 10^{-83.1}.$$

From (6.3)

$$P_{H_2}^2 P_{O_2} = 10^{-83.1}. \tag{6.6}$$

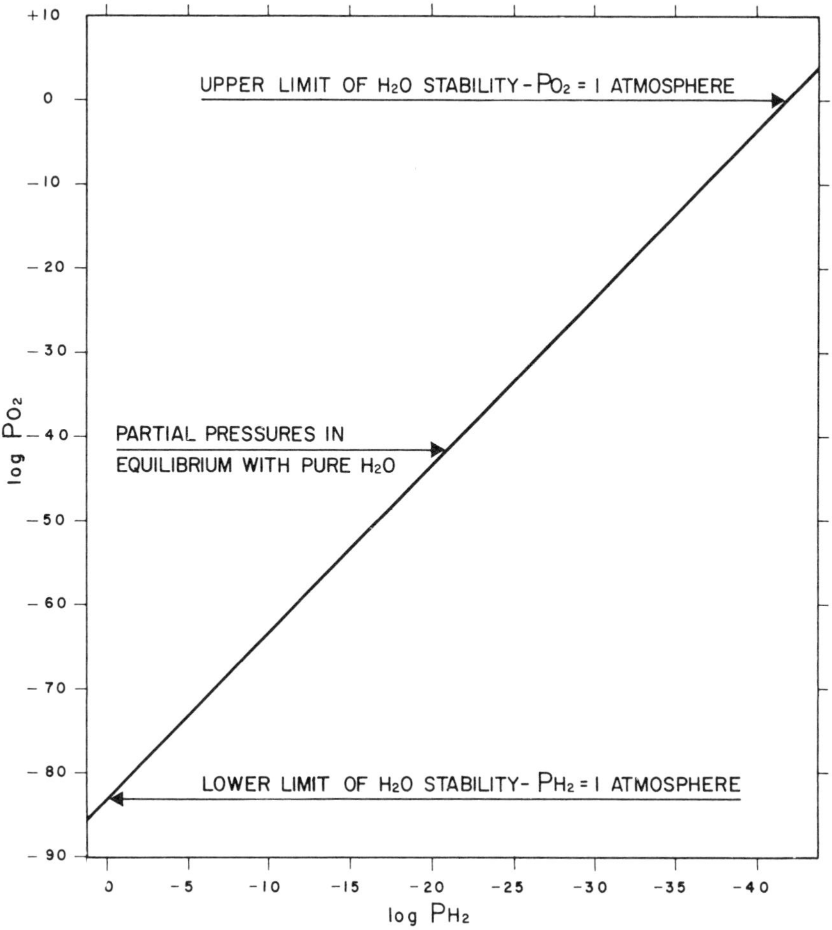

FIG. 6.1. Partial pressures of oxygen and hydrogen in equilibrium with water at 25 °C and approximately 1 atmosphere total pressure.

Thus, the degree of dissociation of water into hydrogen and oxygen at room temperature is extremely small, but nonetheless the partial pressure can be used, assuming equilibrium, to characterize the range of stability of water. If the partial pressure of either hydrogen or oxygen exceeds 1 atmosphere at earth surface conditions, the gas pressure will exceed that of the atmosphere, and oxygen or hydrogen, as the case may be, will be released from the water, which will decompose to yield more oxygen or hydrogen. When $P_{O_2} = 1$, then

$$P_{H_2}^2 = 10^{-83.1}; \quad P_{H_2} = 10^{-41.5}. \tag{6.7}$$

When $P_{H_2} = 1$, then

$$P_{O_2} = \frac{10^{-83.1}}{1} = 10^{-83.1}. \tag{6.8}$$

If an arbitrary value is given to either oxygen or hydrogen partial pressure, the equilibrium is fixed. It is convenient for further reference to make a plot of P_{H_2} against P_{O_2} so that if either gas pressure is known, the other can be obtained without calculation (Figure 6.1).

STABILITY OF IRON OXIDES

The stability relations among the various iron oxides of mineralogical interest can be handled in the same way. The first step is to write the reactions in terms of the iron compounds and gaseous oxygen, being careful to include all permutations and combinations

$$Fe_c + \tfrac{1}{2}O_{2\,g} = FeO_c \tag{6.9}$$

$$3Fe_c + 2O_{2\,g} = Fe_3O_{4\,c} \tag{6.10}$$

$$2Fe_c + \tfrac{3}{2}O_{2\,g} = Fe_2O_{3\,c} \tag{6.11}$$

$$3FeO_c + \tfrac{1}{2}O_{2\,g} = Fe_3O_{4\,c} \tag{6.12}$$

$$2FeO_c + \tfrac{1}{2}O_{2\,g} = Fe_2O_{3\,c} \tag{6.13}$$

$$2Fe_3O_4 + \tfrac{1}{2}O_{2\,g} = 3Fe_2O_{3\,c}. \tag{6.14}$$

Because the activities of the crystalline solids are unity, each reaction is at equilibrium at a particular partial pressure of oxygen. The calculations necessary are illustrated by solving for P_{O_2} for equilibrium in reaction (6.14)

$$\frac{[Fe_2O_3]^3}{[Fe_3O_4]^2 P_{O_2}^{1/2}} = K, \quad \frac{1}{P_{O_2}^{1/2}} = K. \tag{6.15}$$

PARTIAL PRESSURE DIAGRAMS

The standard free energy of the reaction is

$$3\Delta F^\circ_{f\text{Fe}_2\text{O}_3} - 2\Delta F^\circ_{f\text{Fe}_3\text{O}_4} - \tfrac{1}{2}\Delta F^\circ_{f\text{O}_2} = \Delta F^\circ_r. \tag{6.16}$$

Substituting numbers

$$3(-177.1) - 2(-242.4) - \tfrac{1}{2}(0) = \Delta F^\circ_r$$

$$\Delta F^\circ_r = -46.5 \text{ kcal.} \tag{6.17}$$

Then

$$\Delta F^\circ_r = -1.364 \log \frac{1}{P^{1/2}_{\text{O}_2}}$$

$$-46.5 = -1.364 \log \frac{1}{P^{1/2}_{\text{O}_2}}$$

$$P_{\text{O}_2} = 10^{-68.2} \text{ atm.} \tag{6.18}$$

Table 6.1 gives the results of the calculations for the various equations.

TABLE 6.1. Partial Pressure of Oxygen for Equilibrium Between Iron-Oxygen Compounds at 25 °C and 1 Atmosphere Total Pressure

Reaction	P_{O_2}	log P_{O_2}
$\text{Fe}_c + \tfrac{1}{2}\text{O}_{2\,g} = \text{FeO}_c$	$10^{-85.6}$	-85.6
$3\text{Fe}_c + 2\text{O}_{2\,g} = \text{Fe}_3\text{O}_{4\,c}$	$10^{-88.9}$	-88.9
$2\text{Fe}_c + \tfrac{3}{2}\text{O}_{2\,g} = \text{Fe}_2\text{O}_{3\,c}$	$10^{-86.6}$	-86.6
$3\text{FeO}_c + \tfrac{1}{2}\text{O}_{2\,g} = \text{Fe}_3\text{O}_{4\,c}$	$10^{-98.5}$	-98.5
$2\text{FeO}_c + \tfrac{1}{2}\text{O}_{2\,g} = \text{Fe}_2\text{O}_{3\,c}$	$10^{-88.4}$	-88.4
$2\text{Fe}_3\text{O}_{4\,c} + \tfrac{1}{2}\text{O}_{2\,g} = 3\text{Fe}_2\text{O}_{3\,c}$	$10^{-68.2}$	-68.2

The relations among these compounds are better illustrated by showing the P_{O_2} pressure on a bar graph, as in Figure 6.2. Then, starting at the bottom, it is possible to follow the sequence of events that should occur if metallic iron is exposed to increasing partial pressures of oxygen, with pressure increased so slowly that equilibrium is approximated. The actual experiment, at 25 °C, is not possible to perform directly, as indicated by the incredibly low oxygen pressures. The calculated pressures are a useful yardstick, but cannot be translated into a real oxidation experiment.

The first reaction encountered, from FeO to Fe_3O_4, is clearly metastable, inasmuch as FeO cannot oxidize to Fe_3O_4 before iron has oxidized at all. Thus, iron would be stable until a P_{O_2} of 10^{-89} was attained, at which value Fe_c and $\text{Fe}_3\text{O}_{4\,c}$ would be in equilibrium. If $P_{\text{O}_2} = 10^{-88}$, Fe will convert to Fe_3O_4. Consequently, the next reaction, FeO to Fe_2O_3,

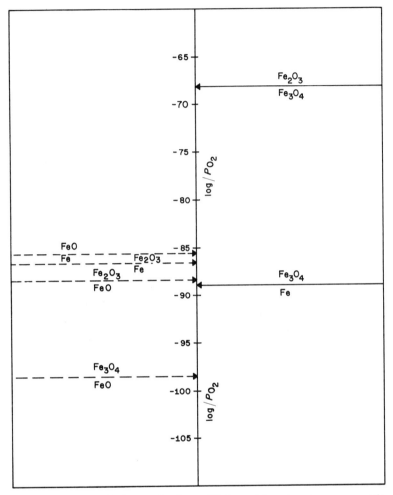

FIG. 6.2. Bar graph showing P_{O_2} in equilibrium with various iron compounds at 25 °C and 1 atmosphere total pressure. Dashed lines indicate metastable reactions.

is metastable because iron converts to magnetite before it goes to FeO. To put it another way, if we write

$$4FeO_c = Fe_3O_{4\,c} + Fe_c \tag{6.19}$$

and thus put the relation in terms of a solid-solid reaction, we find that ΔF_r° is negative, and that FeO will disproportionate into magnetite and metallic iron at room temperature, and hence has no field of stability. There are numerous ways of checking metastable relations, but the use of the P_{O_2} yardstick is as convenient as most.

PARTIAL PRESSURE DIAGRAMS

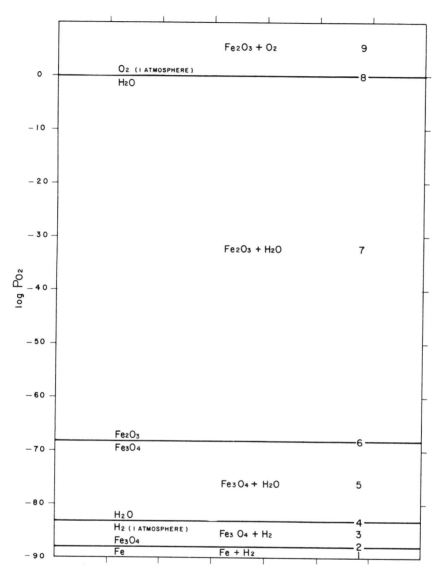

FIG. 6.3. Composite diagram showing interrelation of stability of iron oxides and water at 25 °C and 1 atmosphere total pressure. Numbers correspond to tie lines and three-phase fields on Figure 6.4.

Continuing upward P_{O_2}-wise along the bar graph, it is found that the next reaction to occur is from Fe_3O_4 to Fe_2O_3.

A simplified bar graph of the stable phases Fe, Fe_3O_4, and Fe_2O_3 can now be constructed, and the first dividend accruing from use of P_{O_2} can be realized, for direct comparison can be made with the P_{O_2} relations for water (Figure 6.3). Although a P_{O_2} value of 10^{-89} perhaps has little physical significance by itself, comparison with the stability limits of water shows immediately that metallic iron is not stable in the presence of water at earth surface conditions, and furthermore, that magnetite oxidizes to hematite near the lower limit of water stability. Therefore, hematite, although a compound of ferric iron and oxygen, is clearly a poor environmental indicator, inasmuch as it is stable over such a large range of oxidation conditions. On the other hand, magnetite, which might be thought of as an oxidized compound because two-thirds of its iron is in the ferric state, is stable only under strong reducing conditions, and is not reduced itself even in the presence of water and hydrogen under 1 atmosphere pressure.

Figure 6.3 shows the interrelations of iron, hydrogen, and oxygen, and can be translated into the three-component composition diagram Fe-O_2-H_2. Figure 6.4 shows such a diagram, with the tie lines deduced from Figure 6.3. At the bottom of Figure 6.3, iron and hydrogen

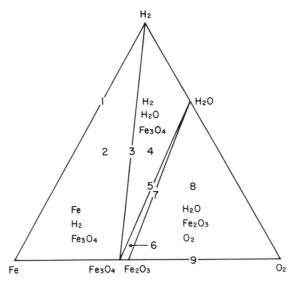

FIG. 6.4. Composition diagram of the three-component system Fe-O_2-H_2 at 25 °C and 1 atmosphere total pressure. Numbers correspond to fields and boundaries on Figure 6.3. Compositions are on atomic percent basis.

coexist, and water is unstable. Thus, this field corresponds to the tie line from iron to hydrogen on Figure 6.4. The field on Figure 6.3 is independent of the amount of iron present, as long as there is any at all, and this is reflected in the range of compositions indicated by the tie line

PARTIAL PRESSURE DIAGRAMS

of Figure 6.4. The three-phase field Fe-H$_2$-Fe$_3$O$_4$ similarly is reflected by the line in Figure 6.3, showing that the pressures of oxygen and hydrogen are invariant during the coexistence of three phases. Note that the combination of Figures 6.3 and 6.4 conveniently provides specific values for gas pressures in the three-phase fields and also gives the ranges of pressures possible for two-phase associations.

P_{O_2}-P_{CO_2} DIAGRAMS

Fe-O$_2$-CO$_2$

The simplicity of representation of the iron-oxygen species, in which a bar graph can be used, suggests immediately the possibility of representing reactions with two gas partial pressures as variables. Of the minerals involving iron, the obvious gases to use are CO_2 and S_2. Using CO_2 first and writing the reactions from native iron and the oxides to the lone carbonate mineral species siderite

$$Fe_c + CO_{2\,g} + \tfrac{1}{2}O_{2\,g} = FeCO_{3\,c} \qquad (6.20)$$

$$Fe_3O_{4\,c} + 3CO_{2\,g} = 3FeCO_{3\,c} + \tfrac{1}{2}O_{2\,g} \qquad (6.21)$$

$$Fe_2O_{3\,c} + 2CO_{2\,g} = 2FeCO_{3\,c} + \tfrac{1}{2}O_{2\,g}. \qquad (6.22)$$

For reaction (6.20)

$$\Delta F°_{f\,FeCO_3} - \Delta F°_{f\,Fe} - \Delta F°_{f\,CO_2} - \tfrac{1}{2}\Delta F°_{f\,O_2} = \Delta F°_r$$

$$-(161.06) - (0) - (-94.26) - (0) = -66.80 \text{ kcal.}$$

Then

$$-66.80 = -1.364 \log K$$

$$\log K = 49.0 = \log \frac{1}{P_{O_2}^{1/2} P_{CO_2}}$$

and

$$-\tfrac{1}{2} \log P_{O_2} - \log P_{CO_2} = 48.2$$

$$\log P_{CO_2} = -49.0 - \tfrac{1}{2} \log P_{O_2}.$$

Similar calculations for reactions (6.21) and (6.22) yield the following relations

Equation	Stability Boundary	Defining Equation in Terms of P_{O_2} and P_{CO_2}
(6.20)	Fe-FeCO$_3$	$\log P_{CO_2} = -49.0 - \tfrac{1}{2} \log P_{O_2}$
(6.21)	Fe$_3$O$_4$-FeCO$_3$	$\log P_{CO_2} = 10.3 + \tfrac{1}{6} \log P_{O_2}$
(6.22)	Fe$_2$O$_3$-FeCO$_3$	$\log P_{CO_2} = 15.9 + \tfrac{1}{4} \log P_{O_2}$

In a plot of $\log P_{O_2}$ vs. $\log P_{CO_2}$, the boundaries between these three pairs are straight lines with slopes of $-\frac{1}{2}$, $+\frac{1}{6}$, and $+\frac{1}{4}$, respectively. Figure 6.5 shows these boundaries, plus those for the iron oxides and water. A simple method of plotting is to substitute an arbitrary value

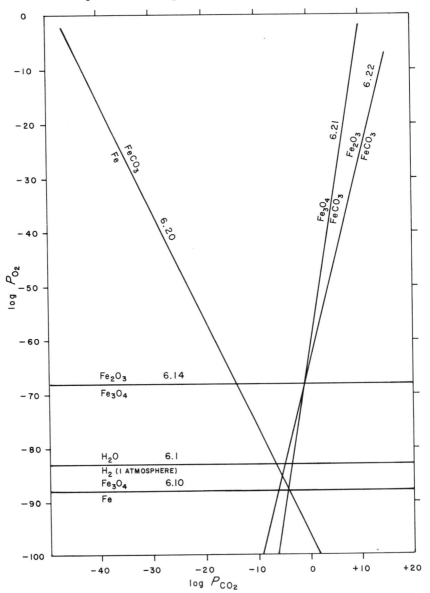

FIG. 6.5. Plot of equations representing stability relations of iron compounds as functions of P_{O_2} and P_{CO_2} at 25 °C.

PARTIAL PRESSURE DIAGRAMS

for P_{CO_2}, obtain a corresponding value for P_{O_2}, and then draw a line of proper slope from the point obtained.

From the relations of Figure 6.5, the metastability of the various reaction pairs in various regions of the diagram can be deduced by

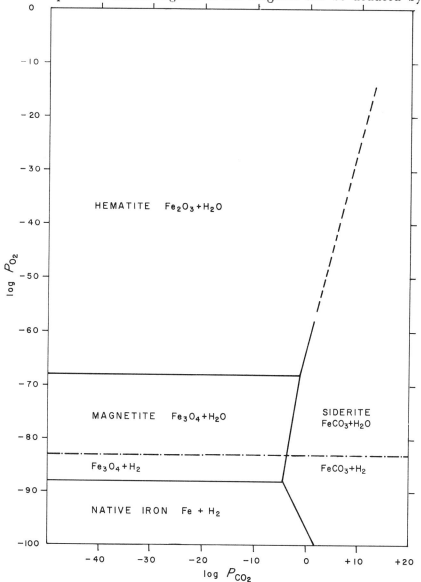

FIG. 6.6. Stability relations of some iron compounds as functions of P_{O_2} and P_{CO_2} at 25 °C and a few atmospheres maximum total pressure. Stability relations for water are superimposed.

inspection. For example, a boundary between Fe and $FeCO_3$ at oxygen pressures above 10^{-88} is clearly metastable, because Fe is unstable relative to the oxides above this oxygen pressure. After elimination of metastable relations, a final diagram is obtained as shown in Figure 6.6. Here the fields of stability can be designated. Comparison of Figures 6.5 and 6.6 shows that the free energy values used are at least internally consistent. Note on Figure 6.5 how equations (6.14), (6.21), and (6.22) intersect, and how (6.21), (6.20), and (6.10) also intersect. A hiatus at either of these points would indicate an error in the free energy values used for one of the equations. On Figure 6.6 the boundary between Fe_2O_3 and $FeCO_3$ is dashed at P_{CO_2} values above 10^0, to indicate the upper limit of the validity of the partial pressure calculated from data appropriate to 25 °C and 1 atmosphere total pressure. In making the calculations to fix the phase boundaries, the activities of the solids have been taken as unity at all total pressures. This procedure is in accordance with the results obtained in Chapter 9, where it is shown that the relative activities of the solids in a given reaction do not change significantly from unity when the total pressure is increased from 1 atmosphere to 50 atmospheres, or so. Also, the assumption is made that $P_g = a_g$, for all values of P_g.

A diagram of this type is valid *only for the species considered*; it does not reveal whether species not considered might not be stable relative to those used. Specifically, the diagram as shown is a superimposition of $Fe-O_2-CO_2$ relations on H_2O-H_2 relations, and it is not impossible that Fe_2O_3 and H_2O might be unstable with respect to the reaction $Fe_2O_{3\,c} + H_2O_l = 2FeOOH_c$. If so, the hematite-water field would disappear in favor of a goethite-water field, and the goethite-water field could be larger than the hematite-water field as shown. Simple superimposition of stability relations, using the same coordinates for representation, should always be followed by an attempt to see if interaction will occur to form new species. *It is difficult to overemphasize that the methods used here provide reasonably satisfactory stability relations among the species examined, but does not guarantee that they are the only ones that should have been considered.* As always, a better answer can be given concerning those relations that cannot occur at equilibrium (i.e., Fe and H_2O), rather than those that can (i.e., Fe_2O_3 and H_2O?). Thus, the safest method of use is to ask questions concerning the equilibrium relations of actual mineral associations.

$Cu-CO_2-O_2$

The number of diagrams of this type that could be constructed is limited only by the investigator's interest in oxide-carbonate relations for

PARTIAL PRESSURE DIAGRAMS

the elements in the periodic table (as well as the availability of free energy data). The copper minerals are of special significance to the geologist because of the economic importance of oxidation and secondary enrichment. Also, they are a convenient demonstration of the results of interaction with water. The major species are native copper (Cu), cuprite (Cu_2O), tenorite (CuO), malachite ($Cu_2(OH)_2CO_3$), and azurite ($Cu_3(OH)_2(CO_3)_2$). If we restrict P_{O_2} values to those within the field of water stability ($P_{O_2} = 10^{-83.1}$ to 10^0), it is possible to interrelate the various copper species by equations containing nearly pure liquid water. In nature the solutions in oxidizing copper deposits are commonly dilute, so that the activity of the water can be assumed to be that of pure water.

The interrelating reactions are (eliminating metastable reactions, as was done for iron)

$$2Cu_c + \tfrac{1}{2}O_{2\,g} = Cu_2O_c; \quad K = \frac{1}{P_{O_2}^{1/2}} \tag{6.23}$$

$$Cu_2O_c + \tfrac{1}{2}O_{2\,g} = 2CuO_c; \quad K = \frac{1}{P_{O_2}^{1/2}} \tag{6.24}$$

$$3Cu_c + \tfrac{3}{2}O_{2\,g} + 2CO_{2\,g} + H_2O_l$$
$$= Cu_3(OH)_2(CO_3)_{2\,c}; \quad K = \frac{1}{P_{O_2}^{3/2}P_{CO_2}^2} \tag{6.25}$$

$$3Cu_2O_c + 4CO_{2\,g} + \tfrac{3}{2}O_{2\,g} + 2H_2O_l$$
$$= 2Cu_3(OH)_2(CO_3)_{2\,c}; \quad K = \frac{1}{P_{CO_2}^4 P_{O_2}^{3/2}} \tag{6.26}$$

$$Cu_2O_c + \tfrac{1}{2}O_{2\,g} + CO_{2\,g} + H_2O_l$$
$$= Cu_2(OH)_2CO_{3\,c}; \quad K = \frac{1}{P_{O_2}^{1/2}P_{CO_2}} \tag{6.27}$$

$$2CuO_c + CO_{2\,g} + H_2O_l = Cu_2(OH)_2CO_{3\,c}; \quad K = \frac{1}{P_{CO_2}} \tag{6.28}$$

$$3Cu_2(OH)_2CO_{3\,c} + CO_{2\,g}$$
$$= 2Cu_3(OH)_2(CO_3)_2 + H_2O_l; \quad K = \frac{1}{P_{CO_2}}. \tag{6.29}$$

The corresponding equations are

Equation	Stability Boundary	Defining Equations in Terms of P_{O_2} and P_{CO_2}
(6.23)	Cu-Cu_2O	$\log P_{O_2} = -51.30$
(6.24)	Cu_2O-CuO	$\log P_{O_2} = -37.6$
(6.25)	Cu-$Cu_3(OH)_2(CO_3)_2$	$\log P_{CO_2} = -36.9 - \tfrac{3}{4}\log P_{O_2}$
(6.26)	Cu_2O-$Cu_3(OH)_2(CO_3)_2$	$\log P_{CO_2} = -17.6 - \tfrac{3}{8}\log P_{O_2}$
(6.27)	Cu_2O-$Cu_2(OH)_2CO_3$	$\log P_{CO_2} = -22.7 - \tfrac{1}{2}\log P_{O_2}$
(6.28)	CuO-$Cu_2(OH)_2CO_3$	$\log P_{CO_2} = -3.8$
(6.29)	$Cu_2(OH)_2CO_3$-$Cu_3(OH)_2(CO_3)_2$	$\log P_{CO_2} = -2.5$

156 SOLUTIONS, MINERALS, AND EQUILIBRIA

The results of the plotting of the equations are shown in Figure 6.7. The general relations are strikingly similar to those of the iron oxides and carbonate, as shown in Figure 6.6, with the added feature of an extra carbonate species. The plot leads to some interesting speculations;

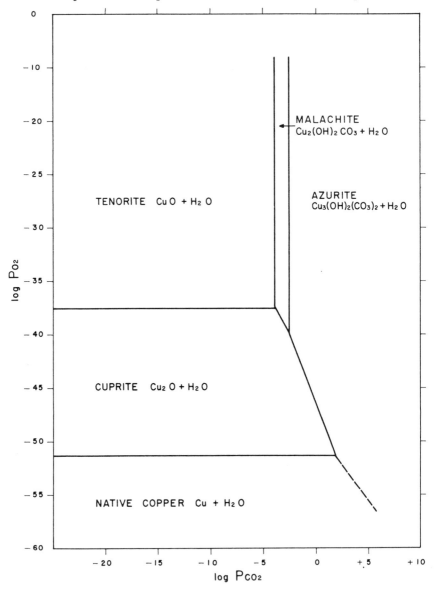

FIG. 6.7. Stability of copper compounds as functions of P_{O_2} and P_{CO_2} at 25 °C and 1 atmosphere, or slightly greater than 1 atmosphere, total pressure. Pure liquid water is assumed present in all cases.

under surface conditions of high P_{O_2} and P_{CO_2} of $10^{-3.5}$ malachite is stable relative to azurite; azurite can exist stably only by increasing P_{CO_2} or by decreasing the activity of water (equation 6.29). Thus it is quite possible that in the dry air of semi-arid climates the relative humidity is low enough to permit azurite to form stably, but when specimens are brought to regions of higher humidity, azurite transforms into malachite. This is the explanation of the green skies of so many pictures removed from a homeland where azurite is a stable pigment.

P_{O_2}-P_{S_2} DIAGRAMS

Fe-O$_2$-S$_2$

The interrelations of metal oxides and sulfides are also handled conveniently by partial pressure diagrams. To depict their relations, diatomic sulfur gas is chosen here. The stable molecular species of sulfur gas is in doubt at 25 °C; it is very likely a more condensed species than S_2, but the species chosen is not important, inasmuch as it is to be used only as a yardstick for comparison of iron mineral reactions with reactions among other minerals. Also, after obtaining stability relations in terms of any convenient pair of variables such as P_{O_2} and P_{S_2}, we can translate the relations into functions of other gases, if we wish, by writing reactions interrelating the original variables to the new ones.

As before, the first step is to write the reactions in terms of the minerals and the gases. A pattern of procedure begins to emerge: first we write the equations for the reactions involving only one variable

$$3Fe_c + 2O_{2\,g} = Fe_3O_{4\,c} \tag{6.10}$$

$$2Fe_3O_{4\,c} + \tfrac{1}{2}O_{2\,g} = 3Fe_2O_{3\,c} \tag{6.14}$$

$$2Fe_c + S_{2\,g} = 2FeS_c \tag{6.30}$$

$$Fe_c + S_{2\,g} = FeS_{2\,c} \tag{6.31}$$

$$FeS_c + \tfrac{1}{2}S_{2\,g} = FeS_{2\,c}. \tag{6.31a}$$

Then the reactions involving both

$$Fe_3O_{4\,c} + \tfrac{3}{2}S_{2\,g} = 3FeS_c + 2O_{2\,g} \tag{6.32}$$

$$Fe_3O_{4\,c} + 3S_{2\,g} = 3FeS_{2\,c} + 2O_{2\,g} \tag{6.33}$$

$$Fe_2O_{3\,c} + 2S_{2\,g} = 2FeS_{2\,c} + \tfrac{3}{2}O_{2\,g}. \tag{6.34}$$

Note that we can use previous experience with the iron oxide relations (see Figure 6.2) to eliminate consideration of all but the stable reactions for iron-oxygen interplay. In writing the iron-sulfur and iron-oxygen-sulfur set for the first time, all possible reactions should be considered as guided by known mineral species. Only one has been left out here, the reaction of Fe_2O_3 with S_2 to produce FeS. If included and plotted, it would have been discovered to be metastable, as might have been surmised from mineral relations. Furthermore, we have restricted our interest to *sulfides*—sulfates and other sulfur-oxygen compounds could appear as a result of the interaction of iron, sulfur, and oxygen. But we have arbitrarily decided to limit our attention to the sulfides, recognizing all the while that we are not attempting to discover all the possible interactions at this juncture.

One final word of caution, which will be repeated from place to place. The calculations are made for FeS_c, which is in fact not FeS, but $Fe_{1-x}S$. Therefore, the free energy value used is for a particular composition ($Fe_{0.95}S$ as given by the free energy tables in Appendix 2), and is not valid for other compositions. The error, however, is *generally* slight for deviations of this order of magnitude from strict stoichiometry, but should always be considered and assessed before taking a calculated boundary too seriously.

Figure 6.8 shows the result of obtaining the equilibrium constants for equations (6.10), (6.14), and (6.30)–(6.34). Even though this is a 25 °C diagram, some of the associations and impossible associations have a familiar ring to the geologist. The three-phase association of pyrrhotite, pyrite, and magnetite is one such, and the incompatibility of pyrrhotite and hematite is another. The overlay of the boundaries of water also tells a story. The equilibria plotted are independent of water, and water is not a necessary companion to the iron minerals, but note that pyrrhotite and water cannot coexist at 25 °C except under a highly restricted set of conditions.

Cu-S$_2$-O$_2$

Again the treatment of iron can be paralleled with that for copper. The reactions are

$$2Cu_c + \tfrac{1}{2}O_{2\,g} = Cu_2O_c \tag{6.23}$$

$$Cu_2O_c + \tfrac{1}{2}O_{2\,g} = 2CuO_c \tag{6.24}$$

$$2Cu_c + \tfrac{1}{2}S_{2\,g} = Cu_2S_c \tag{6.35}$$

$$Cu_2S_c + \tfrac{1}{2}S_{2\,g} = 2CuS_c \tag{6.36}$$

$$Cu_2O_c + \tfrac{1}{2}S_{2\,g} = Cu_2S_c + \tfrac{1}{2}O_{2\,g} \tag{6.37}$$

$$2CuO_c + \tfrac{1}{2}S_{2\,g} = Cu_2S_c + O_2 \tag{6.38}$$

$$CuO_c + \tfrac{1}{2}S_{2\,g} = CuS_c + \tfrac{1}{2}O_{2\,g}. \tag{6.39}$$

PARTIAL PRESSURE DIAGRAMS

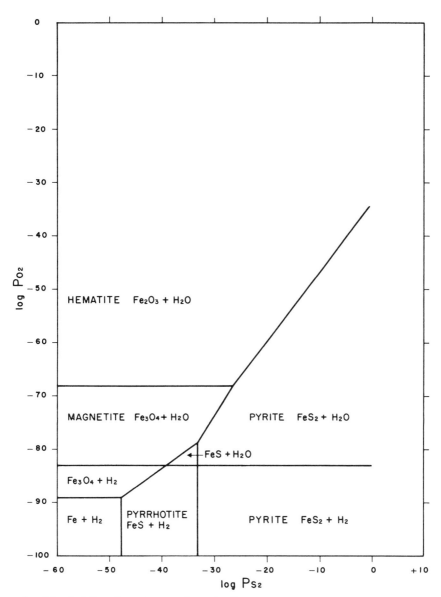

FIG. 6.8. Stability of some iron oxides and sulfides as functions of P_{O_2} and P_{S_2} at 25 °C and 1 atmosphere total pressure. Stability relations for water are superimposed.

This list does not include a compound that would be of interest, Cu_9S_5, the mineral digenite. Free energy values for digenite are not available from standard sources, so it will be omitted with the recognition that we should like to include it. The same caution must be used with

Cu_2S as was used with FeS, for the mineral chalcocite is commonly not stoichiometric.

Figure 6.9 shows the graphical relations resulting from obtaining

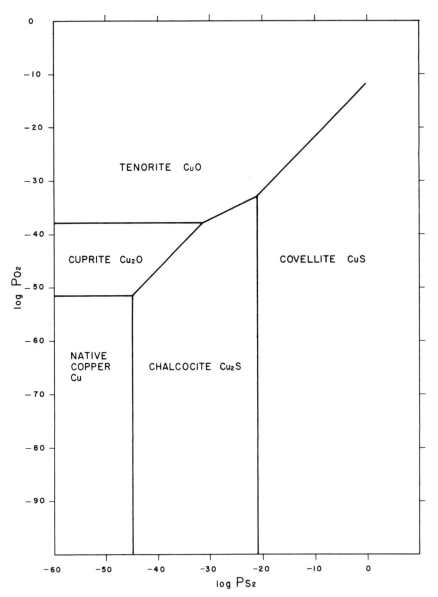

FIG. 6.9. Stability relations among some copper oxides and sulfides at 25 °C and 1 atmosphere total pressure as functions of P_{O_2} and P_{S_2}.

PARTIAL PRESSURE DIAGRAMS

equilibrium constant values for equations (6.23), (6.24), and (6.35)–(6.39), and making a plot similar to that for iron.

P_{O_2}-P_{CO_2}-P_{S_2} DIAGRAMS

Fe-O$_2$-CO$_2$-S$_2$

Relations among iron compounds as functions of P_{O_2} and P_{S_2}, and as functions of P_{O_2} and P_{CO_2}, automatically raise the question of stability as simultaneous functions of all three variables. Figure 6.10 shows the result of combining Figures 6.8 and 6.6. Construction is fairly simple, in that it involves only setting up coordinates for P_{S_2} and P_{CO_2} at right angles from a common P_{O_2} axis, and then tracing out the intersection of the planes. Only one new equation is needed, that for pyrite to siderite

$$\text{FeS}_{2\,c} + \text{CO}_{2\,g} + \tfrac{1}{2}\text{O}_{2\,g} = \text{FeCO}_{3\,c} + \text{S}_{2\,g}. \quad (6.40)$$

This is the only stability boundary that makes an angle with all three axes, that is to say, includes P_{S_2}, P_{O_2}, and P_{CO_2} in its equation.

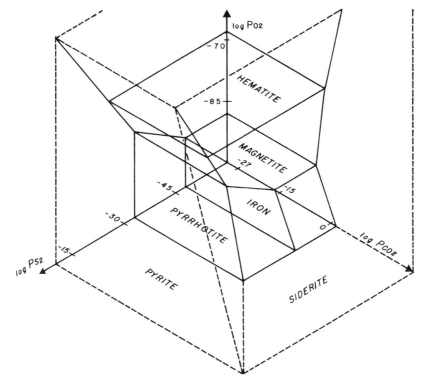

FIG. 6.10. Stability relations of some iron compounds as functions of P_{O_2}, P_{CO_2}, and P_{S_2} at 25 °C and 1 atmosphere, or greater, total pressure.

$Cu-O_2-CO_2-S_2$

The relations among some of the copper oxides, hydroxycarbonates, and sulfides also can be delineated in terms of all three partial pressures, as shown in Figure 6.11. The figure portrays relations at CO_2 pressures far above those for which the data are applicable, but the diagram still serves to show the very high pressures necessary to convert the sulfides to hydrocarbonates at low oxygen pressures. Except for these equilibria, the diagram is well within the range of useful values of the partial pressures.

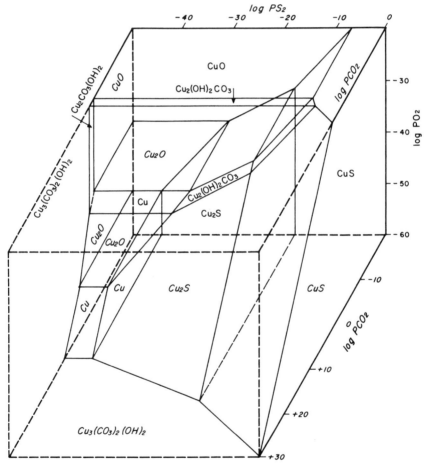

FIG. 6.11. Stability relations of some copper minerals as functions of P_{O_2}, P_{CO_2}, and P_{S_2} in the presence of pure water at 25 °C, and for 1 atmosphere, and greater, total pressure. Values for P_{CO_2} are extrapolated beyond the validity of the calculation. [Courtesy J. Anderson, Department of Geology, Harvard University Graduate School, 1958.]

SUPERIMPOSITION OF DIAGRAMS

Because partial pressure diagrams for minerals of different elements can be expressed in terms of the same variables, it is often of interest to overlay diagrams, to compare stability relations of various metal sulfides, oxides, and carbonates.

Fe-Mn-O$_2$-S$_2$

In Figure 6.12 the P_{O_2}-P_{S_2} diagram for manganese has been overlaid on the iron diagram (Figure 6.8). In addition to the combination plot, a new restraint has been put on the system. The partial pressure of S$_2$ in equilibrium with alabandite, MnS, rises to values in excess of 1 atmosphere at low P_{O_2} pressures, which reminds us that the vapor pressure of solid sulfur at low temperature is probably pretty low, so that it may be worth determining if solid sulfur should be considered as a possible phase. From the relation

$$S_{2\,g} = 2S_{rhombic} \tag{6.41}$$

$$2\Delta F^\circ_{fS_c} - \Delta F^\circ_{fS_{2\,g}} = \Delta F^\circ_r$$

$$0 - 19.13 = -19.13$$

$$-19.13 = -1.364 \log K$$

$$\log K = 14 = \log \frac{1}{P_{S_2}}$$

$$\log P_{S_2} = -14.$$

Thus, the vapor pressure of S$_2$ in equilibrium with native sulfur is 10^{-14}, and this value cannot be exceeded, for the activity of S$_2$ is fixed by the presence of the solid. Consequently, we need not worry about any species stable at P_{S_2} values above 10^{-14}, and we can add sulfur to the associations at this pressure. This relation stresses the use of such diagrams to determine incompatible and compatible associations, and also the danger of using them as a predictive device, inasmuch as a compatible association may be metastable with respect to species not considered. Some important aspects of the diagram are the highly restricted field of alabandite and water, a relation consistent with the rare occurrence of alabandite; the compatibility of Fe$_2$O$_3$ and MnO over a limited range of oxygen pressure; the compatibility of Mn$_3$O$_4$ and FeS$_2$; and the wide range of stability of Fe$_2$O$_3$, which spans the whole sequence of manganese oxides. Also, Mn$_3$O$_4$ and Fe$_3$O$_4$ apparently are incompatible, although the difference in oxygen pressure from that calculated necessary to stabilize them is so small that this conclusion is suspect.

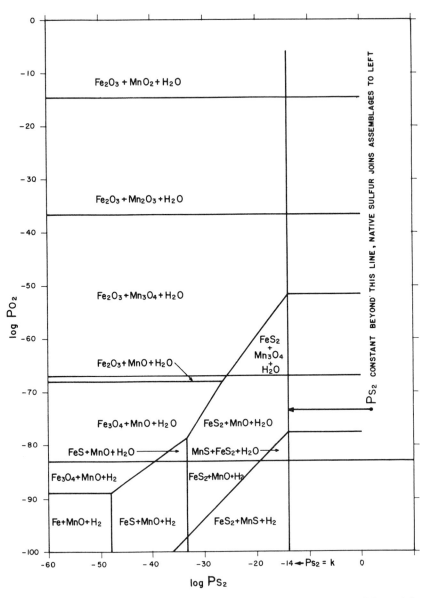

FIG. 6.12. Stability of some iron and manganese compounds as functions of P_{S_2} and P_{O_2} at 25 °C and 1 atmosphere total pressure. Consideration of stability of native sulfur and of water shows limiting effect on P_{O_2} and P_{S_2} values that can be achieved. [Courtesy J. Reitzel, Department of Geology, Harvard University Graduate School, 1958.]

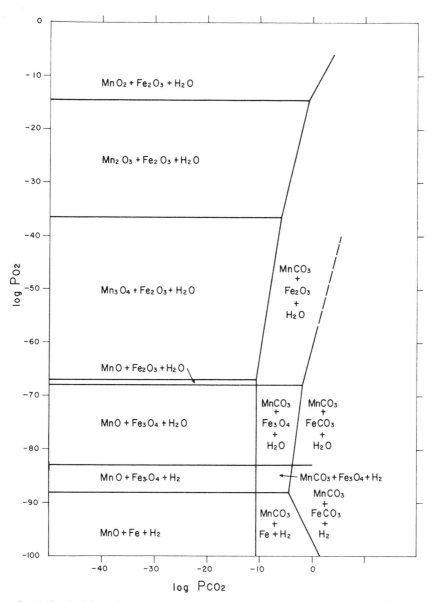

FIG. 6.13. Stability relations among iron and manganese oxides and carbonates as functions of P_{O_2} and P_{CO_2} at 25 °C and a few atmospheres total pressure maximum. Stability range of water also shown. No interactions between Fe and Mn compounds are considered. [Courtesy J. Reitzel, Department of Geology, Harvard University Graduate School, 1958.]

Fe-Mn-O$_2$-CO$_2$

Relations between oxides and carbonates of iron and manganese are shown in Figure 6.13. It illustrates the considerably greater stability of pure manganese carbonate relative to the oxides, compared to that of iron carbonate relative to its oxides. Comparison of iron and manganese behavior in this fashion serves to remind us of the possibility of interaction, such as the formation of $(Fe_xMn_y)CO_3$, and that we are looking at two separate diagrams in which interactions have not been considered. In general, however, solid solution is least under the conditions being considered.

Fe-Cu-O$_2$-S$_2$

The superposition of the Fe-O$_2$-S$_2$ and Cu-O$_2$-S$_2$ diagrams (Figures 6.8 and 6.9) yields Figure 6.14. This figure is eloquently barren in terms of compounds containing both iron and copper, such as chalcopyrite (CuFeS$_2$) and bornite (Cu$_5$FeS$_4$). R. Natarajan (unpublished manuscript, Department of Geology, Harvard University, 1958) has attempted to take into account such interactions, basing his interpretation of the position of the compounds on the diagrams of McKinstry and Kennedy [H. E. McKinstry and G. C. Kennedy, Some suggestions concerning the sequence of certain ore minerals: *Econ. Geol.*, **52**, 379–390 (1957)], and McKinstry [H. E. McKinstry, Phase assemblages in sulfide ore deposits: *N.Y. Acad. Sci.*, Ser. II, **20**, 15–26 (1957)], who have assembled data on the natural associations in this system. For the 25 °C relations, Natarajan studied the mineralogy of the zone of oxidation and secondary sulfide enrichment. The composition diagram for Cu-Fe-S$_2$ at 25 °C, as deduced by Natarajan, is shown in Figure 6.15. The reactions involved in following the heavy arrow through the diagram (increasing sulfur content) can be written as follows, with the equation letters keyed to the tie lines in the composition diagram.

Starting from the Cu-Fe tie line, the three-phase field Cu-Fe-FeS is encountered, then the tie line Cu-FeS. (The Cu-FeS association is not known to occur in nature, probably because of the low S$_2$ pressure required, as well as a pressure of O$_2$ lower than that in equilibrium with water.) The three-phase field is at a constant P_{S_2} and corresponds to a vertical line on the P_{O_2}-P_{S_2} diagram. The line is defined by the reaction

$$Fe_c + Cu_c + \tfrac{1}{2}S_{2\,g} = FeS_c + Cu_c. \tag{A}$$

The next reaction is from Cu + FeS to Cu$_2$S + FeS

$$2Cu_c + FeS_c + \tfrac{1}{2}S_{2\,g} = Cu_2S_c + FeS_c \tag{B}$$

and in sequence

PARTIAL PRESSURE DIAGRAMS

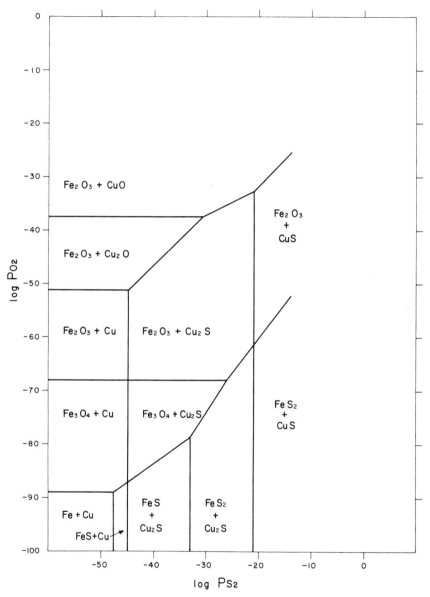

FIG. 6.14. Relations of some compounds in the Fe-O$_2$-S$_2$ system and in the Cu-O$_2$-S$_2$ system plotted simultaneously as functions of P_{O_2} and P_{S_2} at 25 °C and 1 atmosphere total pressure. No interactions are considered.

$$5\text{Cu}_2\text{S}_c + 3\text{FeS}_c + \tfrac{1}{2}\text{S}_{2\,g} = 2\text{Cu}_5\text{FeS}_{4\,c} + \text{FeS}_c \qquad \text{(C)}$$

$$\text{Cu}_5\text{FeS}_{4\,c} + 5\text{FeS}_c + \text{S}_{2\,g} = 5\text{CuFeS}_{2\,c} + \text{FeS}_c \qquad \text{(D)}$$

$$\text{CuFeS}_{2\,c} + \text{FeS}_c + \tfrac{1}{2}\text{S}_{2\,g} = \text{CuFeS}_{2\,c} + \text{FeS}_{2\,c} \qquad \text{(E)}$$

$$5\text{CuFeS}_{2\,c} + \text{FeS}_{2\,c} + \text{S}_{2\,g} = \text{Cu}_5\text{FeS}_{4\,c} + 5\text{FeS}_{2\,c} \qquad \text{(F)}$$

$$2\text{Cu}_5\text{FeS}_{4\,c} + \text{FeS}_{2\,c} + \tfrac{1}{2}\text{S}_{2\,g} = 5\text{Cu}_2\text{S}_c + 3\text{FeS}_{2\,c} \qquad \text{(G)}$$

$$\text{Cu}_2\text{S}_c + \text{FeS}_{2\,c} + \tfrac{1}{2}\text{S}_{2\,g} = 2\text{CuS}_c + \text{FeS}_{2\,c} \qquad \text{(H)}$$

$$2\text{CuS}_c + \text{FeS}_{2\,c} + \tfrac{1}{2}\text{S}_{2\,g} = \text{CuS}_c + \text{FeS}_{2\,c} + \text{S}_c. \qquad \text{(I)}$$

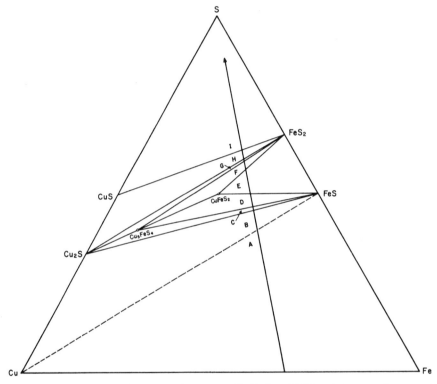

FIG. 6.15. Stability relations in the system Cu-Fe-S at 25 °C and 1 atmosphere total pressure, as deduced by R. Natarajan. There is some uncertainty concerning the composition of Cu₂S and FeS at this temperature.

No free-energy data are available for Cu_5FeS_4, and those for CuFeS_2 are only estimates. Nonetheless, reasonable values for the constants of all the reactions can be obtained. The explanation is best illustrated by reference to Figure 6.16, which shows the stability relations as deduced. The base from which the diagram was started is Figure 6.14 (heavy lines on Figure 6.16), which shows the stability fields for which free energy data are available. Reactions (A)–(H) then give the sequence in which the minerals occur as P_{S_2} increases, and serves to place the light lines in correct positions relative to known equilibria across the lower part of the diagram. By adding information concerning mineral associations from the ternary system $Cu\text{-}Fe\text{-}O_2\text{-}S_2$, it becomes necessary to adjust further the position of the light lines representing the new equilibria. To obtain the mineral associations $CuFeS_2\text{-}Fe_3O_4$ and $CuFeS_2\text{-}Fe_2O_3$, both of

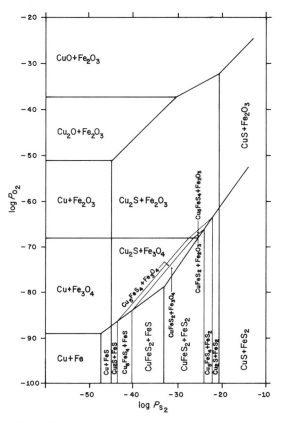

FIG. 6.16. Relations among some iron and copper oxides and sulfides at 25 °C and 1 atmosphere total pressure. Heavy lines are from superposition of iron and copper $P_{O_2}\text{-}P_{S_2}$ diagrams; lighter lines are relations deduced from mineral associations. [R. Natarajan and R. Garrels, unpublished manuscript, 1958.]

which are common, the line for the boundary $Cu_5FeS_4 + FeS = CuFeS_2 + FeS$ (D) cannot be placed on the diagram unless it is within 1 or 2 log units of P_{S_2}. Furthermore, because the *slopes* of all equilibria are known, irrespective of their absolute values, once a position for P_{S_2} of equation (D) is chosen, the position for the boundary $CuFeS_2 + FeS_2 =$

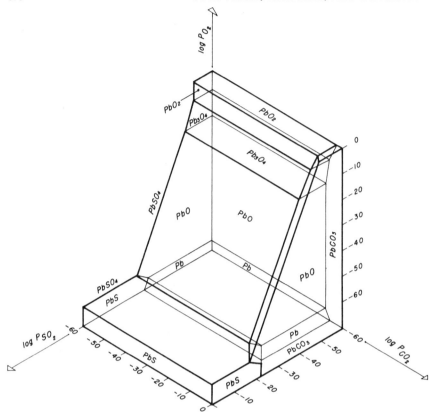

FIG. 6.17. Stability of some lead compounds at 25 °C and 1 atmosphere total pressure as a function of P_{O_2}, P_{CO_2}, and P_{SO_2}. P_{SO_2} was used in this instance to emphasize that various gases can be used in describing mineral equilibria. Note that PbS, PbSO$_4$, and PbCO$_3$ are the phases expected at near-surface conditions. [Courtesy J. Anderson, Department of Geology, Harvard University Graduate School, 1958.]

$Cu_5FeS_4 + FeS_2$ (F) also is fixed. It will thus be noted that any significant change in the boundaries shown will cause an observed mineral association to disappear, or it will cause the presence of an association unknown in nature.

Figure 6.16 is not purported to be an accurate diagram of the mineral relations shown, but it is a useful working model from which deductions can be made and tested against natural occurrences or experiment.

SUMMARY

In this chapter a variety of diagrams has been developed to illustrate the utility of expressing mineral relations in terms of the common denominators of partial pressures of gases. The gases of greatest geological interest are O_2, CO_2, and S_2. Relations among the oxides, carbonates, and sulfides of various metals can be expressed as individual plots, or they can be superimposed for purposes of comparison. Such superimposition permits only comparison of the minerals considered in the individual systems, and is not to be construed as a representation of all stable phases that may occur. There are three chief limitations of such diagrams: (1) Data are not yet available to take into consideration compositional variation of real minerals. (2) Interactions to form new mineral species should be considered to achieve relations of geological interest, and simple superposition of diagrams is not sufficient in many cases to be of great utility. (3) Although partial pressures of gases are convenient for comparison of mineral relations, the calculated values, with the exception of P_{CO_2}, are so low as to have little physical significance. The calculated values are not verifiable by direct experiment, and they are not easy to translate into the environmental conditions that can be observed in the field.

SELECTED REFERENCES

Darken, L. S., and R. W. Gurry, *The Physical Chemistry of Metals*. New York, McGraw-Hill, 1953, 535 pp.
Schmitt, H. H., (ed.), *Equilibrium Diagrams for Minerals*. Cambridge, Mass., The Geological Club of Harvard, 1962.

PROBLEMS

6.1. What is the equilibrium partial pressure of oxygen for the reaction: $SnO_c + \frac{1}{2}O_{2\,g} = SnO_{2\,c}$, at 25 °C and a total pressure of 1 atmosphere?
Ans. $P_{O_2} = 10^{-90.5}$ atm.

6.2. What is the partial pressure of $S_{2\,g}$ in equilibrium with rhombic sulfur, at 25 °C and a total pressure of 1 atmosphere? *Ans.* $P_{S_2} = 10^{-14.0}$ atm.

6.3. Construct a diagram showing the stability relations of $UO_{2\,c}$, $UO_{3\,c}$, and UO_2CO_3 as functions of $\log P_{O_2}$ and $\log P_{CO_2}$.

CHAPTER 7

Eh-pH Diagrams

The current rapid increase in use of Eh-pH diagrams to show mineral stability relations is a little surprising because the ingredients necessary to manufacture them have been known for a long time. The development of the concept of pH and its extensive application to the interpretation of the chemistry of soils and natural waters dates back several decades, and the Nernst equation relating Eh to $E°$ and to the activities of reactants and products of half-cells is of even more ancient derivation.

It would be impossible to attempt a complete bibliography of all papers of geological interest that have made significant use of Eh or pH or both, but the answer to the enigma concerning the late development of mineral stability diagrams perhaps can be given in part by citing some key publications in chronological sequence. As early as 1923, Clark (cf. W. M. Clark and Barnett Cohen, An analysis of the theoretical relations between reduction potentials and pH: *Public Health Reports, Reprint 826*, 1923) discussed at length the interrelation of Eh and pH, but his interests were to a large extent in the field of biochemistry, and his attention was directed to equilibria among dissolved species rather than among solids. Thus, his work created little stir in geochemistry.

One of the first papers in geology devoted to the stability relations of minerals, as well as those of dissolved ions, was Scerbina's discussion of the role of oxidation potential as applied to mineral paragenesis [V. V. Scerbina, Oxidation-reduction potentials as applied to the study of the paragenesis of minerals. *Compte. rend. acad. sci. U.R.S.S.*, 22, 503–506 (1939)]. He considered the effect of pH in his discussion, but made no plots involving the two variables. On the other hand, he suggested groups of minerals that could coexist under given Eh conditions.

The next milestone was the publication of ZoBell's classic work on Eh values for marine sediments [Claude E. ZoBell, 1946, Studies on redox potentials of marine sediments: *Bull. Am. Assoc. Petrol. Geol., 30*, 477–513 (1946)]. He made available in a widely circulated geological journal a comprehensive treatment of the theory and practice of Eh and pH measurement, and the interpretation of such measurements in terms of the natural environment. The scope of the article is greater than indicated by the title, for it is a critical résumé of most of the previous work. The one aspect missing is the depiction of the stability fields of solids as functions of Eh and pH. His paper, plus one a year older, by Starkey and Wight (R. L. Starkey and K. M. Wight, *Anaerobic Corrosion of Iron in Soil.* New York, American Gas Association, 1945, 108 pp.) stand as basic references for anyone interested in practicing the profession of Eh and pH.

Attention was further directed toward the use of pH and potential by Mason [Brian Mason, Oxidation and reduction in geo-chemistry. *J. Geol.*, 57, 62–72 (1949)], who discussed mineral stability relations as functions of Eh and pH, but did not use any clear-cut mineral stability-field diagrams.

In 1946, M. Pourbaix published a book on the thermodynamics of dilute solutions; in 1949 it was translated into English. (M. J. N. Pourbaix, *Thermodynamics of Dilute Aqueous Solutions.* London, Edward Arnold & Co., 1949, 136 pp.) Because of the title, it was missed by geochemists at first, although the methods are ideally suited to the treatment of mineral equilibria. Pourbaix's book is the summary of a decade of development of methods of representing the stability of solids and dissolved species as functions of pH and Eh. Much of the following development is based on Pourbaix's work, with slight adaptation and extension of some of his methods.

Since the publication of his book, Pourbaix, as director of the Belgian Institute for the Study of Corrosion, has published, with his coworkers, numerous technical reports, each concerned with equilibria of a given element in oxygenated water. A complete list of the reports of interest is given at the end of this chapter. A compilation of most of these reports is now available in an atlas (Pourbaix *et al.*, *Atlas d'Équilibres, etc.*, listed in references at end of chapter). The reports, although aimed primarily at metallurgists, can be used without modification by anyone interested in mineral stability relations, insofar as pure chemical compounds resemble minerals.

Largely through the impact of Pourbaix's work, use of Eh-pH stability diagrams has ramified rapidly through the geochemical literature. A bibliography and discussion of some of the publications of geological interest is given in Chapter 11.

CONSTRUCTION OF Eh-pH DIAGRAMS

The mechanics of constructing Eh-pH diagrams is illustrated in this text by a detailed development of diagrams showing relations among iron minerals. The calculations and equations are presented with a degree of elaboration not entirely necessary if the text were limited to a discussion of iron alone, in the anticipation that readers may want to make similar calculations for other elements. The complexities that may be encountered in such attempts are most likely to be demonstrated if all aspects of the calculations for iron are treated with comparable thoroughness.

The definition and measurement of Eh and pH were discussed in some detail in Chapter 5. It was shown there that for every oxidation-reduction system, we can write the half-reaction

$$\text{reduced state} = \text{oxidized state} + ne, \tag{5.23}$$

where n denotes the number of electrons involved in the half-reaction. The corresponding half-cell potential is given by

$$\text{Eh} = E° + \frac{RT}{n\mathscr{F}} \ln \frac{[\text{oxidized state}]}{[\text{reduced state}]}. \tag{5.24}$$

Here Eh is the half-cell potential relative to the standard hydrogen electrode; $E°$ is the standard half-cell potential, i.e., the voltage of the half-cell when the activities are unity for all species entering into the half-reaction.

For the half-reaction

$$a\text{A} + b\text{B} = c\text{C} + d\text{D} + ne$$

(5.24) yields, for 25 °C,

$$\text{Eh} = E° + \frac{0.0592}{n} \log \frac{[\text{C}]^c[\text{D}]^d}{[\text{A}]^a[\text{B}]^b}.$$

This equation together with the fundamental relation given by equation (1.22), i.e., $E° = (\Delta F_r°)/(n\mathscr{F})$, form the basis for the calculations developed in the following discussion.

All calculations in the following are made for 25 °C and 1 atmosphere total pressure; activities of pure solids and pure liquids are taken to be unity, and the activity of a gas phase is taken to be equal to its pressure. Finally, the effect of temperature and pressure variations not too far removed from 25 °C and 1 atmosphere will be considered.

THE STABILITY OF WATER

In Chapter 6 the stability of water was expressed as a function of hydrogen and oxygen partial pressures by obtaining the equilibrium constant for the reaction

$$2\text{H}_2\text{O}_\text{l} = 2\text{H}_{2\,\text{g}} + \text{O}_{2\,\text{g}}.$$

Eh-pH DIAGRAMS

The upper limit of water stability was determined as the equilibrium between water and oxygen at 1 atmosphere pressure. These relations are shown in Figure 6.1.

If we wish to show these same limits as functions of Eh and/or pH, the method is to write a reaction between water and oxygen in terms of hydrogen ions and/or electrons, so that representation of the relation can be plotted with Eh and pH as ordinate and abscissa.

For the upper limit of water stability ($P_{O_2} = 1$ atmosphere)

$$2H_2O_l = O_{2\,g} + 4H^+_{aq} + 4e. \tag{7.1}$$

Then from equation (5.24)

$$Eh = E° + \frac{0.059}{4} \log \frac{P_{O_2}[H^+]^4}{[H_2O]^2}. \tag{7.2}$$

Under the conditions chosen, P_{O_2} is unity, as is the activity of pure liquid water, so

$$Eh = E° + \frac{0.059}{4} \log [H^+]^4.$$

Substituting $-$pH for [H$^+$]

$$Eh = E° - 0.059 \text{ pH}. \tag{7.3}$$

Therefore, the equilibrium between water and oxygen at a partial pressure of 1 atmosphere is a straight line in an Eh-pH plot, with a slope of -0.059 volt per pH unit, and has an intercept of E°. To obtain a numerical value of E° for the reaction, the standard free energy is obtained and is substituted in equation (1.22)

$$E° = \frac{\Delta F°_r}{n\mathscr{F}}. \tag{1.22}$$

First, $\Delta F°_r$ is obtained

$$2H_2O_l = O_{2\,g} + 4H^+_{aq} + 4e \tag{7.1}$$

$$\Delta F°_{fO_2} + 4\Delta F°_{fH^+} - 2\Delta F°_{fH_2O} = \Delta F°_r$$

$$0 + (4 \times 0) - (2 \times -56.69) = +113.4 \text{ kcal.}$$

Substituting in (1.22)

$$E° = \frac{113.4}{4 \times 23.06} = 1.23 \text{ volts.}$$

The final equation is

$$Eh = 1.23 - 0.059 \text{ pH}. \tag{7.4}$$

The line representing this relation is shown on Figure 7.1. Note that this line must represent a fixed oxygen pressure, so that in a sense a change from a representation in terms of P_{O_2} to one involving Eh and

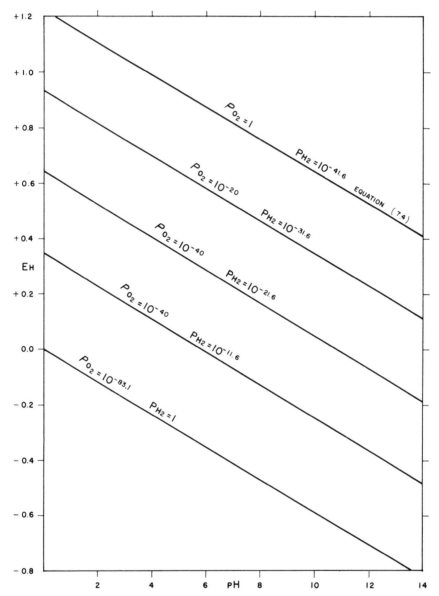

FIG. 7.1. Stability limits of water as a function of Eh and pH at 25 °C and 1 atmosphere total pressure. Contours show partial pressures of hydrogen and oxygen at intermediate Eh values.

The upper limit of water stability was determined as the equilibrium between water and oxygen at 1 atmosphere pressure. These relations are shown in Figure 6.1.

If we wish to show these same limits as functions of Eh and/or pH, the method is to write a reaction between water and oxygen in terms of hydrogen ions and/or electrons, so that representation of the relation can be plotted with Eh and pH as ordinate and abscissa.

For the upper limit of water stability (P_{O_2} = 1 atmosphere)

$$2H_2O_l = O_{2\,g} + 4H^+_{aq} + 4e. \tag{7.1}$$

Then from equation (5.24)

$$Eh = E° + \frac{0.059}{4} \log \frac{P_{O_2}[H^+]^4}{[H_2O]^2}. \tag{7.2}$$

Under the conditions chosen, P_{O_2} is unity, as is the activity of pure liquid water, so

$$Eh = E° + \frac{0.059}{4} \log [H^+]^4.$$

Substituting $-$pH for [H$^+$]

$$Eh = E° - 0.059\ pH. \tag{7.3}$$

Therefore, the equilibrium between water and oxygen at a partial pressure of 1 atmosphere is a straight line in an Eh-pH plot, with a slope of -0.059 volt per pH unit, and has an intercept of E°. To obtain a numerical value of E° for the reaction, the standard free energy is obtained and is substituted in equation (1.22)

$$E° = \frac{\Delta F°_r}{n\mathscr{F}}. \tag{1.22}$$

First, $\Delta F°_r$ is obtained

$$2H_2O_l = O_{2\,g} + 4H^+_{aq} + 4e \tag{7.1}$$

$$\Delta F°_{fO_2} + 4\Delta F°_{fH^+} - 2\Delta F°_{fH_2O} = \Delta F°_r$$

$$0 + (4 \times 0) - (2 \times -56.69) = +113.4\ \text{kcal.}$$

Substituting in (1.22)

$$E° = \frac{113.4}{4 \times 23.06} = 1.23\ \text{volts.}$$

The final equation is

$$Eh = 1.23 - 0.059\ pH. \tag{7.4}$$

The line representing this relation is shown on Figure 7.1. Note that this line must represent a fixed oxygen pressure, so that in a sense a change from a representation in terms of P_{O_2} to one involving Eh and

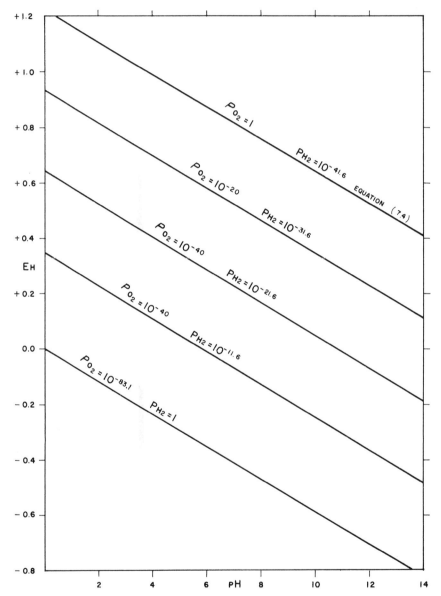

FIG. 7.1. Stability limits of water as a function of Eh and pH at 25 °C and 1 atmosphere total pressure. Contours show partial pressures of hydrogen and oxygen at intermediate Eh values.

Eh-pH DIAGRAMS

pH has resulted in the splitting of a single variable into two. In general this is the reverse of the usual procedure, which is to use as few variables as possible, but because Eh and pH are both easily measurable, and because much of our thinking is in terms of acidity, alkalinity, reduction, and oxidation, the representation is useful.

If we re-examine equation (7.2), we see that it can be rewritten with the numerical value obtained for E°

$$\text{Eh} = 1.23 + \frac{0.059}{4} \log P_{O_2} - 0.059 \text{ pH}. \tag{7.5}$$

At a partial pressure of oxygen of 1 atmosphere, the P_{O_2} term disappears, but evidently the equation can be used to plot a line on an Eh-pH diagram for any chosen pressure of oxygen. Therefore isobaric oxygen contours on an Eh-pH diagram (in equilibrium with water!) are parallel straight lines with a slope of -0.059 volt/pH.

Also, because of the relation shown in equation (6.1)

$$2H_2O_l = 2H_{2\,g} + O_{2\,g}$$

the partial pressure of hydrogen is also fixed if that of oxygen is stipulated, and P_{H_2} contours must also be straight lines with the same slope. However, values for P_{H_2} can also be derived by writing the half-cell reaction

$$H_{2\,g} = 2H^+_{aq} + 2e. \tag{7.6}$$

This half-cell reaction does not include liquid water, but its presence is implicit in the term H^+_{aq}. The Eh equation is

$$\text{Eh} = E° + \frac{0.059}{2} \log \frac{[H^+]^2}{P_{H_2}}.$$

Substituting $-\text{pH}$ for $\log [H^+]$ and rearranging

$$\text{Eh} = E° - \frac{0.059}{2} \log P_{H_2} - 0.059 \text{ pH}. \tag{7.7}$$

As before, the numerical value of E° is obtained by calculating $\Delta F°_r$ and substituting into the relation $E° = \Delta F°_r / n\mathscr{F}$

$$2\Delta F°_{f\,H^+} - \Delta F°_{f\,H_2} = \Delta F°_r$$
$$0 \quad\quad -0 \quad = 0$$

$$E° = \frac{0}{2 \times 23.06} = 0.$$

The final equation, since $E° = 0$, is

$$\text{Eh} = -\frac{0.059}{2} \log P_{H_2} - 0.059 \text{ pH}. \tag{7.8}$$

Figure 7.1 is the result of plotting equations (7.5) and (7.8). The lines at top and bottom are for $P_{O_2} = 1$ atmosphere and $P_{H_2} = 1$ atmosphere respectively, and show the equilibrium limits of the existence of water under earth surface or near-surface conditions.

Under experimental conditions, water can exist for long periods of time at higher and lower Eh values than those shown. Overvoltages of about 0.5 volt are generally necessary to achieve water decomposition at rates easily observed in the laboratory. However, the reaction can be catalyzed, and inasmuch as bacteria function well in this regard, it is probably unlikely that natural waters maintain for appreciable time-intervals Eh values much above or below the limits shown on Figure 7.1.

THE STABILITY OF IRON OXIDES

As derived in Chapter 6, the stable iron phases are native iron, magnetite, and hematite. The stability relations were determined as functions of the partial pressure of oxygen. These can be converted to reactions expressible as functions of Eh and pH by adding the water dissociation half-cell to the reactions expressed in terms of oxygen partial pressure. For the oxidation of iron to magnetite

$$3Fe_c + 2O_{2\,g} = Fe_3O_{4\,c} \qquad (6.10)$$

$$4H_2O_l = 8H^+_{aq} + 2O_{2\,g} + 8e \qquad (7.1)$$

$$\overline{3Fe_c + 4H_2O_l = Fe_3O_{4\,c} + 8H^+_{aq} + 8e.} \qquad (7.9)$$

The addition serves to eliminate oxygen gas and to substitute for it H^+ and electrons as variables.

The standard free energy of reaction 7.9 is

$$\Delta F^\circ_{fFe_3O_4} + 8\Delta F^\circ_{fH^+} - 3\Delta F^\circ_{fFe} - 4\Delta F^\circ_{fH_2O} = \Delta F^\circ_r$$

$$-242.4 + (8 \times 0) - (3 \times 0) - (4 \times -56.69) = -15.6 \text{ kcal.}$$

From the relation $E^\circ = \dfrac{\Delta F^\circ_r}{n\mathscr{F}}$

$$E^\circ = \frac{-15.6}{8 \times 23.06} = -0.084 \text{ volt.}$$

The half-cell represented by reaction 7.9 can then be expressed by the Eh equation

$$Eh = -0.084 + \frac{0.059}{8} \log \frac{[Fe_3O_4][H^+]^8}{[Fe]^3[H_2O]^4}. \qquad (7.10)$$

Remembering that the activities of $Fe_3O_{4\ c}$, Fe_c, and H_2O_l are unity (if the water is nearly pure!)

$$Eh = -0.084 + \frac{0.059}{8} \log [H^+]^8.$$

Substituting $-pH$ for $\log [H^+]$

$$Eh = -0.084 - 0.059\ pH. \tag{7.11}$$

The reaction between iron and magnetite is therefore a straight line on the Eh-pH plot, and has the same slope as the water stability boundary lines.

For the oxidation of magnetite to hematite, the derivation of the Eh-pH equation is similar and is shown without explanatory steps.

$$2Fe_3O_{4\ c} + \tfrac{1}{2}O_{2\ g} = 3Fe_2O_{3\ c} \tag{6.14}$$

$$H_2O_l \qquad\qquad = 2H^+_{aq} + \tfrac{1}{2}O_{2\ g} + 2e \tag{7.1}$$

$$\overline{2Fe_3O_{4\ c} + H_2O_l = 3Fe_2O_{3\ c} + 2H^+ + 2e} \tag{7.12}$$

$$3\Delta F°_{fFe_2O_3} + 2\Delta F°_{fH^+} - 2\Delta F°_{fFe_3O_4} - \Delta F°_{fH_2O} = \Delta F°_r$$

$$(3 \times -177.1) + (2 \times 0) - (2 \times -242.4) - (-56.69) = +10.2\ \text{kcal}$$

$$E° = \frac{+10.2}{2 \times 23.06} = 0.221\ \text{volt}$$

$$Eh = 0.221 + \frac{0.059}{2} \log \frac{[Fe_2O_3]^3[H^+]^2}{[Fe_3O_4]^2[H_2O]}$$

$$Eh = 0.221 + \frac{0.059}{2} \log [H^+]^2$$

$$Eh = 0.221 - 0.059\ pH. \tag{7.13}$$

Equation (7.13) shows that the boundary between magnetite and hematite is parallel to that for iron-magnetite. Figure 7.2 shows equations (7.11) and (7.13), as well as the stability limits of water. The reaction Fe-Fe_3O_4 lies below the lower limit of water stability, and therefore the reaction cannot take place stably in the presence of water. In other words, considering the relations from a reduction standpoint, magnetite exists stably down to the potentials at which water dissociates to release hydrogen gas at 1 atmosphere pressure, and the stability field of iron cannot be reached in the presence of water, if equilibrium is maintained. The position of the line as drawn, then, is the boundary between iron and magnetite under the condition that liquid water is present metastably.

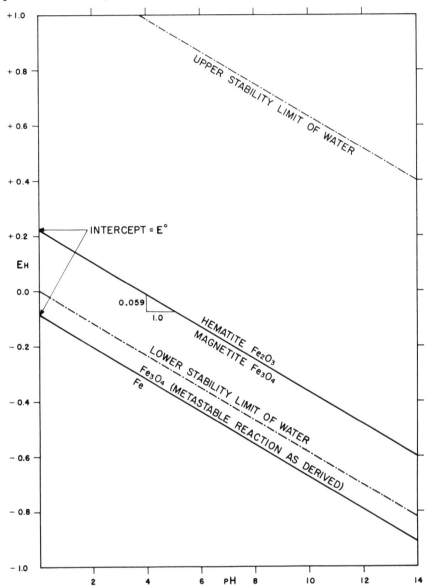

FIG. 7.2. Boundary for iron-magnetite and for magnetite-hematite as a function of Eh and pH at 25 °C and 1 atmosphere total pressure. Because the Fe-Fe_3O_4 boundary is below the lower stability limit for water, the reaction is metastable, and metallic iron cannot exist at equilibrium in the presence of water at any pH.

Eh-pH DIAGRAMS

Figure 7.3 summarizes the stability relations of the iron oxides and water as functions of Eh and pH and will be used as the base for determining ionic activities.

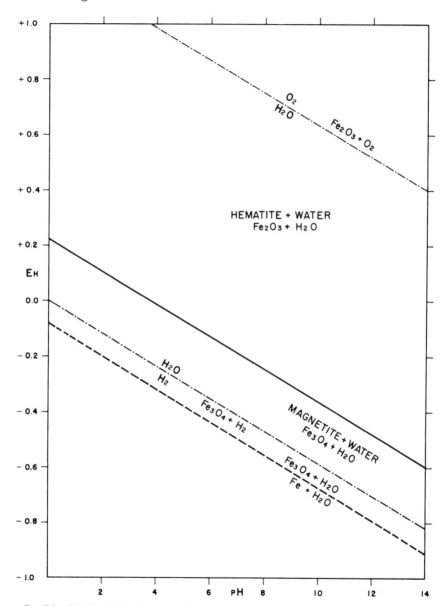

FIG. 7.3. Stability fields of iron oxides as functions of Eh and pH at 25 °C and 1 atmosphere total pressure. Dashed line indicates lower stability limit of magnetite in presence of metastable water.

STABILITY RELATIONS OF IRON HYDROXIDES

When a ferric salt in solution is precipitated by addition of hydroxide, the first precipitate is a hydrous oxide of indeterminate water content, often referred to as ferric hydroxide. The first precipitate is unstable with respect to both hematite and goethite.

For the relation between freshly precipitated "ferric hydroxide" and hematite we can write:

$$2Fe(OH)_{3 \text{ pptd}} \rightleftharpoons Fe_2O_3{}_c + 3H_2O_1 \qquad (7.14)$$

The activities of reactants and products are all unity, so that this is a go-no-go relation.

From the free energies of formation

$$\Delta F°_{fFe_2O_3} + 3\Delta F°_{fH_2O} - 2\Delta F°_{fFe(OH)_3} = \Delta F°_r$$

$$-177.1 + (3 \times -56.69) - (2 \times -166.0) = -15.17 \text{ kcal.}$$

Thus, with sufficient time, freshly precipitated ferric hydroxide will convert to the much more stable hematite (or goethite).

Similarly, addition of hydroxide to a ferrous salt in the absence of oxygen yields ferrous hydroxide—in this case a fairly well crystallized compound. But if we consider the reaction

$$3Fe(OH)_{2 \text{ c}} = Fe_3O_4{}_c + H_{2\text{ g}} + 2H_2O_1 \qquad (7.15)$$

then

$$\Delta F°_{fFe_3O_4} + \Delta F°_{fH_2g} + 2\Delta F°_{fH_2O} - 3\Delta F°_{fFe(OH)_2} = \Delta F°_r$$

$$-242.4 + 0 + (2 \times -56.69) - (3 \times -115.57) = -11.1 \text{ kcal.}$$

$$K = P_{H_2} = \frac{-11.1}{-1.364} = 10^{8.1}.$$

Therefore, $Fe(OH)_2$ is unstable with respect to decomposition into magnetite, water, and hydrogen gas. Under a total pressure of 1 atmosphere, a precipitate of ferrous hydroxide should eventually decompose to yield magnetite, and hydrogen would bubble from the system until the conversion was complete.

Yet both ferric and ferrous hydroxides are compounds of more than transitory existence. Although they clearly are not stable relative to hematite and magnetite, we can show their fields of temporary stability just as well as if they were permanent products, if it suits our purpose to examine short-lived equilibria.

For the reaction

$$Fe_c + 2H_2O_1 = Fe(OH)_{2\text{ c}} + 2H^+_{aq} + 2e \qquad (7.16)$$

$$Eh = -0.047 - 0.059 \text{ pH}. \qquad (7.17)$$

Similarly
$$Fe(OH)_{2\,c} + H_2O_l = Fe(OH)_{3\,c} + H^+_{aq} + e \quad (7.18)$$
$$Eh = 0.271 - 0.059\,pH. \quad (7.19)$$

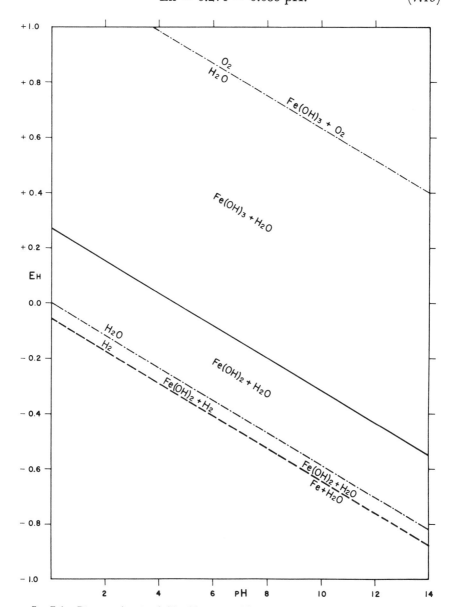

FIG. 7.4. Diagram showing fields of ferrous and ferric hydroxides at 25 °C and 1 atmosphere total pressure. Stability limits of water also shown. Dashed line is metastable boundary of Fe and Fe(OH)$_2$ in water.

The equations, plus the stability limit for water, are shown in Figure 7.4. The similarity to the relations among the oxides shown in Figure 7.3 is obvious; the chief difference is that $Fe(OH)_2$ oxidizes to $Fe(OH)_3$ at a higher potential than Fe_3O_4 oxidizes to Fe_2O_3.

This diagram of the relations among metastable iron compounds serves to illustrate again that the calculations throughout are valid for the species considered, but have no guarantee attached that the species considered are the truly stable ones.

ACTIVITIES OF IONS IN EQUILIBRIUM WITH IRON OXIDES

After diagrams showing the stability fields of the iron oxides have been prepared, the aspect of chief utility of Eh-pH diagrams can be developed. Because free energy of formation values are available for numerous ionic species, it is possible to calculate the activities of these species in equilibrium with the solid oxides. These ionic activities provide a useful picture of the major contributions to the solubility of the various solids and serve to delineate the conditions of acidity or alkalinity, or of oxidation potential, in which the solids are probably least soluble in dilute solution.

The dissolved species involving iron, oxygen, and water, for which free energy values are available, are Fe^{3+}_{aq}, $Fe(OH)^{++}_{aq}$, $Fe(OH)^{+}_{2\,aq}$, $FeO^{-}_{2\,aq}$, Fe^{++}_{aq}, $Fe(OH)^{+}_{aq}$, and $HFeO^{-}_{2\,aq}$. The first four are ferric species; the final three are ferrous. Although these are the species for which data are available, they are not necessarily the only important species that should be considered in a relatively complete treatment of the iron-water relations. For example, $Fe(OH)^{\circ}_{3\,aq}$, or undissociated dissolved ferric hydroxide, may be an important contributor to iron solubility (J. Winchester, personal communication). On the other hand, we progress if we prepare diagrams showing the relations of known species to the stable solids.

Reference to Figure 7.3 shows that the relations needed to show the activities of the dissolved species are reactions expressing relations between magnetite and hematite, the stable solids in water, and the dissolved species, written in such a way as to be expressible in terms of Eh and pH. There are certain general relations that are helpful in calculation if recognized prior to the attempt to write the reactions.

For example, reactions involving hematite and any dissolved species containing ferric iron will be Eh-independent, inasmuch as no oxidation or reduction is required. Similarly, reactions involving hematite and dissolved species containing ferrous iron will certainly be Eh-dependent, and may be pH-dependent as well. Consequently, as a general rule, derivation of relations between dissolved species and solids is easier if reactions involving only one valence state are considered first.

The pattern of procedure used here is to calculate the activities of each dissolved species in equilibrium with hematite and magnetite, and to contour the activities within the Eh-pH framework of solid stability. After individual diagrams have been prepared, the final step is to superimpose them all, and emerge with a composite diagram showing all information concerning dissolved species for which data are available. Thus, the individual diagrams are reasonably permanent records, whereas the composite diagram can be considered a progress report in the sense that it can be added to at any time that more data become available on additional species.

FERRIC ION. Let us first calculate in detail the relations among magnetite, hematite, and the activity of ferric ion. Observing the rule previously stated, the first step is to write the reaction of ferric ion with ferric oxide, always trying to keep it in terms of water and hydrogen ions

$$Fe_2O_{3\,c} + 6H^+_{aq} = 2Fe^{3+}_{aq} + 3H_2O_l. \qquad (7.20)$$

The standard free energy of the reaction is

$$2\Delta F°_{fFe^{3+}} + 3\Delta F°_{fH_2O} - \Delta F°_{fFe_2O_3} - 6\Delta F°_{fH^+} = \Delta F°_r$$

$$(2 \times -2.53) + (3 \times -56.69) - (-177.1) - 6(0) = +2.0 \text{ kcal.}$$

The equilibrium constant is

$$\log K = \frac{\Delta F°_r}{-1.364} = \frac{2.0}{-1.364} = -1.45.$$

Eliminating $Fe_2O_{3\,c}$ and H_2O_l from the constant because their activity is unity

$$\log \frac{[Fe^{3+}]^2}{[H^+]^6} = -1.45.$$

Rearranging and substituting $-$pH for $\log [H^+]$

$$2 \log [Fe^{3+}] = -1.45 - 6 \text{ pH}$$

$$\log [Fe^{3+}] = -0.72 - 3 \text{ pH}. \qquad (7.21)$$

The log of the activity of ferric ion in equilibrium with $Fe_2O_{3\,c}$ is seen to be a linear function of pH. If pH is stipulated, $[Fe^{3+}]$ is fixed, and vice versa. The usual method is to assume convenient values of $[Fe^{3+}]$ and solve for pH. Because Eh is not involved, such contours of $[Fe^{3+}]$ will lie parallel to the Eh axis. Table 7.1 shows pairs of values of $\log [Fe^{3+}]$ and pH. The values chosen for $[Fe^{3+}]$ or other ions usually range from about 10^{-1} to 10^{-8} or 10^{-10}. Values higher than

10^{-1} fall in such a high concentration range that activities can be expected to depart markedly from molalities, for one thing; also we are rarely interested geologically in activities greater than 10^{-1} for a given species. On the other hand, when values for a given ion become less than 10^{-8} (about 1 part Fe^{3+} per 10 billion), the activity or molality of the ion can be considered insignificant.

TABLE 7.1. Solution of Equation (7.21) to Yield Values of pH for Activities of Fe^{3+} in Equilibrium with Hematite

log [Fe^{3+}]	pH
−1	0.09
−2	0.43
−3	0.76
−4	1.09
−5	1.43
−6	1.76
−7	2.09
−8	2.43

Figure 7.5a shows contours of log [Fe^{3+}] plotted from the table. Next we can consider the equilibrium between ferric ions and magnetite

$$Fe_3O_{4\,c} + 8H^+_{aq} = 3Fe^{3+}_{aq} + 4H_2O_l + e. \qquad (7.22)$$

In this case an oxidation is involved, so that this half-cell reaction is described by (eliminating substances of unit activity)

$$Eh = E° + \frac{0.059}{1} \log \frac{[Fe^{3+}]^3}{[H^+]^8}.$$

Obtaining $\Delta F°_r$ as before, and from it a numerical value of E°, and substituting −pH for log [H^+]

$$Eh = 0.337 + 0.177 \log [Fe^{3+}] + 0.472\, pH. \qquad (7.23)$$

Here we are faced with a new problem—it is not enough to stipulate [Fe^{3+}] to define the equation—a pH or an Eh value must also be chosen. However, this does in fact provide us with a test of the consistency of the thermodynamic data. If we choose a value such as 10^{-6} for [Fe^{3+}] and the corresponding pH value for [Fe^{3+}] in equilibrium with *hematite*, then the Eh we obtain should be that of the hematite-magnetite boundary at the appropriate [Fe^{3+}] and pH, inasmuch as at a given point on the boundary the two solids are in equilibrium, which means in turn that all ionic activities in equilibrium with one must also be in equilibrium with the other. If the Eh of the point chosen checks, then the contour can be drawn through the magnetite field, because at a fixed value of [Fe^{3+}] the *slope* of the contour in the magnetite field is fixed. Alternatively, Eh values within the magnetite field can be chosen arbitrarily and the pH calculated for the given activity of Fe^{3+}. Specifically, let us calculate Eh at the hematite-magnetite boundary

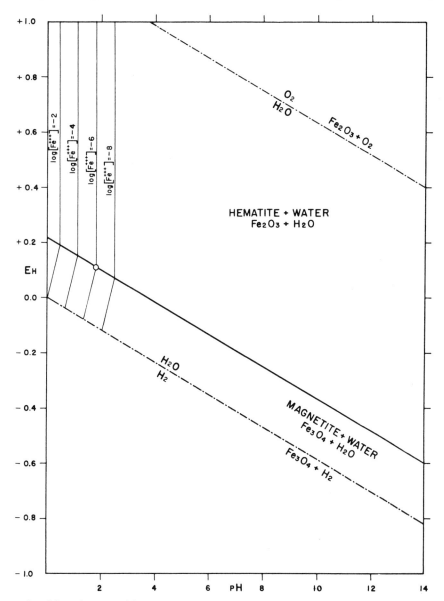

FIG. 7.5a. Activity of ferric iron in equilibrium with hematite and magnetite at 25 °C and 1 atmosphere total pressure. Contours are log $[Fe^{3+}]$. Note that values become vanishingly small at low pH in the fields of both solids.

for the pair of values log $[Fe^{3+}] = -6$; pH $= 1.76$ from Table 7.1. Substituting these values into equation (7.23)

$$Eh = 0.377 + 0.177 \times -6 + 0.472 \times 1.76 = +0.106.$$

This point is circled on Figure 7.5a, and shows that the check is within the limits of plotting. The contours in the magnetite field have been drawn in this instance with a slope of 0.472 volt per pH unit. If we translate activity of ferric ion into molality of ferric ion, we see that $m_{Fe^{3+}}$ is entirely negligible above pH values of about 2, and is low at pH values of 1. In passing, we note that the role of ferric ion as an important species in nature may have been overplayed, for it is a rare water that has a pH of 2 or less.

The contours are arbitrarily ended at the lower limit of stability of water, because Fe^{3+}_{aq} cannot exist at equilibrium below that point. In some later diagrams the contours will be continued into the metastable region to show the pattern of behavior expected if water persists beyond its stable limits.

OTHER IONIC SPECIES. For the rest of the ionic species, the reactions necessary to express their relations to hematite and magnetite, and the corresponding Eh-pH equations, as well as those already developed for Fe^{3+} are given in Table 7.2. The mechanics of calculation are parallel to those for ferric ion. Diagrams illustrating the distribution of each ion in the fields of the two solids constitute Figure 7.5a–f.

Composite Diagram

Figure 7.6 is a composite diagram drawn according to conventions set up by Pourbaix. The limit of the stability field of a given solid is arbitrarily drawn where the sum of the activities of the ions in equilibrium with the solid exceeds some chosen value. For geological purposes, a value of 10^{-6} is chosen, on the premise that if the sum of the activities of known dissolved species in equilibrium with a solid is less than 10^{-6}, that solid will behave as an immobile constituent in its environment. This rule, developed largely from experience, seems to correlate well with the observed behavior of minerals. A contour representing Σ activity ions $= 10^{-4}$ is also drawn to show the slope of the "solubility" as a function of pH and Eh. Boundaries are drawn between the fields of ions. A given field is labeled with the ion that is preponderant within it, and a boundary is placed where the ion becomes equal to an adjacent preponderant ion. Reference to Figure 7.5a–f shows that the distribution of ions is such that the fields labeled Fe^{++}_{aq} and Fe^{3+}_{aq} on Figure 7.6 in fact are overwhelmingly populated by ferrous and ferric ions, and

TABLE 7.2. Reactions and Equations Relating Ionic Activities to Magnetite and Hematite

Reaction		Equation	
A. Fe^{3+}_{aq}			
$Fe_2O_{3c} + 6H^+_{aq} = 2Fe^{3+}_{aq} + 3H_2O_l$	(7.20)	$\log[Fe^{3+}] = -0.72 - 3\,pH$	(7.21)
$Fe_3O_{4c} + 8H^+_{aq} = 3Fe^{3+}_{aq} + 4H_2O_l + e$	(7.22)	$Eh = 0.337 + 0.177\log[Fe^{3+}] + 0.472\,pH$	(7.23)
B. $Fe(OH)^{++}_{aq}$			
$2Fe(OH)^{++}_{aq} + H_2O_l = Fe_2O_{3c} + 4H^+_{aq}$	(7.24)	$\log[FeOH^{++}] = -3.151 - 2\,pH$	(7.25)
$Fe_3O_{4c} + 5H^+_{aq} = 3Fe(OH)^{++}_{aq} + H_2O_l + e$	(7.26)	$Eh = 0.780 + 0.177\log[FeOH^{++}] + 0.295\,pH$	(7.27)
C. $Fe(OH)^+_{2\,aq}$			
$Fe_2O_{3c} + H_2O_l + 2H^+_{aq} = 2Fe(OH)^+_{2\,aq}$	(7.28)	$\log[Fe(OH)^+_2] = -7.84 - pH$	(7.29)
$Fe_3O_{4c} + 2H_2O_l + 2H^+_{aq} = 3Fe(OH)^+_{2\,aq} + e$	(7.30)	$Eh = 1.61 + 0.177\log[Fe(OH)^+_2] + 0.118\,pH$	(7.31)
D. Fe^{++}_{aq}			
$2Fe^{++}_{aq} + 3H_2O_l = Fe_2O_{3c} + 6H^+_{aq} + 2e$	(7.32)	$Eh = 0.728 - 0.059\log[Fe^{++}] - 0.177\,pH$	(7.33)
$3Fe^{++}_{aq} + 4H_2O_l = Fe_3O_{4c} + 8H^+_{aq} + 2e$	(7.34)	$Eh = 0.980 - 0.0885\log[Fe^{++}] - 0.236\,pH$	(7.35)
E. $Fe(OH)^+_{aq}$			
$2Fe(OH)^+_{aq} + H_2O_l = Fe_2O_{3c} + 4H^+_{aq} + 2e$	(7.36)	$Eh = 0.217 - 0.059\log[Fe(OH)^+] - 0.118\,pH$	(7.37)
$3Fe(OH)^+_{aq} + H_2O_l = Fe_3O_{4c} + 5H^+_{aq} + 2e$	(7.38)	$Eh = 0.214 - 0.0885\log[Fe(OH)^+] - 0.148\,pH$	(7.39)
F. $HFeO^-_{2\,aq}$			
$2HFeO^-_{2\,aq} = Fe_2O_{3c} + H_2O_l + 2e$	(7.40)	$Eh = -1.139 - 0.059\log[HFeO^-_2]$	(7.41)
$3HFeO^-_{2\,aq} + H^+_{aq} = Fe_3O_{4c} + 2H_2O_l + 2e$	(7.42)	$Eh = -1.819 - 0.0885\log[HFeO^-_2] + 0.0295\,pH$	(7.43)

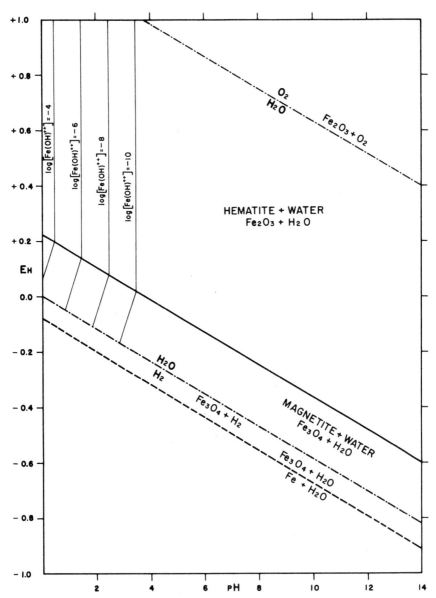

FIG. 7.5b. Activity of Fe(OH)$^{++}$ ion in equilibrium with hematite and magnetite at 25 °C and 1 atmosphere total pressure, within the stability field of water.

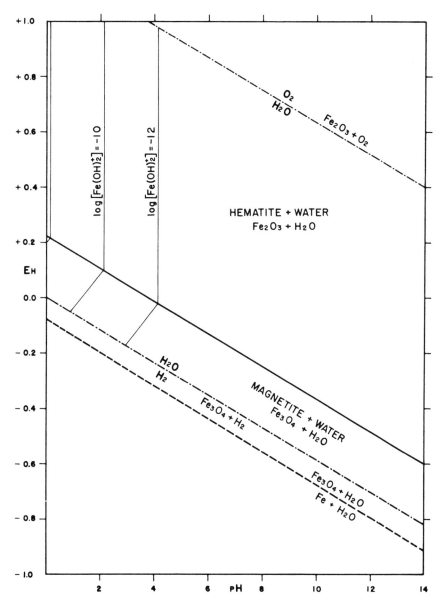

FIG. 7.5c. Activity of Fe(OH)$_2^+$ ion in equilibrium with hematite and magnetite at 25 °C and 1 atmosphere total pressure, within the stability field of water.

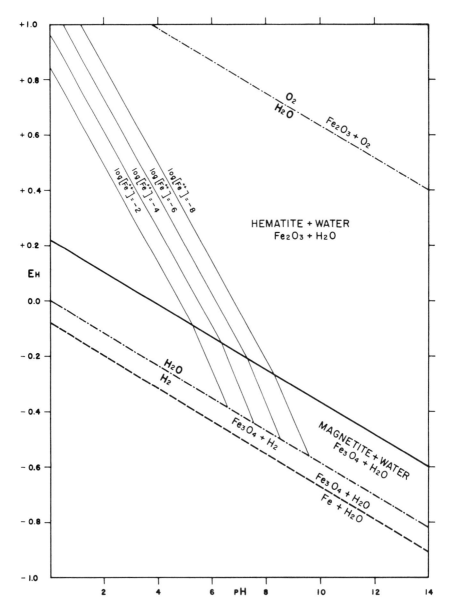

FIG. 7.5d. Activity of Fe^{++} ion in equilibrium with hematite and magnetite at 25 °C and 1 atmosphere total pressure, within the stability field of water.

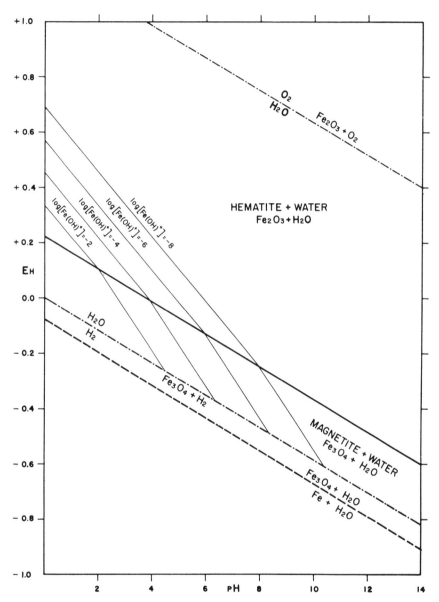

FIG. 7.5e. Activity of Fe(OH)$^+$ ion in equilibrium with hematite and magnetite at 25 °C and 1 atmosphere total pressure, within the stability field of water.

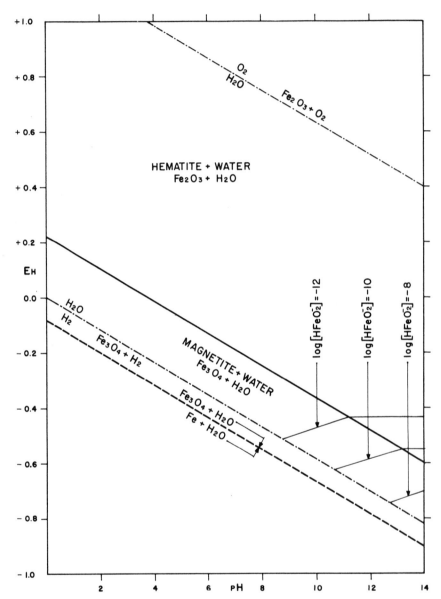

FIG. 7.5f. Activity of $HFeO_2^-$ ion in equilibrium with hematite and magnetite at 25 °C and 1 atmosphere total pressure, within the stability field of water.

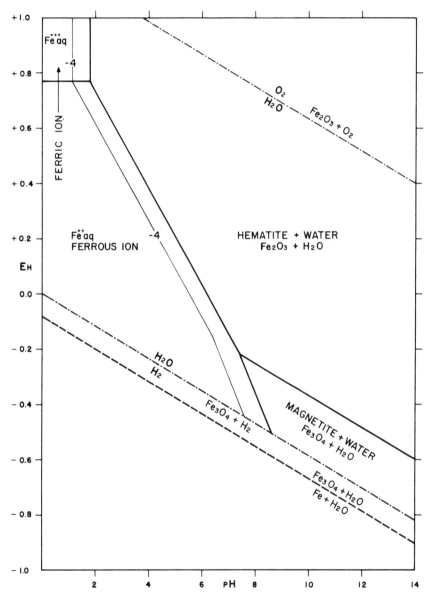

FIG. 7.6. Composite diagram showing stability fields of hematite and magnetite in water. Fields of ions are designated where total activity of dissolved species are $> 10^{-6}$. Fields of ions are labeled with dominant species. Contour of log [dissolved species] $= -4$ is included to show slope of activity change. Plot at 25 °C and 1 atmosphere total pressure.

except in local areas, the hidden contribution from the other ions is almost negligible.

Because ionic activities in equilibrium with solids change so rapidly with change in Eh and pH, it is relatively rare that any region on an Eh-pH diagram cannot be ascribed satisfactorily to a single ion that makes up 99 percent or more of the total ionic activity.

There are several methods for obtaining composite diagrams. One that appeals to geologists, because of their experience with topographic maps, is to make tracings of the Eh-pH diagrams showing contours of individual ions, and then to add the individual ion contributions graphically by overlaying the tracings. A boundary, such as that between Fe^{++}_{aq} and Fe^{3+}_{aq} on Figure 7.6, is quickly determined by connecting points of intersection of equal contours.

The composite diagram also can be assembled analytically by writing reactions between the ions and solving for the Eh-pH conditions at which they are equal. For example, if we write

$$Fe^{++}_{aq} = Fe^{3+}_{aq} + e \qquad (7.44)$$

then

$$Eh = E° + 0.059 \log \frac{[Fe^{3+}]}{[Fe^{++}]}. \qquad (7.45)$$

And, under the condition that $Fe^{++}_{aq} = Fe^{3+}_{aq}$, $Eh = E°$. Consequently, the boundary between the ions is at $E°$ for the half-cell, or at 0.771 volt.

Similarly, for the pair Fe^{++}_{aq}-$Fe(OH)^+_{aq}$

$$Fe^{++}_{aq} + H_2O_1 = Fe(OH)^+_{aq} + H^+_{aq} \qquad (7.46)$$

and

$$\log \frac{[Fe(OH)^+]}{[Fe^{++}]} = \log K + pH. \qquad (7.47)$$

When the ions are equal, $pH = -\log K$. In this instance it is found that the ionic activities become equal at such a high pH that neither ion is as great as 10^{-6}; in other words, $Fe(OH)^+$ does not become a dominant species within the field we have arbitrarily designated as a field of "solubility."

In the example of the iron oxides, and in many following diagrams, it is gratifying to discover that in dilute solution, the major ions that have to be considered are chiefly old friends. In other words, the ions we think of as "common" tend to be discovered under ordinary laboratory conditions and to be described first, and only by greater and greater refinement, or by extension of chemical work to unusual conditions, are fields entered in which they are displaced by other species.

Among the most striking aspects of the composite diagram for iron is the highly restricted field of predominance of ferric ion. Only under strongly acid and oxidizing conditions does the activity of the ion exceed 10^{-6}. The great stability of hematite is evidenced by the size of its field, which ranges from moderately acid oxidizing conditions to strongly reducing neutral and alkaline environments. "Solubility" in this system is achieved almost exclusively through the contribution of the ferrous ion, which strikes deepest into the fields of the solids in reducing environments. Neither hematite nor magnetite shows appreciable amphoteric behavior in the pH range 0–14; the activity of $HFeO_2^-$ does not exceed 10^{-6}, so that it does not even appear on the final diagram.

EFFECT OF CO_2 ON IRON-WATER-OXYGEN RELATIONS

In natural environments, iron occurs as magnetite, hematite, and as siderite, the carbonate, as well as in the form of iron sulfides and silicates. Thus, the next step is to consider the influence of CO_2 on the stability relations of magnetite and hematite. From the earlier development of partial pressure diagrams, the most obvious attack is to try to develop relations among Eh, pH, and P_{CO_2}. Inasmuch as Eh and pH can be considered as a two-variable representation of P_{O_2}, it should be possible to consider siderite in relation to magnetite and hematite by using a three-dimensional representation with Eh, pH, and P_{CO_2} as the axes.

Stability as a Function of P_{CO_2}

The foregoing treatment requires writing reactions between magnetite and siderite, and between hematite and siderite. Considering hematite first

$$2FeCO_{3\ c} + H_2O_l = Fe_2O_{3\ c} + 2CO_{2\ g} + 2H^+_{aq} + 2e. \quad (7.48)$$

The corresponding Eh equation is

$$Eh = E° + 0.059 \log P_{CO_2} - 0.059\ pH \quad (7.49)$$

$$Eh = 0.286 + 0.059 \log P_{CO_2} - 0.059\ pH.$$

For the reaction between siderite and magnetite

$$3FeCO_{3\ c} + H_2O_l = Fe_3O_{4\ c} + 3CO_{2\ g} + 2H^+_{aq} + 2e \quad (7.50)$$

and

$$Eh = 0.319 + 0.0885 \log P_{CO_2} - 0.059\ pH. \quad (7.51)$$

Then, as shown in Figure 7.7, it is possible to plot magnetite-hematite relations as previously developed, and to show them on the front face of the figure. Then, plotting P_{CO_2} as a third dimension, the magnetite-hematite boundaries, which are independent of P_{CO_2}, can be extended as planes parallel to the P_{CO_2} axis. Equations (7.49) and (7.51), on

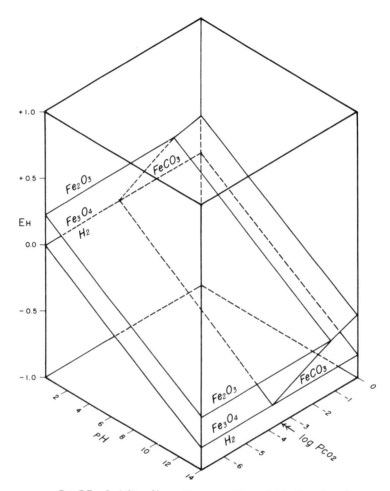

FIG. 7.7. Stability of hematite, magnetite, and siderite as function of Eh, pH, and P_{CO_2} at 25 °C and 1 atmosphere total pressure. Relations shown are terminated at lower stability limit of water. Double arrow on log P_{CO_2} axis shows partial pressure of CO_2 in present earth atmosphere.

the other hand, are functions of all three variables, so they plot as sloping planes in three dimensions, and serve to show how the fields of magnetite and hematite are encroached upon by siderite as P_{CO_2} increases. The double arrow at a P_{CO_2} of $10^{-3.5}$ atmosphere indicates the partial pressure of CO_2 in the earth's atmosphere, and it is of interest that under such conditions siderite has only a small field of stability. On the other hand, magnetite is entirely displaced by siderite when P_{CO_2} reaches a value of about $10^{-1.4}$ atmosphere.

The three-dimensional diagram is useful for showing gross relations in the system, but for many purposes iso-P_{CO_2} sections are more convenient. Such sections can be used to show the equilibrium activities of dissolved ions more easily than can the three-dimensional diagrams.

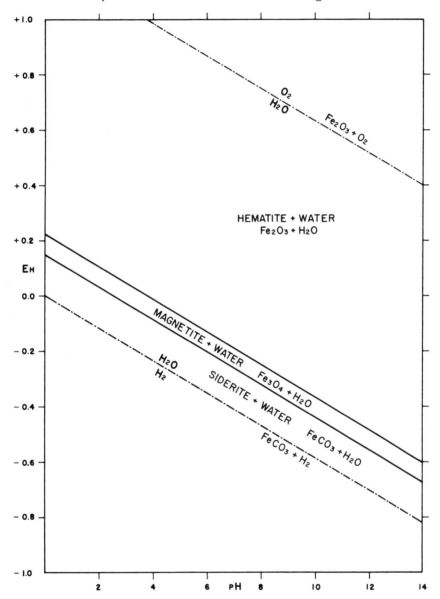

FIG. 7.8a. Stability of Fe_2O_3, Fe_3O_4, and $FeCO_3$ at 25 °C and 1 atmosphere total pressure, with $P_{CO_2} = 10^{-2.0}$ atmosphere. Relations shown are for stability range of water.

Figure 7.8a is a section of Figure 7.7 at $P_{CO_2} = 10^{-2.0}$ atmosphere. The boundary between siderite and magnetite is plotted directly from equation (7.51), rather than attempting to transfer it graphically from

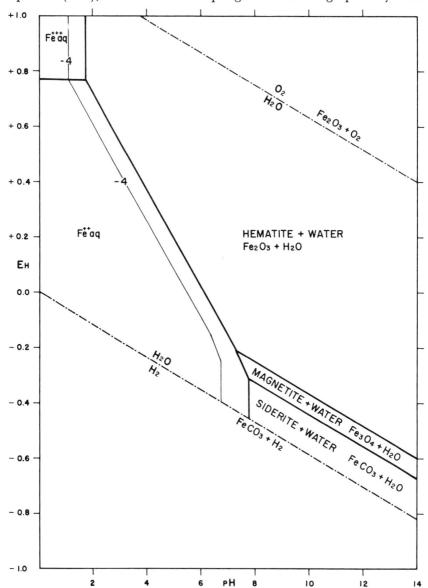

FIG. 7.8b. Stability of Fe_2O_3, Fe_3O_4, and $FeCO_3$ at 25 °C and 1 atmosphere total pressure, with $P_{CO_2} = 10^{-2.0}$ atmosphere. Contour is for Σactivity ions $= 10^{-4}$. Boundary of solids is at Σactivity ions $= 10^{-6}$. Relations shown only within boundaries of water stability.

Figure 7.7. In Figure 7.8b the activity of dissolved ionic species is shown for the new system, including siderite. Because the field of stability of siderite relative to hematite and magnetite has already been defined by equations (7.49) and (7.51), it is necessary to alter the hematite-magnetite ion activity diagram (Figure 7.6) only within the field of this new stable solid. At first glance it might appear necessary to calculate equilibrium between siderite and Fe^{3+}, $Fe(OH)^{++}$, $Fe(OH)_2^+$, $HFeO_2^-$, $Fe(OH)^+$, and Fe^{++}, but inasmuch as only Fe^{++} was important ($> 10^{-6}$) in the field of magnetite now occupied by siderite, and inasmuch as siderite is more stable than magnetite in the region in which it has displaced magnetite, the activities of all ions in equilibrium with siderite will be less than they were for magnetite. Consequently, only the activity of Fe^{++} need be considered.

Only one reaction need be written

$$FeCO_{3\ c} + 2H^+_{aq} = Fe^{++}_{aq} + CO_{2\ g} + H_2O_l. \qquad (7.52)$$

The equation is

$$\log [Fe^{++}] = 7.47 - \log P_{CO_2} - 2\ pH. \qquad (7.53)$$

P_{CO_2} has been stipulated as fixed at $10^{-2.0}$, so that $\log [Fe^{++}]$ is a function only of pH and must be plotted as contours parallel to the Eh axis. A check of the validity of the free energy values used, as well as of the arithmetic and plotting, is obtained if the Fe^{++} contours in the field of $FeCO_3$ join appropriately those carried forward from the magnetite-hematite diagram.

At a $P_{CO_2} = 10^{-3.5}$, that of the earth's atmosphere, the field of $FeCO_3$ is so small that it barely gets above the stability limit of water. Thus, it was necessary to choose a larger value to show the behavior of the dissolved Fe^{++}.

Stability as a Function of $\Sigma\ CO_2$.

Siderite is an important primary sedimentary mineral of iron ores, but the preceding discussion shows that it is not an equilibrium species in the presence of the atmosphere. Its occurrence is indicative of strongly reducing conditions and the presence of CO_2 in more than atmospheric amounts. Let us now consider siderite stability in terms of the total dissolved carbonate. In a given solution, free or nearly free from oxygen, how much carbonate or bicarbonate ion is necessary to stabilize siderite relative to magnetite? Does the required amount fall within the range of ordinary ground water analyses? Consider a natural water with total dissolved carbonate ($H_2CO_3 + HCO_3^- + CO_3^{--}$) of 10^{-2} (roughly 600 ppm) molal. Under what conditions, if any, is siderite stable?

The problem to be solved is essentially identical to Case 3 considered in Chapter 3 on carbonate equilibria. Total carbonate is known. What happens at arbitrarily selected values of Eh and pH?

One approach is to determine the distribution of $[CO_3^{--}]$ as a function of pH, and then to draw contours of $[Fe^{++}]$ from the relation

$$[Fe^{++}][CO_3^{--}] = K_{FeCO_3}. \qquad (7.54)$$

Then, by superimposing these contours for $[Fe^{++}]$ on the diagram for $[Fe^{++}]$ in equilibrium with Fe_2O_3 and Fe_3O_4 (Figure 7.5d), the field of stability of $FeCO_3$ is delineated as that area in which $[Fe^{++}]$ is smaller in equilibrium with $FeCO_3$ than with the other two solids.

Both methods will be illustrated; the first has perhaps the more general utility, because it can be used no matter how complex the system becomes. If the activity of a given ion in equilibrium with a given solid is less than that in equilibrium with any other solid considered, the region of the minimum is a stability field of the given solid.

For a total carbonate of 10^{-2}, we can write, without serious error (Chapter 3)

$$[H_2CO_3] + [HCO_3^-] + [CO_3^{--}] = 10^{-2} \qquad (7.55)$$

$$\frac{[H^+][CO_3^{--}]}{[HCO_3^-]} = 10^{-10.3} \qquad (7.56)$$

$$\frac{[H^+][HCO_3^-]}{[H_2CO_3]} = 10^{-6.4}. \qquad (7.57)$$

Because we know that carbonate species are not reduced or oxidized under natural conditions, except perhaps under special circumstances,[1] Eh need not be considered, and contours of $[CO_3^{--}]$ will lie parallel to the Eh axis. Equations (7.55)–(7.57) can be solved in terms of $[H^+]$ and $[CO_3^{--}]$, first by obtaining $[HCO_3^-]$ in terms of $[H^+]$ and $[CO_3^{--}]$ from equation (7.56), and then by using this value to obtain $[H_2CO_3]$ in terms of $[H^+]$ and $[CO_3^{--}]$ from equation (7.57). Then, substituting these values in (7.55)

$$\frac{[H^+]^2[CO_3^{--}]}{10^{-16.7}} + \frac{[H^+][CO_3^{--}]}{10^{-10.3}} + [CO_3^{--}] = 10^{-2}.$$

Simplifying

$$[H^+]^2 + 10^{-6.4}[H^+] + 10^{-16.7} = \frac{10^{-18.7}}{[CO_3^{--}]}. \qquad (7.58)$$

From this equation, substitution of various arbitrary values of $[CO_3^{--}]$

---

[1] The various carbonate species are in fact thermodynamically unstable with respect to carbon near the lower oxidation potential limit of water, a fact that may have considerable geological significance, but is ignored in the present treatment.

allows calculation of the corresponding pH. The result is Figure 7.9. Then, substituting the numerical value of K into equation (7.54) and rearranging

$$\log [\text{Fe}^{++}] = -10.67 - \log [\text{CO}_3^{--}]. \tag{7.59}$$

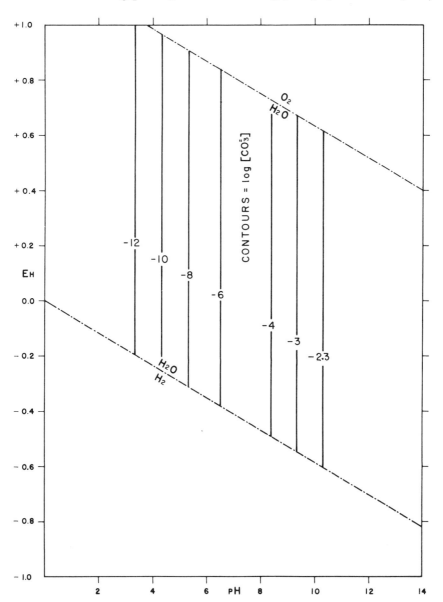

FIG. 7.9. Contours of $\log [\text{CO}_3^{--}]$ shown on an Eh-pH plot for the condition that $[\text{H}_2\text{CO}_3] + [\text{HCO}_3^-] + [\text{CO}_3^{--}] = 10^{-2}$ at 25 °C and 1 atmosphere total pressure.

Thus, for every contour of log $[CO_3^{--}]$ a corresponding contour of log $[Fe^{++}]$ can be obtained. In Figure 7.10 these contours for $[Fe^{++}]$ in equilibrium with $FeCO_3$ are superimposed on the contours of $[Fe^{++}]$

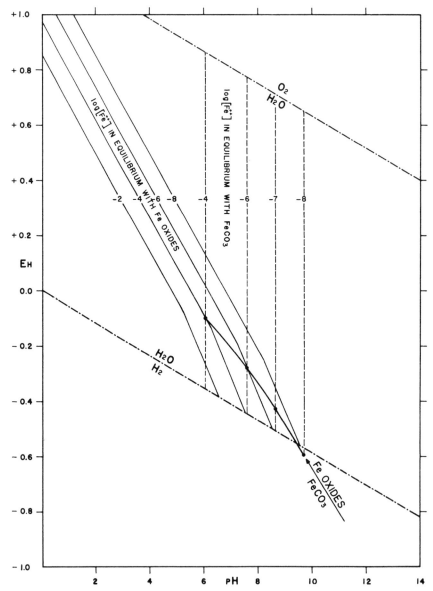

FIG. 7.10. Method of establishing stability field of $FeCO_3$ relative to Fe oxides by delineating area in which $[Fe^{++}]$ is a minimum in equilibrium with $FeCO_3$. Solution contains $10^{-2}m$ $H_2CO_3 + HCO_3^- + CO_3^{--}$. Dashed lines are log $[Fe^{++}]$ in equilibrium with $FeCO_3$; solid lines in equilibrium with Fe oxides. Black dots are points of equal $[Fe^{++}]$. System at 25 °C and 1 atmosphere total pressure.

in equilibrium with Fe_2O_3 and Fe_3O_4 (Figure 7.5d), and the boundary between $FeCO_3$ and the oxides is indicated. Figure 7.11 shows a composite diagram for the oxides and siderite, obtained by overlaying

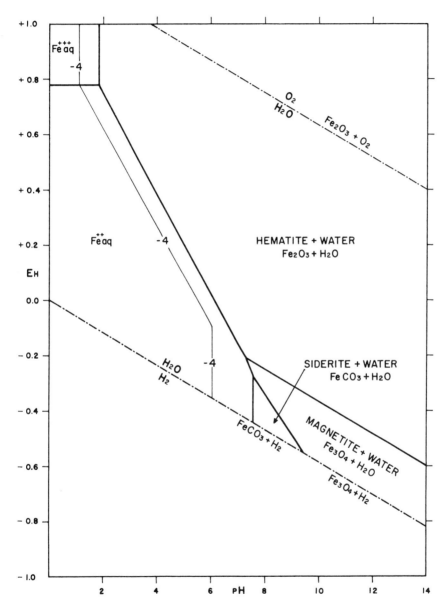

FIG. 7.11. Stability of hematite, magnetite, and siderite in aqueous solution containing total dissolved carbonate species of $10^{-2}\ m$, at 25 °C and 1 atmosphere total pressure. Contour is for log [dissolved iron] $= -4$. Boundary between solids and ions at log [dissolved iron] $= -6$.

Figure 7.10 on Figure 7.6. Note the similarities and differences of Figure 7.8b, which is at constant P_{CO_2}, and Figure 7.11, with fixed total dissolved carbonate. When total dissolved carbonate is fixed, $[CO_3^{--}]$ rises to a maximum at a pH of about 10.5 and remains constant, whereas at a constant P_{CO_2}, $[CO_3^{--}]$ rises continuously with increasing pH. This is reflected in a restricted field of siderite at high pH at constant total carbonate, for the oxides become continuously more stable as pH is increased (they are precipitated by OH^-), whereas the stability of siderite remains constant above a pH of about 10.5. In other words, fluctuations of pH without a change in oxidation conditions can cause an alternation of siderite and magnetite in a sediment in a system closed to CO_2, but in the open system, where P_{CO_2} is constant, the change from siderite to magnetite cannot take place without a change in P_{O_2}.

In the second method of obtaining the siderite stability field, the first step is to delineate the pH-Eh areas in which a given dissolved carbonate species is preponderant. From the equations

$$\frac{[CO_3^{--}]}{[HCO_3^-]} = \frac{10^{-10.3}}{[H^+]} \qquad (7.56)$$

$$\frac{[HCO_3^-]}{[H_2CO_3]} = \frac{10^{-6.4}}{[H^+]} \qquad (7.57)$$

it can be seen that carbonate ion equals bicarbonate ion activity at pH 10.3, and bicarbonate ion equals carbonic acid activity at pH 6.4. Thus, below pH 6.4, H_2CO_3 is the dominant dissolved species; between pH 6.4 and pH 10.3, HCO_3^- takes over; and above pH 10.3, CO_3^{--} is king. Moreover, these boundaries, being based on ratios, are valid for any given total dissolved carbonate. In the dilute range (ionic strength ≤ 0.2), we can also assign all the dissolved CO_2 to the preponderant species within its field of dominance without much error, as long as we do not approach the pH boundaries too closely.

Consequently, when total dissolved CO_2 is 10^{-2} m, we can write for pH values lower than 6.4

$$3FeCO_{3\,c} + 4H_2O_1 = Fe_3O_{4\,c} + 3H_2CO_{3\,aq} + 2H^+_{aq} + 2e, \qquad (7.60)$$

and we can express the result entirely in terms of Eh and pH because the activity of $H_2CO_{3\,aq}$ is fixed at 10^{-2}. In the field of HCO_3^-, the corresponding reaction is

$$3FeCO_{3\,c} + 4H_2O_1 = Fe_3O_{4\,c} + 3HCO_{3\,aq}^- + 5H^+_{aq} + 2e. \qquad (7.61)$$

In the CO_3^{--} field, the reaction is

$$3FeCO_{3\,c} + 4H_2O_1 = Fe_3O_{4\,c} + 3CO_{3\,aq}^{--} + 8H^+_{aq} + 2e. \qquad (7.62)$$

The Eh-pH equations are

$$Eh = 0.445 - 0.059\, pH + 0.0885 \log [H_2CO_3] \quad (7.63)$$
$$Eh = 1.010 - 0.148\, pH + 0.0885 \log [HCO_3^-] \quad (7.64)$$
$$Eh = 1.920 - 0.236\, pH + 0.0885 \log [CO_3^{--}]. \quad (7.65)$$

Also, the relations between $FeCO_3$ and Fe_2O_3 must be considered

$$2FeCO_{3\,c} + 3H_2O_l = Fe_2O_{3\,c} + 2H_2CO_{3\,aq} + 2H^+_{aq} + 2e \quad (7.66)$$
$$2FeCO_{3\,c} + 3H_2O_l = Fe_2O_{3\,c} + 2HCO_{3\,aq}^- + 4H^+_{aq} + 2e \quad (7.67)$$
$$2FeCO_{3\,c} + 3H_2O_l = Fe_2O_{3\,c} + 2CO_{3\,aq}^{--} + 6H^+_{aq} + 2e. \quad (7.68)$$

And the Eh-pH equations are

$$Eh = 0.370 - 0.059\, pH + 0.059 \log [H_2CO_3] \quad (7.69)$$
$$Eh = 0.747 - 0.118\, pH + 0.059 \log [HCO_3^-] \quad (7.70)$$
$$Eh = 1.359 - 0.177\, pH + 0.059 \log [CO_3^{--}]. \quad (7.71)$$

All six Eh-pH equations, (7.63)–(7.65) and (7.69)–(7.71), are plotted in Figure 7.12 and labeled, as well as the boundaries between the carbonate species (7.56) and (7.57). Note that the reactions are plotted only in the field of the carbonate ion concerned, i.e., the reaction from $FeCO_3$ to Fe_2O_3 is shown only in the range of pH from 0 to 6.4, in which H_2CO_3 constitutes essentially all the dissolved carbonate. The stable and metastable reactions are determined by working from the bottom of the diagram upwards; in the field of H_2CO_3, Fe_3O_4 cannot go to Fe_2O_3 because $FeCO_3$ is not yet oxidized to Fe_3O_4; and $FeCO_3$ oxidizes to Fe_2O_3 before it oxidizes to Fe_3O_4. Thus, in this region the stable reaction is from $FeCO_3$ directly to Fe_2O_3.

As shown, the lines change slope abruptly at pH 6.4 and 10.3; in fact they must bend, inasmuch as the carbonate species do not persist as 10^{-2} all the way to the pH boundaries. In Figure 7.13 the stable boundaries are traced from Figure 7.12 to show the relations of $FeCO_3$, Fe_2O_3, and Fe_3O_4 at a total dissolved carbonate of 10^{-2} molal.

Once the fields of the solids are delineated in this fashion, the contour for $[Fe^{++}] = 10^{-6}$ can be found from the relation

$$FeCO_{3\,c} + H^+_{aq} = Fe^{++}_{aq} + HCO^-_{3\,aq}. \quad (7.72)$$

In the field of dominance of HCO_3^- (pH 6.4–10.3), $[HCO_3^-] = 10^{-2}$, and the equilibrium relations can be expressed

$$\frac{[Fe^{++}][HCO_3^-]}{[H^+]} = K$$

or

$$\log [Fe^{++}] - 2 + pH = \log K. \quad (7.73)$$

If K is obtained from standard free energy values, the equation becomes

$$\log [Fe^{++}] = -0.33 - pH + 2. \qquad (7.74)$$

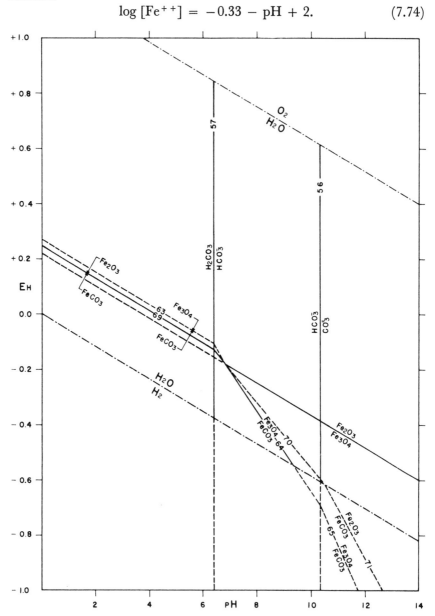

FIG. 7.12. Eh-pH plot of equations for reactions of siderite to magnetite, and siderite to hematite at a total dissolved carbonate of 10^{-2}. Solid lines are stable boundaries; short dashed lines are metastable boundaries. Numbers on lines refer to equations in text. Conditions are 25 °C and 1 atmosphere total pressure.

The Eh-pH equations are

$$Eh = 0.445 - 0.059 \, pH + 0.0885 \log [H_2CO_3] \quad (7.63)$$
$$Eh = 1.010 - 0.148 \, pH + 0.0885 \log [HCO_3^-] \quad (7.64)$$
$$Eh = 1.920 - 0.236 \, pH + 0.0885 \log [CO_3^{--}]. \quad (7.65)$$

Also, the relations between $FeCO_3$ and Fe_2O_3 must be considered

$$2FeCO_{3\,c} + 3H_2O_l = Fe_2O_{3\,c} + 2H_2CO_{3\,aq} + 2H_{aq}^+ + 2e \quad (7.66)$$
$$2FeCO_{3\,c} + 3H_2O_l = Fe_2O_{3\,c} + 2HCO_{3\,aq}^- + 4H_{aq}^+ + 2e \quad (7.67)$$
$$2FeCO_{3\,c} + 3H_2O_l = Fe_2O_{3\,c} + 2CO_{3\,aq}^{--} + 6H_{aq}^+ + 2e. \quad (7.68)$$

And the Eh-pH equations are

$$Eh = 0.370 - 0.059 \, pH + 0.059 \log [H_2CO_3] \quad (7.69)$$
$$Eh = 0.747 - 0.118 \, pH + 0.059 \log [HCO_3^-] \quad (7.70)$$
$$Eh = 1.359 - 0.177 \, pH + 0.059 \log [CO_3^{--}]. \quad (7.71)$$

All six Eh-pH equations, (7.63)–(7.65) and (7.69)–(7.71), are plotted in Figure 7.12 and labeled, as well as the boundaries between the carbonate species (7.56) and (7.57). Note that the reactions are plotted only in the field of the carbonate ion concerned, i.e., the reaction from $FeCO_3$ to Fe_2O_3 is shown only in the range of pH from 0 to 6.4, in which H_2CO_3 constitutes essentially all the dissolved carbonate. The stable and metastable reactions are determined by working from the bottom of the diagram upwards; in the field of H_2CO_3, Fe_3O_4 cannot go to Fe_2O_3 because $FeCO_3$ is not yet oxidized to Fe_3O_4; and $FeCO_3$ oxidizes to Fe_2O_3 before it oxidizes to Fe_3O_4. Thus, in this region the stable reaction is from $FeCO_3$ directly to Fe_2O_3.

As shown, the lines change slope abruptly at pH 6.4 and 10.3; in fact they must bend, inasmuch as the carbonate species do not persist as 10^{-2} all the way to the pH boundaries. In Figure 7.13 the stable boundaries are traced from Figure 7.12 to show the relations of $FeCO_3$, Fe_2O_3, and Fe_3O_4 at a total dissolved carbonate of 10^{-2} molal.

Once the fields of the solids are delineated in this fashion, the contour for $[Fe^{++}] = 10^{-6}$ can be found from the relation

$$FeCO_{3\,c} + H_{aq}^+ = Fe_{aq}^{++} + HCO_{3\,aq}^-. \quad (7.72)$$

In the field of dominance of HCO_3^- (pH 6.4–10.3), $[HCO_3^-] = 10^{-2}$, and the equilibrium relations can be expressed

$$\frac{[Fe^{++}][HCO_3^-]}{[H^+]} = K$$

or

$$\log [Fe^{++}] - 2 + pH = \log K. \quad (7.73)$$

If K is obtained from standard free energy values, the equation becomes

$$\log [Fe^{++}] = -0.33 - pH + 2. \qquad (7.74)$$

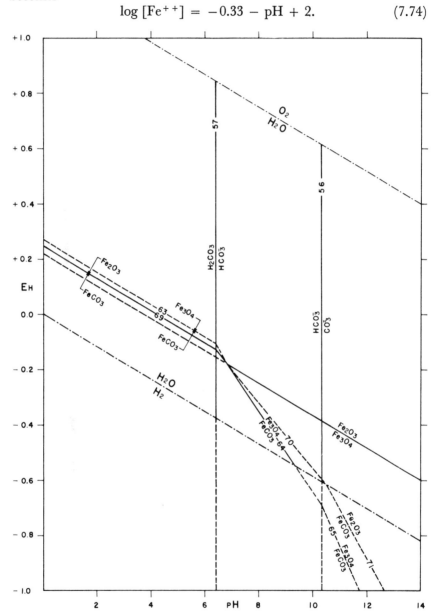

FIG. 7.12. Eh-pH plot of equations for reactions of siderite to magnetite, and siderite to hematite at a total dissolved carbonate of 10^{-2}. Solid lines are stable boundaries; short dashed lines are metastable boundaries. Numbers on lines refer to equations in text. Conditions are 25 °C and 1 atmosphere total pressure.

Substituting -6 for $\log [Fe^{++}]$, and solving for pH, we obtain a value of 7.67, which agrees well with that obtained graphically (Figure 7.11). This contour of $[Fe^{++}]$ is shown as a dotted line on Figure 7.13.

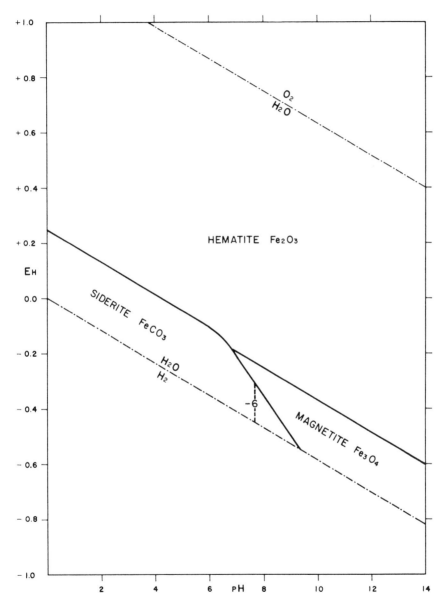

FIG. 7.13. Stability fields of hematite, magnetite, and siderite in water at 25 °C, 1 atmosphere total pressure, and total dissolved carbonate of 10^{-2}. Dotted line is for $\log [Fe^{++}] = -6$.

The field of $FeCO_3$ delineated by this contour and the boundaries between $FeCO_3$ and Fe_3O_4, and H_2-H_2O, is identical to that obtained by the method of contouring used before (Figure 7.10).

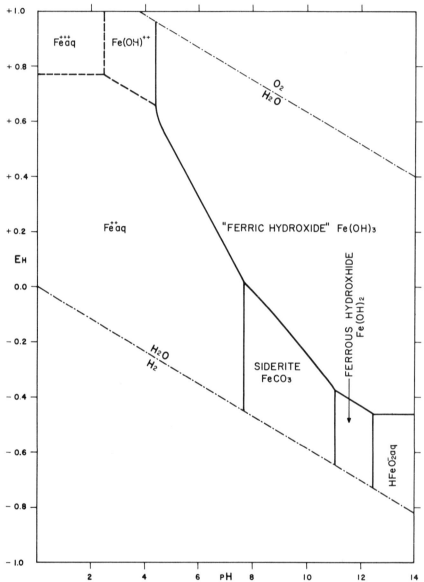

FIG. 7.14. Diagram showing the relations among the metastable iron hydroxides and siderite at 25 °C and 1 atmosphere total pressure. Boundary between solids and ions at total activity of dissolved species = 10^{-6}. Total dissolved carbonate species = 10^{-2}. Dashed lines are boundaries between fields dominated by the labeled ion.

If the total dissolved carbonate is increased, the field of siderite expands; if it is decreased, it contracts. If total dissolved carbonate drops below about $10^{-3}m$, the field of siderite disappears, if defined as the area in which the activity of dissolved iron-containing species is less than 10^{-6}.

On the other hand, at the same total dissolved carbonate considered here ($10^{-2}m$), the field of siderite relative to the freshly precipitated iron hydroxides is much larger, as shown in Figure 7.14. This figure can be considered the "experimental" representation, inasmuch as precipitation of iron in the laboratory follows the relations shown here fairly well. Because of the lesser stability of the iron hydroxides, note the appearance of fields of $Fe(OH)^{++}$ and $HFeO_2^-$.

EFFECT OF SULFUR ON IRON-WATER-OXYGEN RELATIONS

Adding CO_2 to the iron-water-oxygen system permitted assessment of the stability of siderite relative to the oxides; addition of sulfur should give information on pyrite and pyrrhotite. As in the development of CO_2 relations, the most obvious type of diagram to construct is a three-variable plot of Eh, pH, and P_{S_2}. Again we can start with the relations of magnetite and hematite in water and write reactions from these species to pyrrhotite and pyrite with addition of sulfur

$$3FeS_c + 4H_2O_l = Fe_3O_{4\,c} + \tfrac{3}{2}S_{2\,g} + 8H^+_{aq} + 8e \quad (7.75)$$

$$2FeS_c + 3H_2O_l = Fe_2O_{3\,c} + S_{2\,g} + 6H^+_{aq} + 6e \quad (7.76)$$

$$3FeS_{2\,c} + 4H_2O_l = Fe_3O_{4\,c} + 3S_{2\,g} + 8H^+_{aq} + 8e \quad (7.77)$$

$$2FeS_{2\,c} + 3H_2O_l = Fe_2O_{3\,c} + 2S_{2\,g} + 6H^+_{aq} + 6e. \quad (7.78)$$

The corresponding Eh-pH-P_{S_2} equations are

$$Eh = 0.444 + 0.011 \log P_{S_2} - 0.059\, pH \quad (7.79)$$

$$Eh = 0.419 + 0.0098 \log P_{S_2} - 0.059\, pH \quad (7.80)$$

$$Eh = 0.811 + 0.0221 \log P_{S_2} - 0.059\, pH \quad (7.81)$$

$$Eh = 0.746 + 0.0197 \log P_{S_2} - 0.059\, pH. \quad (7.82)$$

We must also consider the boundary between pyrrhotite and pyrite

$$2FeS_c + S_{2\,g} = 2FeS_{2\,c}. \quad (7.83)$$

This boundary is at a fixed P_{S_2} and is independent of Eh and pH. The value of P_{S_2} is 10^{-33} atmosphere.

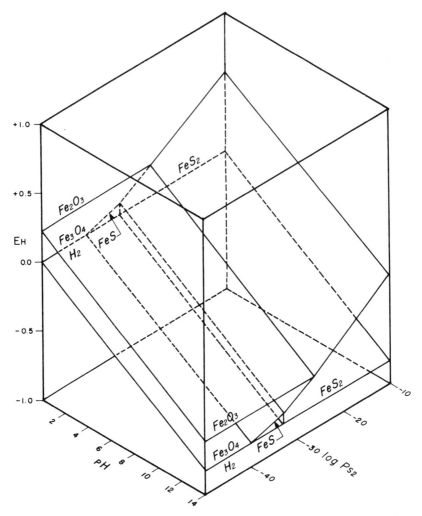

FIG. 7.15. Stability relations of Fe_2O_3, Fe_3O_4, FeS, and FeS_2 as functions of Eh, pH, and log P_{S_2} at 25 °C and 1 atmosphere total pressure in the presence of water. Note the small field of stability of FeS and the small values of P_{S_2} for the association pyrite, pyrrhotite, and magnetite.

The easiest method of plotting is to make first an Eh-log P_{S_2} section at pH = 0, plotting the lines for the various equilibria. Then, after eliminating metastable relations, the planes can be drawn to obtain the three-dimensional diagram. Construction is easy after elimination of metastable boundaries, inasmuch as all the lines have the same slope, except for reaction (7.83). The results are shown in Figure 7.15. The similarity to the diagram for Eh-pH-log P_{CO_2} (Figure 7.7) is striking.

Stability as a Function of Σ S

As shown in the diagrams concerning CO_2, it is frequently useful to construct constant P_{S_2} sections of Eh-pH diagrams. The procedure is identical to that already discussed for CO_2-bearing systems, so that none is illustrated here. On the other hand, new complexities arise in systems with a fixed amount of dissolved sulfur. There are but three important dissolved carbonate species, H_2CO_3, HCO_3^-, and CO_3^{--}, and they are not Eh-sensitive. More than 40 ionic and molecular species containing sulfur have been isolated and studied, and their relative stabilities are functions of both Eh and pH. Fortunately, it has been shown by Valensi (G. Valensi, Contribution au diagramme potential-pH du soufre. *Compt. rend. 2ème Réunion, Comité intern. thermo. kinetics électrochim.*, Milan, 1950, 51–68) that only those familiar in nature are thermodynamically stable in appreciable quantity at room temperature. Valensi's diagrams of major species shows only sulfate ion, bisulfate ion, native sulfur, hydrogen sulfide, bisulfide ion, and sulfide ion. As for the carbonates, it is possible to write reactions involving dissolved species by limiting the reaction to the field in which the sulfur species involved makes up by far the greater part of the dissolved sulfur.

The construction of the sulfur species distribution diagram is fairly straightforward. Rather than attempt a "formal" presentation, which might consist of plotting all possible equilibria and then eliminating metastable reactions, the development will be based on the kind of procedure that in practice speeds up construction of most stability diagrams.

The completed diagram is shown in Figure 7.16. The following discussion is best followed by continuous reference to it. First, we can be sure that the stable sulfur species under acid reducing conditions is H_2S, and that it will change to HS^- and to S^{--} as the pH is raised, according to the relations

$$\frac{[H^+][HS^-]}{[H_2S]} = K_{H_2S}; \frac{[HS^-]}{[H_2S]} = \frac{K_{H_2S}}{[H^+]} \quad (7.84)$$

$$\frac{[H^+][S^{--}]}{[HS^-]} = K_{HS^-}; \frac{[S^{--}]}{[HS^-]} = \frac{K_{HS^-}}{[H^+]}. \quad (7.85)$$

Thus, $[H_2S] = [HS^-]$ when $K_{H_2S} = [H^+]$, and $[HS^-] = [S^{--}]$ when $K_{HS^-} = [H^+]$. The constants K_{H_2S} and K_{HS^-} are 10^{-7} and 10^{-14}, respectively, so that the boundaries among these species are vertical lines at pH 7 and 14.

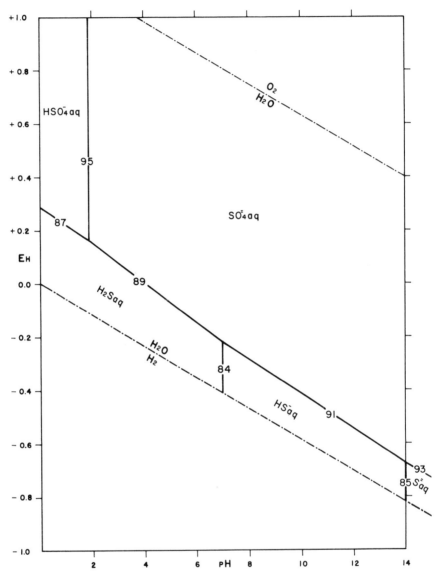

FIG. 7.16. Equilibrium distribution of sulfur species in water at 25 °C and 1 atmosphere total pressure. Numbers on lines refer to equations in text used to plot boundaries between ions.

Eh-pH DIAGRAMS

Similarly, the oxidation of H_2S at constant pH will yield HSO_4^- or SO_4^{--}, and HS^- and S^{--} should oxidize to SO_4^{--}. The reactions and the corresponding Eh equations are

$$H_2S_{aq} + 4H_2O_l = HSO_{4\,aq}^- + 9H_{aq}^+ + 8e \quad (7.86)$$

$$Eh = 0.290 - 0.066\,pH + 0.0074 \log \frac{[HSO_4^-]}{[H_2S]} \quad (7.87)$$

$$H_2S_{aq} + 4H_2O_l = SO_{4\,aq}^{--} + 10H_{aq}^+ + 8e \quad (7.88)$$

$$Eh = 0.303 - 0.074\,pH + 0.0074 \log \frac{[SO_4^{--}]}{[H_2S]} \quad (7.89)$$

$$HS_{aq}^- + 4H_2O_l = SO_{4\,aq}^{--} + 9H_{aq}^+ + 8e \quad (7.90)$$

$$Eh = 0.252 - 0.066\,pH + 0.0074 \log \frac{[SO_4^{--}]}{[HS^-]} \quad (7.91)$$

$$S_{aq}^{--} + 4H_2O_l = SO_{4\,aq}^{--} + 8H_{aq}^+ + 8e \quad (7.92)$$

$$Eh = 0.148 - 0.059\,pH + 0.0074 \log \frac{[SO_4^{--}]}{[S^{--}]}. \quad (7.93)$$

When the sulfur species are equal, i.e., if $[S^{--}] = [SO_4^{--}]$, the term containing sulfur species in the Eh equation becomes zero, so that lines can be drawn solely as functions of Eh and pH which are boundaries between the dominant species.

Finally, the boundary between HSO_4^- and SO_4^{--} is obtained from the reaction

$$HSO_{4\,aq}^- = H_{aq}^+ + SO_{4\,aq}^{--} \quad (7.94)$$

$$\frac{[SO_4^{--}]}{[HSO_4^-]} = \frac{K_{HSO_4^-}}{[H^+]} = \frac{10^{-1.9}}{[H^+]} \quad (7.95)$$

The relations shown in Figure 7.16 are independent of the total dissolved sulfur considered; they show only where the ratios of sulfur species are unity.

If we consider a given value for the total activity of dissolved sulfur, such as 10^{-1}, then it can be considered that within any given field, the activity of the species shown is very nearly 10^{-1}, and this relation can be used to calculate the abundance of other sulfur species within the field. For example, within the H_2S field the activity of sulfide ion is

$$H_2S_{aq} = S_{aq}^{--} + 2H_{aq}^+ \quad (7.96)$$

$$\frac{[S^{--}][H^+]^2}{[H_2S]} = K$$

$$\log [S^{--}] = -22 + 2\,pH \quad (7.97)$$

Once a given total sulfur (Σ S) is chosen, it becomes of interest to see if the amount chosen can stay in solution in equilibrium with native sulfur, which is a possible solid phase. To ascertain this, we write the reactions for the various species in equilibrium with sulfur

$$H_2S_{aq} = S_c + 2H^+_{aq} + 2e \qquad (7.98)$$

$$HS^-_{aq} = S_c + H^+_{aq} + 2e \qquad (7.99)$$

$$S^{--}_{aq} = S_c + 2e \qquad (7.100)$$

$$S_c + 4H_2O_l = HSO^-_{4\,aq} + 7H^+_{aq} + 6e \qquad (7.101)$$

$$S_c + 4H_2O_l = SO^{--}_{4\,aq} + 8H^+_{aq} + 6e \qquad (7.102)$$

In each of these equations, inasmuch as the activity of crystalline sulfur is unity, there is but one line representing the Eh-pH relation at which the activity of the dissolved species is the chosen value (10^{-1}). Thus, these lines define the stability field of native sulfur. Figure 7.17 shows the stable sulfur species at the specific value of total activity of dissolved sulfur = 10^{-1}. (10^{-1} is chosen because it is approximately the activity of H_2S_{aq} in a solution saturated with H_2S gas at 1 atmosphere pressure at 25 °C.)

Stability Relations of Pyrrhotite and Pyrite

The stability fields of pyrrhotite and pyrite relative to magnetite and hematite in solutions with given total dissolved sulfur can now be determined. The methods are entirely similar to those used for the relation of siderite to magnetite and hematite. $[Fe^{++}]$ in equilibrium with each sulfide can be determined and the field of stability of each sulfide in relation to the oxides found graphically by superimposing upon the $[Fe^{++}]$ in equilibrium with the oxides. Alternatively, the reactions between the sulfides, and between the sulfides and oxides, can be written for each domain of a given dissolved sulfur species. This second method will be illustrated here.

Assuming a total dissolved sulfur activity of 10^{-1}, let us calculate the boundaries between the solids. The development can be followed on Figure 7.18.

In the H_2S field

$$FeS_c + H_2S_{aq} = FeS_{2\,c} + 2H^+_{aq} + 2e \qquad (7.103)$$

$$Eh = -0.133 - 0.059 \, pH - 0.0295 \log 10^{-1}. \qquad (7.104)$$

The plot of the boundary shows that pyrrhotite oxidizes to pyrite in the presence of H_2S before the water boundary is reached. Thus, pyrrhotite is not stable in water in the presence of 10^{-1} H$_2$S. Next

$$3\text{FeS}_c + 4\text{H}_2\text{O}_l = \text{Fe}_3\text{O}_{4\,c} + 3\text{H}_2\text{S}_{aq} + 2\text{H}^+_{aq} + 2e \quad (7.105)$$

$$\text{Eh} = 0.754 - 0.059\,\text{pH} + 0.088 \log 10^{-1}. \quad (7.106)$$

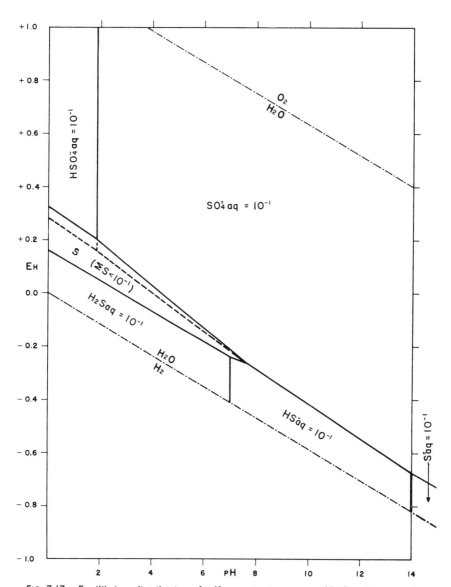

FIG. 7.17. Equilibrium distribution of sulfur species in water at 25 °C and 1 atmosphere total pressure for activity dissolved sulfur = 10^{-1}. Under these conditions, native sulfur is a stable phase. Dashed line indicates equal values of dissolved species within sulfur field.

The line indicated by equation (7.106) does not even plot in the H_2S field; in other words, pyrrhotite is stable relative to magnetite within the entire H_2S field. To put it another way, $[H_2S]$ would have to be

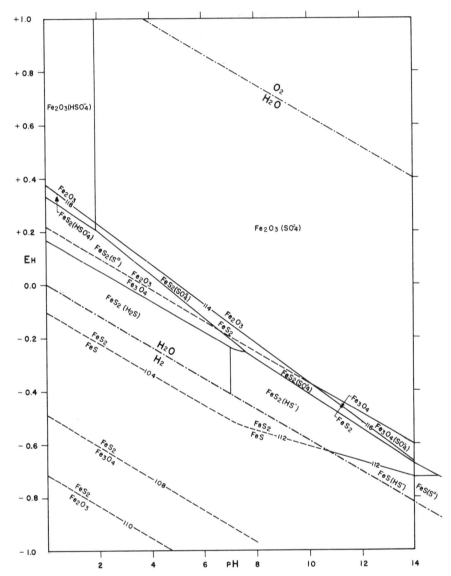

FIG. 7.18. Stability relations of iron sulfides and oxides in water at 25 °C and 1 atmosphere total pressure, and total dissolved sulfur = 10^{-1}. Heavy solid lines are stable boundaries of oxides and sulfides or boundaries of dissolved sulfur species; dashed lines are metastable boundaries. Numbers on lines refer to equations in text.

extremely low before a field of magnetite could appear within the H_2S field of dominance.

Now consider the reaction from magnetite to pyrite

$$Fe_3O_{4\ c} + 6H_2S_{aq} = 3FeS_{2\ c} + 4H_2O_l + 4H^+_{aq} + 4e \quad (7.107)$$

$$Eh = -0.577 - 0.059\ pH - 0.088 \log 10^{-1}. \quad (7.108)$$

Like the pyrrhotite-pyrite boundary (7.104), it plots far below the water stability boundary, showing that magnetite does not exist in the H_2S field. Finally, we write

$$Fe_2O_{3\ c} + 4H_2S_{aq} = 2FeS_{2\ c} + 3H_2O_l + 2H^+_{aq} + 2e \quad (7.109)$$

$$Eh = -0.831 - 0.059\ pH - 0.118 \log 10^{-1}. \quad (7.110)$$

The plot of this reaction shows that it, too, is metastable in the field of H_2S. It is of interest in passing that the reactions of H_2S with hematite and magnetite to yield pyrite (reactions 7.107 and 7.109) are oxidation reactions—the reduction of iron is more than overbalanced by the oxidation of the sulfide ion of the H_2S to the disulfide ion in pyrite.

In summary, for the field where $[H_2S] = 10^{-1}$, pyrite is the stable species. Inasmuch as the reactions in the HS^- field are similar, except for a slight change in slope, only the pyrrhotite-pyrite reaction need be considered, for it is the only one for which the change in slope might intersect the lower water stability boundary. Accordingly

$$FeS_c + HS^-_{aq} = FeS_{2\ c} + H^+_{aq} + 2e \quad (7.111)$$

$$Eh = -0.340 - 0.0295\ pH - 0.0295 \log 10^{-1}. \quad (7.112)$$

The reaction, as shown on Figure 7.18 indicates that the decreased slope is sufficient to pull the pyrrhotite-pyrite boundary into the field of water stability at high pH ($\cong 10.6$).

At this stage of development we can guess that the field of magnetite is going to be obliterated by that of pyrite, and try to find the boundary between pyrite and hematite in the SO_4^{--} field. If it is found to be above the magnetite-hematite boundary, then this assumption is justified. So we write

$$2FeS_{2\ c} + 19H_2O_l = Fe_2O_{3\ c} + 4SO^{--}_{4\ aq} + 38H^+_{aq} + 30e \quad (7.113)$$

$$Eh = 0.380 - 0.075\ pH + 0.0079 \log 10^{-1}. \quad (7.114)$$

When equation (7.114) is plotted, it is found to lie on or above the hematite-magnetite boundary up to pH = 10, showing that pyrite is stable relative to magnetite in that range if $SO_4^{--} = 10^{-1}$, but that above pH = 10, the reaction from pyrite to magnetite must be considered

$$3FeS_{2\,c} + 28H_2O_l = Fe_3O_{4\,c} + 6SO_{4\,aq}^{--} + 56H_{aq}^+ + 44e \quad (7.115)$$

$$Eh = 0.384 - 0.075\,pH + 0.0080 \log 10^{-1} \quad (7.116)$$

When plotted, equation (7.116) joins nicely onto equation (7.114). The final step is to obtain the FeS_2-Fe_2O_3 boundary in the HSO_4^- field. The equation is

$$2FeS_{2\,c} + 19H_2O_l = 4HSO_{4\,aq}^- + Fe_2O_{3\,c} + 34H_{aq}^+ + 30e \quad (7.117)$$

$$Eh = 0.366 - 0.067\,pH + 0.0079 \log 10^{-1}. \quad (7.118)$$

Now that the fields of stability of the various solids are known, the activity of ions in equilibrium with them can be calculated, and a final diagram for iron sulfides and oxides obtained. Where magnetite and hematite are stable, contours of $[Fe^{++}]$ can be transferred directly from Figure 7.5d. For pyrite we can still take advantage of the fields of dissolved sulfur species. In the H_2S field

$$2H_2S_{aq} + Fe_{aq}^{++} = FeS_{2\,c} + 4H_{aq}^+ + 2e \quad (7.119)$$

$$Eh = 0.057 - 0.118\,pH - 0.059 \log 10^{-1} - 0.0295 \log [Fe^{++}]. \quad (7.120)$$

In the HSO_4^- field

$$FeS_{2\,c} + 8H_2O_l = 2HSO_{4\,aq}^- + Fe_{aq}^{++} + 14H_{aq}^+ + 14e \quad (7.121)$$

$$Eh = 0.339 - 0.059\,pH + 0.0084 \log 10^{-1} + 0.0042 \log [Fe^{++}]. \quad (7.122)$$

In the SO_4^{--} field

$$FeS_{2\,c} + 8H_2O_l = 2SO_{4\,aq}^{--} + Fe_{aq}^{++} + 16H_{aq}^+ + 14e \quad (7.123)$$

$$Eh = 0.354 - 0.067\,pH + 0.0084 \log 10^{-1} + 0.0042 \log [Fe^{++}]. \quad (7.124)$$

In the HS^- field

$$2HS_{aq}^- + Fe_{aq}^{++} = FeS_{2\,c} + 2H_{aq}^+ + 2e \quad (7.125)$$

$$Eh = -0.470 - 0.059\,pH - 0.059 \log 10^{-1} - 0.029 \log [Fe^{++}]. \quad (7.126)$$

In the native sulfur field

$$FeS_{2\,c} = Fe_{aq}^{++} + 2S_c + 2e \quad (7.127)$$

$$Eh = 0.340 + 0.0295 \log [Fe^{++}]. \quad (7.128)$$

Because the only variables in the equations are Eh, pH, and log $[Fe^{++}]$ it is possible to substitute arbitrary values of $[Fe^{++}]$ and obtain contours for their values within the field of pyrite. If all relations have been calculated correctly, the contours in the pyrite field should join those in the oxide fields without a break. Figure 7.19 shows the

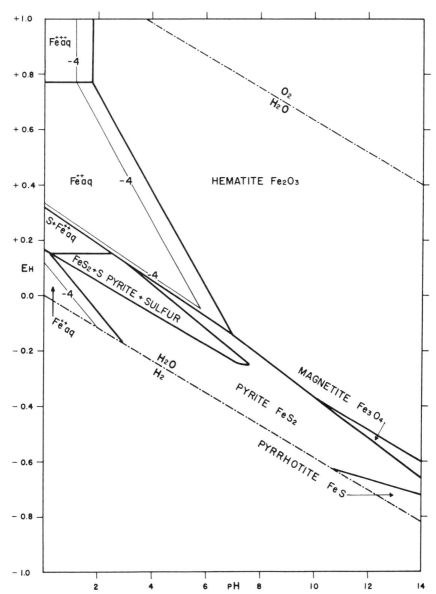

FIG. 7.19. Stability relations of iron oxides and sulfides in water at 25 °C and 1 atmosphere total pressure at an activity of dissolved sulfur of 10^{-1}. Boundaries between ions and solids are at an activity of 10^{-6} of dissolved iron species. The numeral -4 is the log of the iron activity used to show the rate of change of "solubility."

"solubility" diagram thus obtained. Contours for the small field of pyrrhotite can be obtained from the single reaction

$$FeS_c + H^+_{aq} = Fe^{++}_{aq} + HS^-_{aq} \qquad (7.129)$$

$$\log [Fe^{++}] = -4.4 - pH - \log 10^{-1}. \qquad (7.130)$$

The diagram has some interesting and perhaps unsuspected relations. First, it shows that if an activity of dissolved iron of 10^{-6} is used as a criterion of stability, pyrite cannot oxidize to yield sulfur at pH values higher than about 3. Above that pH, sulfur is not a stable phase at the potential at which pyrite oxidizes. Marcasite, on the other hand, which is less stable than pyrite, would be expected to yield sulfur at higher pH values. This pH relation probably explains why marcasite yields sulfur on oxidation by ferric salts, whereas pyrite does not, except when the ferric salt is highly concentrated. Also, the bulge acidward of the pyrite stability field at intermediate Eh values shows why inorganic nonoxidizing acids have no effect on pyrite, whereas oxidizing acids do. Also, there is a field of "solubility" under acid-reducing conditions, showing that pyrite can be decomposed by reducing agents in acid solutions. This conclusion is verifiable experimentally. Finally, the relations among pyrite, pyrrhotite, and magnetite are of geological interest. The association of these three minerals is characteristic of high-temperature deposits. At room temperature, as shown here, the three-phase association can occur only at pH values higher than 14 when total sulfur is as high as 10^{-1}. This relation implies that the conditions of stability of the assemblage move toward lower pH values at higher temperatures.

Figure 7.20 shows the changes resulting from lowering total sulfur activity to 10^{-6}. As might be expected, the fields of the sulfides shrink markedly. Pyrrhotite appears as a stable phase under strong reducing conditions at pH $\cong$ 8, and the area of "solubility" increases markedly on the acid side. From Figures 7.20 and 7.19, it can be seen that the first appearance of stable sulfides, where minute quantities of divalent sulfur appear in the system (perhaps generated by organisms), is at nearly neutral pH values.

RELATIONS OF IRON MINERALS IN WATER CONTAINING CO_2 AND SULFUR

The natural system of greatest geological interest is one that contains both dissolved carbonate and dissolved sulfur. The sedimentary iron ores are characterized by the presence of iron oxides, carbonates, sulfides, and silicates. Omitting silicates for the moment, it is clear that

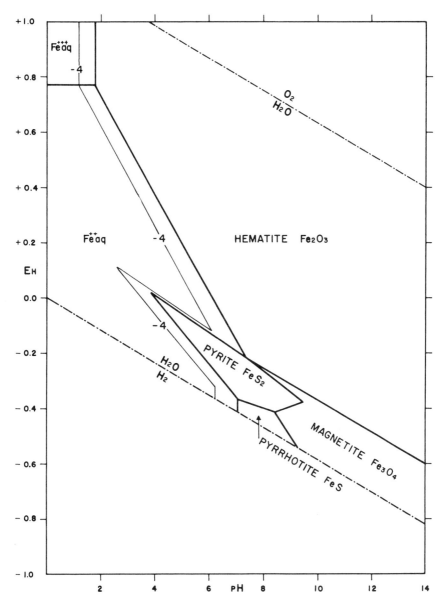

FIG. 7.20. Stability relations of iron oxides and sulfides in water at 25 °C and 1 atmosphere total pressure, when $\Sigma S = 10^{-6}$. Note shrinkage of sulfide boundaries and appearance of FeS as a stable phase at intermediate pH under strongly reducing conditions, as well as marked increase in the area of "acid solubility" over a wide range of Eh.

224 **SOLUTIONS, MINERALS, AND EQUILIBRIA**

if we are to use an Eh-pH framework there are too many variables to represent conveniently if we want to assess the effects of continuous changes in activity of sulfur and carbonate species. On the other hand,

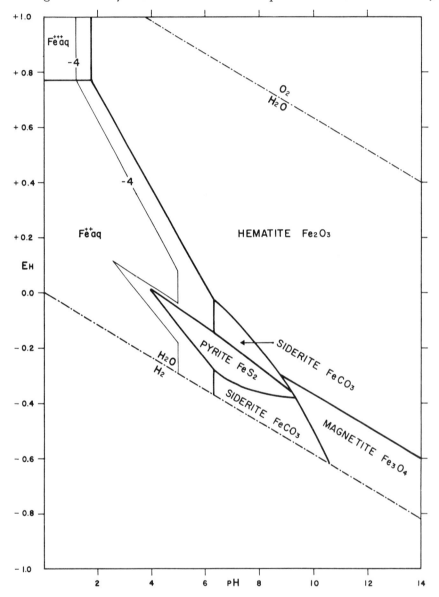

FIG. 7.21. Stability relations of iron oxides, sulfides, and carbonate in water at 25 °C and 1 atmosphere total pressure. Total dissolved sulfur = 10^{-6}. Total dissolved carbonate = 10^0. Note elimination of FeS field by $FeCO_3$ under strongly reducing conditions, and remarkable stability of pyrite in presence of small amount dissolved sulfur.

any given environment, in terms of a fixed $\Sigma\ CO_2$ and $\Sigma\ S$, can be represented on a two-dimensional Eh-pH diagram. Also, if individual diagrams have been prepared, one for an arbitrary $\Sigma\ CO_2$ and another for an arbitrary $\Sigma\ S$, they can be overlain, and the areas of stability of minerals of the total system delineated.

For example, Figure 7.11 shows stability relations among hematite, magnetite, and siderite at a $\Sigma\ CO_2$ of 10^{-2}, and Figure 7.20 shows relations among hematite, magnetite, pyrite, and pyrrhotite at $\Sigma\ S = 10^{-6}$. If the sulfide diagram is overlain on the carbonate diagram, it is immediately apparent that the activity of dissolved iron in equilibrium with the compounds on the $\Sigma\ S$ diagram is everywhere less than on the $\Sigma\ CO_2$ diagram. In other words, when $\Sigma\ S$ is 10^{-6} and $\Sigma\ CO_2$ is 10^{-2}, only oxides and sulfides are stable, and siderite does not appear at equilibrium. It must be emphasized that this is the *equilibrium* situation —it is well known that sulfate ion is extremely slow to reduce at low temperature in the absence of organic intervention.

Figure 7.21 shows the relations when $\Sigma\ CO_2 = 10^0$, and $\Sigma\ S = 10^{-6}$, and serves to illustrate that if siderite is to have an important field of stability, dissolved carbonate must be very high and reduced sulfur extremely low. Under these conditions the field of pyrrhotite is eliminated, but a considerable field of pyrite remains. Note that under these conditions siderite may be a criterion of very strong reducing conditions, or of moderate reducing conditions, and its presence in many iron ores apparently indicates the essential absence of appreciable divalent sulfur and the presence of relatively large amounts of dissolved carbonate.

THE INFLUENCE OF OTHER CONSTITUENTS IN THE IRON SYSTEM: SILICA AS AN EXAMPLE

Unfortunately, thermochemical data at present are not sufficient to permit handling of the total sedimentary iron mineral environment, which includes important quantities of silicates. Free-energy values are available only for iron metasilicate, $FeSiO_3$, which is a poor substitute for the actual compounds, iron chlorites and chamosites. Yet the procedures necessary to handle silicates can be illustrated by using $FeSiO_3$ as an example.

An aspect of stability diagrams that has been neglected is the representation of environments in which the number of variables is large—the tendency has been to restrict to two or three variables and to show their effects as continuous functions. But there is no real difficulty in working with five or six variables, if one is willing to choose arbitrary values of each for a given diagram. The chore of making a sufficient number of diagrams to encompass the range of variables in which one might be interested is large, but there is no present barrier to such treatment except for the labor of preparing the requisite number of diagrams.

To return to the problem of iron silicates, we can see the relation of iron silicate to the iron oxides, if it can be assumed that solid silica is present in the environment. Because there is abundant evidence that opaline silica has been a primary associate of all the other iron minerals, one is justified in assuming that amorphous silica (silica glass) is an invariable associate of the other iron minerals of sedimentary iron ores. Then, if we recognize the limitations of substituting ferrous metasilicate for the real silicates found in sedimentary iron ores, we can write

$$3FeSiO_{3\ c} + H_2O_l = Fe_3O_{4\ c} + 3SiO_{2\ glass} + 2H^+_{aq} + 2e \quad (7.131)$$

$$Eh = 0.272 - 0.059\ pH. \quad (7.132)$$

Equation (7.132) is plotted on Figure 7.22, and the boundary between $FeSiO_3$ and Fe_3O_4 is found to be above that for Fe_3O_4-Fe_2O_3. In other words, in the presence of silica glass, or of a system saturated with respect to amorphous silica, magnetite is unstable relative to iron metasilicate. The next step is to obtain the boundary between $FeSiO_3$ and Fe_2O_3

$$2FeSiO_{3\ c} + H_2O_l = Fe_2O_{3\ c} + 2SiO_{2\ glass} + 2H^+_{aq} + 2e \quad (7.133)$$

$$Eh = 0.258 - 0.059\ pH. \quad (7.134)$$

This equation also plots above the Fe_3O_4-Fe_2O_3 line, showing that in the presence of silica glass, the phases expected are iron silicate and hematite. We can then write for the field occupied by $FeSiO_3$

$$FeSiO_{3\ c} + 2H^+_{aq} = Fe^{++}_{aq} + SiO_{2\ glass} + H_2O_l \quad (7.135)$$

$$\log[Fe^{++}] = 8.03 - 2\ pH. \quad (7.136)$$

Substituting values of 10^{-6} and 10^{-4} for $[Fe^{++}]$, we obtain a "solubility" diagram for the iron silicate-iron oxide system. Therefore, if we consider only iron oxides and iron silicate in the presence of amorphous silica, magnetite disappears as a solid phase. Undoubtedly the actual iron silicates found as primary sedimentary minerals in iron ores are somewhat more stable than pure $FeSiO_3$, but the relations using $FeSiO_3$ show that if bottom waters are saturated with amorphous silica, and if there is enough silica to satisfy all the iron present, iron silicate will form in preference to magnetite. On the other hand, the general conditions of stability of ferrous metasilicate are remarkably similar to those of magnetite.

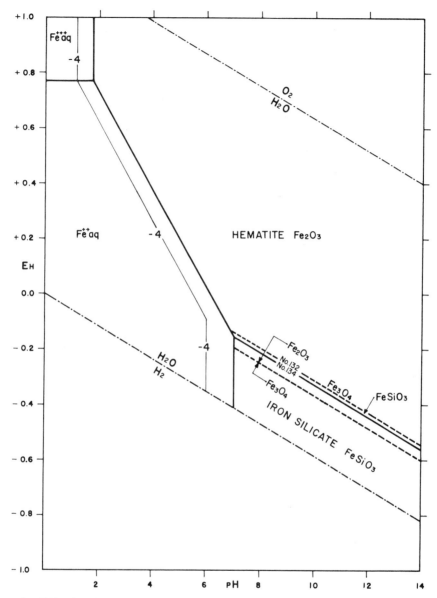

FIG. 7.22. Stability relations of iron oxides and iron metasilicate at 25 °C and 1 atmosphere total pressure in the presence of water. Solid silica glass assumed present. Numbers on lines refer to equations in text. −4 on thin solid lines is log total dissolved iron in equilibrium with solids.

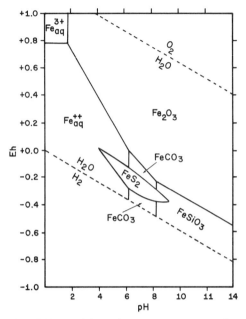

FIG. 7.23. Stability relations among iron oxides, carbonates, sulfides, and silicates at 25 °C, and 1 atmosphere total pressure in the presence of water. Other conditions: total $CO_2 = 10^0$; total sulfur $= 10^{-6}$; amorphous silica is present. This diagram is intended to suggest the innumerable components that can be considered in pH-Eh representation, if arbitrary values are selected for each constituent.

All kinds of fascinating games can be played by determining stability relations for various values of total carbonate, total sulfur, and total dissolved silica, and the conditions can be endlessly varied. Figure 7.23 is a summary diagram for a particular set of conditions, namely, $\Sigma\, CO_2 = 10^0$; $\Sigma\, S = 10^{-6}$, $\Sigma\, SiO_2$ equal that in equilibrium with silica glass (amorphous silica). This diagram was obtained by imposing Figures 7.22 and 7.21.

In terms of the primary sedimentary iron ores, it is obvious that a siderite facies can be obtained by having a high dissolved CO_2 and by removing sulfide sulfur from the system; a silicate facies, by removing sulfur and high CO_2 content, while preserving enough silica to yield chert; a magnetite facies, by reducing sulfur, CO_2, and maintaining silica at a value undersaturated with respect to amorphous silica.[2]

---

[2] The word "chert" is used here on the assumption that the siliceous layers of the iron formations, now cryptocrystalline quartz, were originally polymerized amorphous silica, i.e., opaline silica.

THE TRANSITION FROM PARTIAL PRESSURE DIAGRAMS TO Eh-pH DIAGRAMS

In Chapter 6 stability diagrams of various oxides, sulfides, and carbonates were shown as functions of P_{O_2}, P_{CO_2}, and P_{S_2}. One of the most interesting geologically is that relating copper and iron oxides and sulfides (Figure 6.16). R. Natarajan (unpublished manuscript, Harvard University, 1958) has transposed such diagrams into Eh-pH diagrams. As shown before, P_{O_2} values can be shown as contours on an Eh-pH diagram (Figure 7.1). Also, at a given total sulfur, contours of P_{S_2} can be drawn on Eh-pH diagrams. For example, if $\Sigma S = 10^{-1}$, we can write, for the field in which H_2S is the dominant species (Figure 7.17)

$$2H_2S_{aq} = S_{2\,g} + 4H^+_{aq} + 4e. \tag{7.137}$$

The Eh-pH equation, under the condition that $H_2S = 10^{-1}$, is

$$Eh = E° + \frac{0.059}{4} \log P_{S_2} - 0.059 \,pH - \frac{0.059}{2} \log 10^{-1}. \tag{7.138}$$

Thus, for a selected value of P_{S_2}, a line can be drawn in the field of dominance of H_2S to represent this selected value. Because all relations on Figure 6.16 are shown in terms of P_{S_2} and P_{O_2}, it is possible to transpose them to an Eh-pH diagram, if total dissolved sulfur is constant. Figures 7.24 and 7.25 show relations among the iron and copper sulfides at $\Sigma S = 10^{-1}$ and $\Sigma S = 10^{-4}$ and serve to show expected stability fields of a large number of compounds under conditions corresponding to those encountered in nature during the processes of oxidation and secondary enrichment of copper ores.

SUMMARY OF IRON MINERAL RELATIONS

The foregoing detailed development of relations among the iron minerals has been documented in the hope that the procedures illustrated are sufficient to permit anyone who follows them in detail to handle similar relations for any other element. Although the application to geologic relations has not been emphasized in this chapter, because of the risk of destroying the thread of the development of the mechanical aspects of stability portrayal, the diagrams of interplay among iron oxides, carbonates, sulfides, and silicates fit natural occurrences so beautifully that words may be superfluous. In addition, it is shown that the kinds of diagrams that could be made for iron compounds alone are limitless, and that anyone interested in specific conditions can examine them in the kinds of frameworks suggested.

Eh-pH DIAGRAMS FOR OTHER ELEMENTS

In the following pages Eh-pH diagrams for various elements are presented, with no more than a caption to show the conditions of calculation. In a following chapter some geologic applications that have been made of some of these diagrams are given, but for most of them the significance lies in the experience of the reader. It is obviously impossible, without tremendous increase in the length of this text, to develop such diagrams. Yet it is hoped that the details of the development of the iron diagrams is sufficient to show that the diagrams are valid only under the conditions for which they have been calculated. There is a tendency, for example, to represent relations among iron carbonates, oxides, and sulfides, without regard to the values of Σ S and Σ CO_2 used for calculation. Such disregard is sufficient to invalidate the use of the diagram.

The diagrams showing relations for metals in oxygenated water have been individually calculated by the persons credited, but it should be re-emphasized that a detailed development of each one, with many additional diagrams covering various conditions, can be found in the publications of "Cebelcor," given at the end of this chapter. These publications are the core from which all the diagrams shown here have been developed.

Figures 7.24 and 7.25

The diagrams show the best information currently available on the copper-iron-water-sulfur mineral relations at 25 °C and 1 atmosphere total pressure, in the presence of dissolved sulfur at fixed activity. The diagrams have been prepared as an aid in the understanding of relations in the zones of oxidation and secondary enrichment of ore deposits. The relations appear complex at first glance, but a pattern begins to emerge after a little study. The various iron, copper-iron, and iron sulfides are separated from the iron and copper oxides by a fairly definite Eh-pH band running from an Eh of about 0.3 at pH 0 downward to the right with a slope of about 60 millivolts per pH unit. Only chalcocite (Cu_2S) projects appreciably above this boundary, which is in accord with the occurrence of chalcocite blankets as a result of secondary enrichment of copper ores. Note that oxidation under acid conditions should produce, at equilibrium, either appreciable Cu^{++} and Fe^{++} in solution, or, with slightly lower acidity, Cu^{++} and solid Fe_2O_3 (or $Fe_2O_3 \cdot H_2O$). The reason for the separation of iron and copper on oxidation is thus represented. Native copper easily can be a secondary product if the descending copper solutions are neutralized. Cuprite

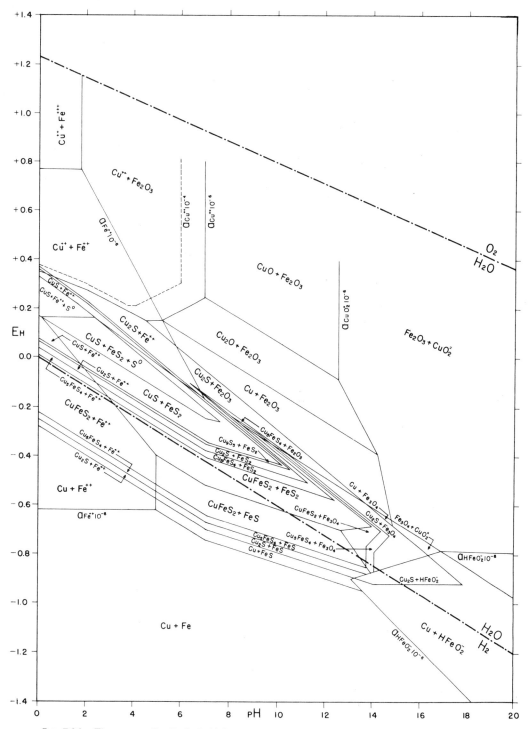

FIG. 7.24. The system Cu–Fe–S–O–H (in part) at 25 °C and 1 atmosphere total pressure. Total dissolved sulfur = 10^{-1} m. [Courtesy R. Natarajan and R. Garrels.]

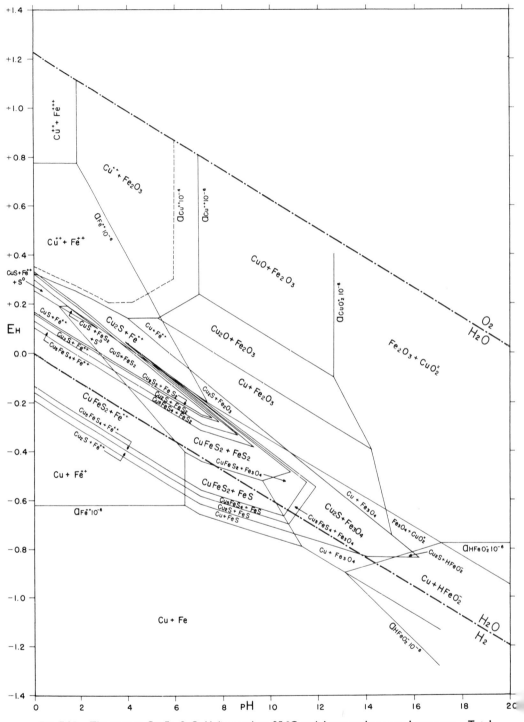

FIG. 7.25. The system Cu–Fe–S–O–H (in part) at 25 °C and 1 atmosphere total pressure. Total dissolved sulfur = 10^{-4} m. [Courtesy R. Natarajan and R. Garrels.]

and tenorite, on the other hand, show a requirement of fairly high pH before their precipitation will remove copper from solution quantitatively. Many other such relations are implicit in the interrelations shown. The diagrams were prepared by R. Natarajan and R. Garrels, under sponsorship of the Committee on Experimental Geology, Harvard University, 1957.

Figures 7.26a–e

This set of figures shows equilibria among lead compounds in water at 25 °C and 1 atmosphere total pressure in the presence of CO_2 or sulfur. Various combinations are shown.

In Figure 7.26a, the presence of a small partial pressure of CO_2 removes the fields of stability of the various lead oxides, except for a small field of PbO_2, suggesting that as a mineral (plattnerite) PbO_2 is a fine indicator of alkaline oxidizing conditions.

Figure 7.26b shows the difference resulting from considering a fixed total CO_2; $Pb_3(OH)_2(CO_3)_2$ appears as a stable compound. Contours of ionic activity exceeding 10^{-4} are shown to illustrate the conditions of occurrence of Pb_3O_4 and PbO in this system. $Pb_3(OH)_2(CO_3)_2$ (hydrocerussite) should be a good indicator of alkaline conditions. Note the field of stability of native lead in these carbonated waters.

Figure 7.26c illustrates the influence of adding sulfur to the Pb-oxygenated water system. Even if dissolved sulfur is 10^{-5}, as shown here, galena has a large stability field, and anglesite also has a large area of occurrence. Note that in this system, the fields of ions, as opposed to solids, have been drawn at an activity of 10^{-2}. If 10^{-6} had been used, both $PbSO_4$ and PbO would have just disappeared, inasmuch as the contours for 10^{-6} Pb^{++} and 10^{-6} $HPbO_2^-$ coincide at pH 9.3. Also, with as little as 10^{-5} total dissolved sulfur, galena has almost squeezed native lead out of the picture.

Figure 7.26d combines the effects of CO_2 and sulfur, and illustrates a system resembling conditions of surface oxidation. Total sulfur is high (10^{-1}), and P_{CO_2} is approximately that of the atmosphere (10^{-4}). Under such conditions, the stability diagram indicates that lead is insoluble; the activity of the ions exceeds 10^{-6} only under very alkaline or very acid conditions. Also, the diagram simplifies to the chief minerals actually observed: galena, cerussite, and anglesite, with a small field of plattnerite under unusual conditions.

Figure 7.26e is a variant on the theme of Figure 7.26d, showing conditions that might be met in an oxidizing galena deposit isolated from the atmosphere, but with fairly high total dissolved carbonate, as well as high dissolved sulfur. Hydrocerussite appears again, owing to

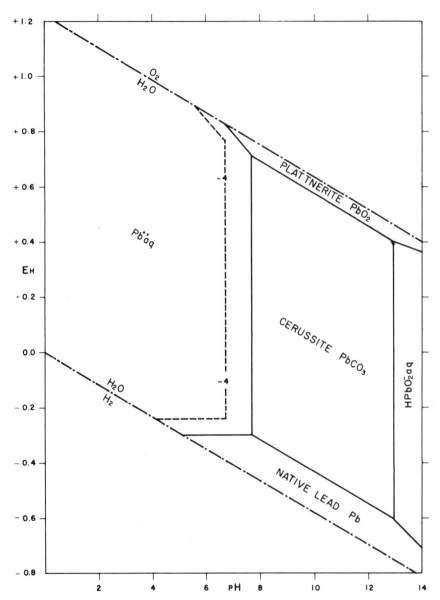

FIG. 7.26a. Stability relations among lead oxides and carbonates in water at 25 °C and 1 atmosphere total pressure. $P_{CO_2} = 10^{-4}$. Boundaries of solids at total ion activity $= 10^{-6}$; contour is at activity $= 10^{-4}$. [Courtesy W. McIntyre.]

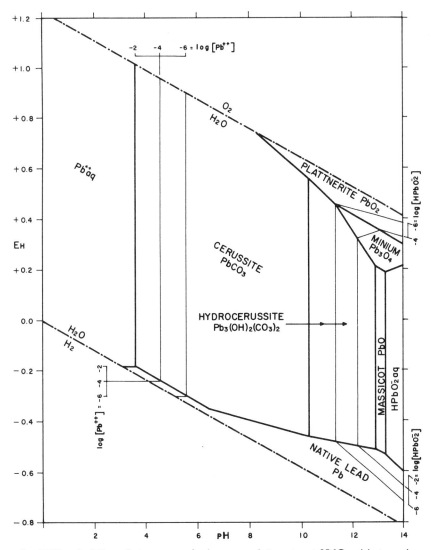

FIG. 7.26b. Stability relations among lead compounds in water at 25 °C and 1 atmosphere total pressure. Total dissolved carbonate species = $10^{-1.5}$. [Courtesy W. McIntyre.]

the fact that the system has fixed total CO_2, and at high pH values the hydrocarbonate can compete with the normal carbonate. When P_{CO_2} is constant, total carbonate rises with pH, and $PbCO_3$ persists as the stable phase. Both Figures 7.26d and e show another noteworthy relation: galena cannot oxidize to yield sulfur at pH values above about 2, but if the acidity exceeds this value, the field of native sulfur rises above that of PbS, and a small area of $PbSO_4$ + S results.

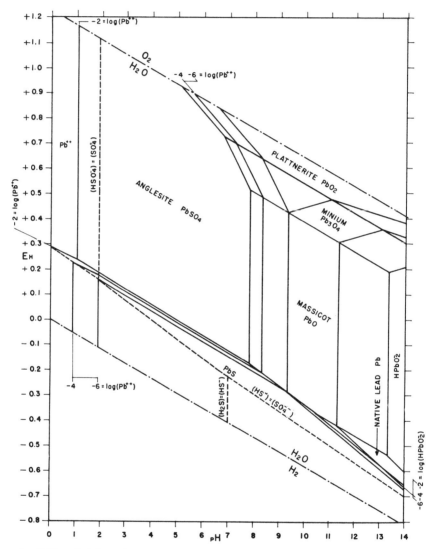

FIG. 7.26c. Stability relations among lead compounds in water at 25 °C and 1 atmosphere total pressure. Total dissolved sulfur species = 10^{-5}. [Courtesy W. McIntyre.]

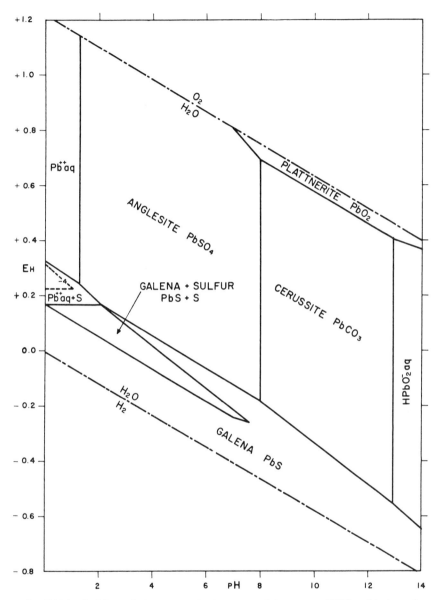

FIG. 7.26d. Stability relations among lead compounds in water at 25 °C and 1 atmosphere total pressure. Total dissolved sulfur = 10^{-1}, $P_{CO_2} = 10^{-4}$. Boundaries of solids at total ionic activity of 10^{-6}. Dashed line is contour at activity of dissolved lead species of 10^{-4}. [Courtesy W. McIntyre.]

In summary, the behavior of lead minerals is described usefully by the diagrams, and several of the minerals can be used as environmental indicators. The diagrams were prepared by W. McIntyre, graduate student of the Department of Geology at Massachusetts Institute of Technology, in 1957, and have been only slightly altered from the originals.

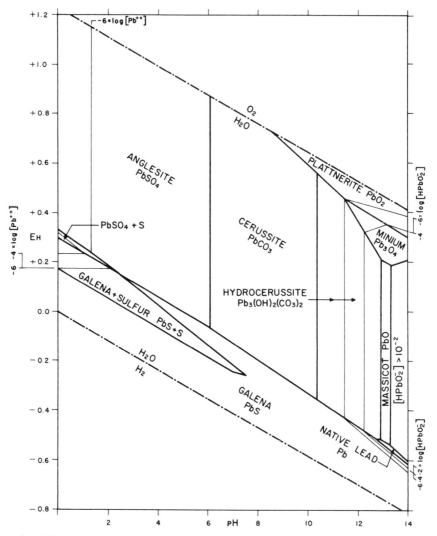

FIG. 7.26e. Stability relations among lead compounds in water at 25 °C and 1 atmosphere total pressure. Total dissolved sulfur species = $10^{-1.5}$, total dissolved carbonate species = 10^{-1}. [Courtesy W. McIntyre.]

Figures 7.27a, b

Relations among copper minerals at 25 °C and 1 atmosphere total pressure in the presence of water are shown here.

Figure 7.27a shows the copper-oxygen-water relations and emphasizes

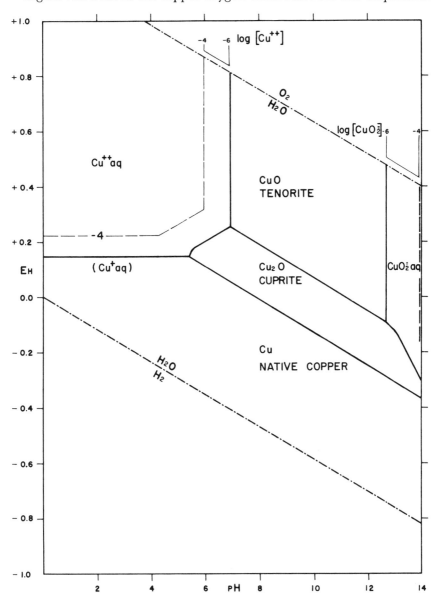

FIG. 7.27a. Stability relations among copper compounds in the system $Cu-H_2O-O_2$ at 25 °C and 1 atmosphere total pressure. [Courtesy J. Anderson.]

the large field of stability of native copper, as well as the field of acid solubility of copper as the cupric ion. If 10^{-7} had been used as the activity cut-off for the fields of ions, Cu^+ would have appeared in the

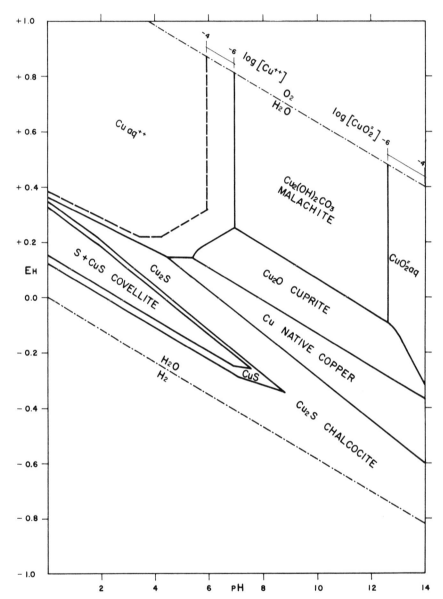

FIG. 7.27b. Stability relations among some copper compounds in the system $Cu-H_2O-O_2-S-CO_2$ at 25 °C and 1 atmosphere total pressure. $P_{CO_2} = 10^{-3.5}$, total dissolved sulfur species = 10^{-1}. [Courtesy J. Anderson.]

Eh-pH DIAGRAMS

Figures 7.27a, b

Relations among copper minerals at 25 °C and 1 atmosphere total pressure in the presence of water are shown here.

Figure 7.27a shows the copper-oxygen-water relations and emphasizes

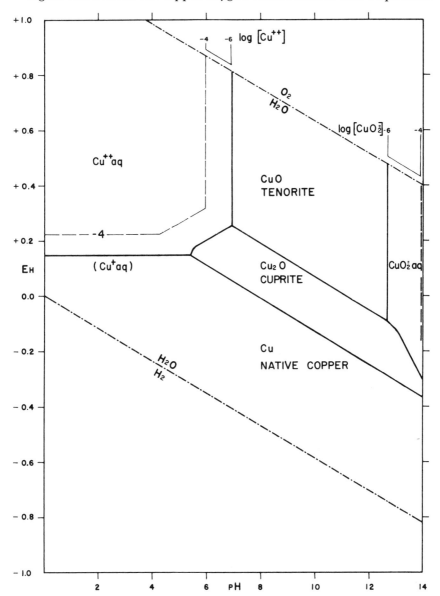

FIG. 7.27a. Stability relations among copper compounds in the system $Cu-H_2O-O_2$ at 25 °C and 1 atmosphere total pressure. [Courtesy J. Anderson.]

the large field of stability of native copper, as well as the field of acid solubility of copper as the cupric ion. If 10^{-7} had been used as the activity cut-off for the fields of ions, Cu^+ would have appeared in the

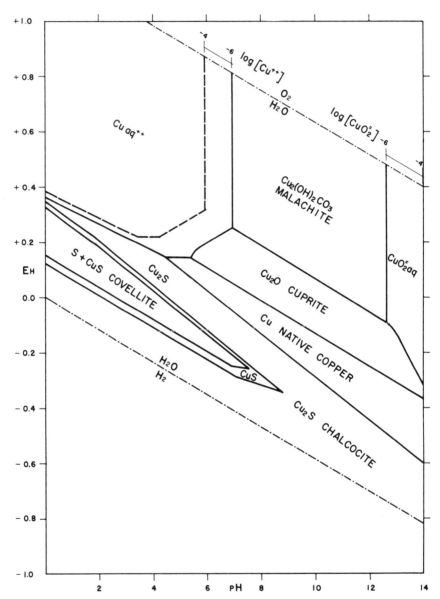

FIG. 7.27b. Stability relations among some copper compounds in the system $Cu-H_2O-O_2-S-CO_2$ at 25 °C and 1 atmosphere total pressure. $P_{CO_2} = 10^{-3.5}$, total dissolved sulfur species = 10^{-1}. [Courtesy J. Anderson.]

position shown in brackets. The peculiar kink in the boundary between Cu_2O and the field of Cu^{++} results from a significant contribution of Cu^+ to the total ionic activity. At this temperature the amphoteric behavior of copper is not important; the existence of important amounts of CuO_2^{--} occurs at pH values above those found naturally. If the nobility of copper is compared to that of lead, cobalt, and nickel, it is apparent why native copper occurs so much more frequently.

Figure 7.27b shows the results of adding CO_2 ($P_{CO_2} = 10^{-3.5}$) and sulfur ($\Sigma S = 10^{-1}$) to the system. Note that malachite has occupied the field of tenorite as shown in Figure 7.27a; at higher P_{CO_2} values, the malachite field would expand still farther and tend to squeeze cuprite out. Even in the presence of this much sulfur, native copper has a good-sized stability field. The sulfides project deeply into the acid range under reducing conditions—this diagram shows clearly why chalcocite precipitates from acid cupriferous waters when they encounter sulfides under reducing conditions. Note also that chalcocite cannot oxidize to yield native sulfur—sulfur does not coexist with chalcocite.

These diagrams were prepared by James Anderson, Department of Geology, Harvard University Graduate School, 1958.

Figures 7.28a, b

Relations among manganese minerals in water containing CO_2 and sulfur at 25 °C and 1 atmosphere total pressure are shown here.

Figure 7.28a shows stability relations among the oxides and the carbonate when total dissolved $CO_2 = 10^{-1.4}$; this is approximately a system saturated with CO_2 under 1 atmosphere pressure, then closed to CO_2. The large field of the carbonate (rhodochrosite) under these conditions is in marked contrast to the much smaller field of siderite under the same conditions. In the absence of CO_2, the higher valence oxides and the manganous hydroxide show increasing stability with increasing oxidation state. No wonder the hydroxide pyrochroite is an extremely rare mineral; it requires alkaline reducing conditions and the practical absence of CO_2. As for iron, the amphoteric behavior of manganese is of no consequence under natural conditions. Both MnO_4^{--} and MnO_4^{-} ions are unstable with respect to water; MnO_4^{--} appears only in a small area under alkaline oxidizing conditions. This relation recalls to mind that potassium permanganate solution, over a period of months, gradually decomposes to give MnO_2. On the other hand, it also illustrates that metastability may be of considerable duration.

Figure 7.28b shows two major new relations. In this instance the system is open to CO_2 ($P_{CO_2} = 10^{-4}$ atmosphere), and as usual the carbonate has wiped out the field of the divalent oxide or hydroxide, owing to the increase in CO_3^{--} as well as OH^- with increasing pH. Also, note the small field of alabandite (MnS) even in the presence of 10^{-1} total dissolved sulfur. As most marine environments become

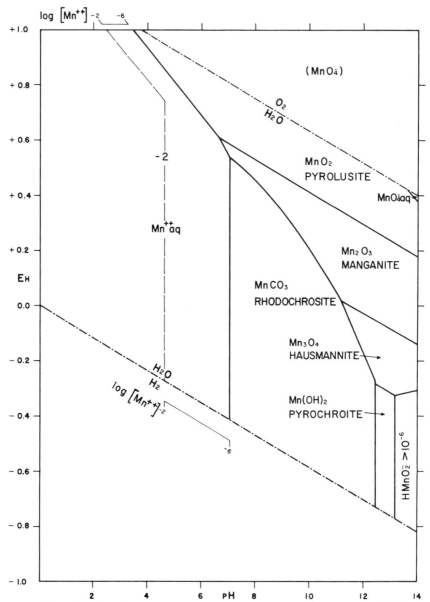

FIG. 7.28a. Stability relations among some manganese compounds in water at 25 °C and 1 atmosphere total pressure. Total dissolved carbonate species = $10^{-1.4}$. [Courtesy E. Gaucher.]

Eh-pH DIAGRAMS

reducing and develop divalent sulfur, the pH drops to 6.5–7.0. Under such conditions manganese would tend to dissolve, or perhaps the carbonate would precipitate, but some unusual OH$^-$-producing reaction

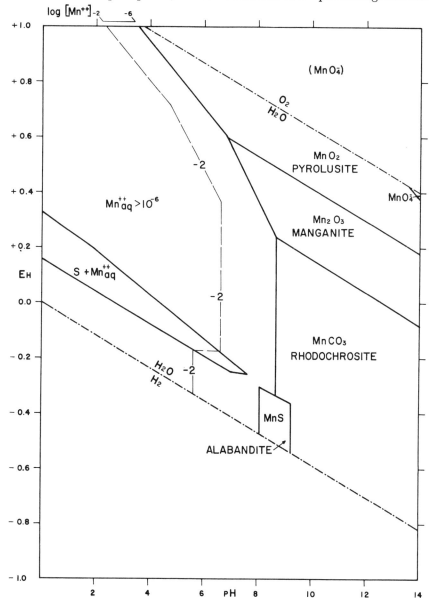

FIG. 7.28b. Stability relations among some manganese compounds in water at 25 °C and 1 atmosphere total pressure. Total dissolved sulfur species = 10^{-1}, $P_{CO_2} = 10^{-4}$. [Courtesy E. Gaucher.]

would be required to produce alabandite. This conclusion is in accord with the sparse occurrence of alabandite in nature.

These diagrams were prepared by Edwin Gaucher, Department of Geology, Harvard University Graduate School, 1957.

Figures 7.29a, b, c

Relations in systems involving nickel and water at 25 °C and 1 atmosphere pressure, with sulfur or CO_2 added, are given here.

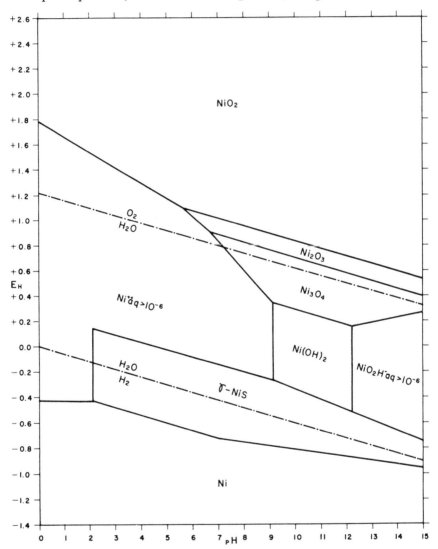

FIG. 7.29a. Stability relations among some nickel compounds in water at 25 °C and 1 atmosphere total pressure. Total dissolved sulfur species = 10^{-5}. [Courtesy J. Anthony.]

Eh-pH DIAGRAMS

Figure 7.29a shows that with total dissolved sulfur activity of 10^{-5}, nickel sulfide is stable over a large pH range under reducing conditions. Nickel oxides and hydroxides, on the other hand, are relatively soluble, as indicated by the large fields of Ni^{++} under acid conditions and NiO_2H^- under alkaline conditions. The general similarity to iron behavior is clear, with the difference of a respectable field of stability of $Ni(OH)_2$ as opposed to the entire lack of such a field for $Fe(OH)_2$.

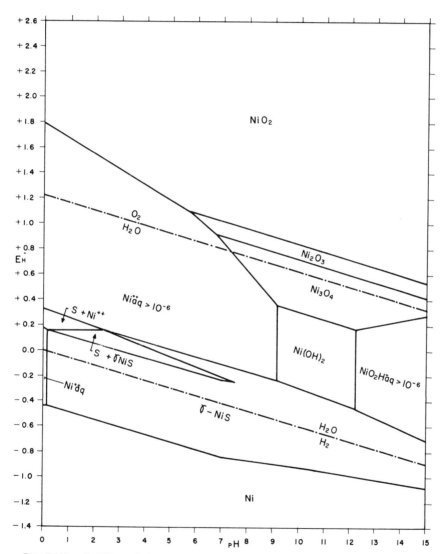

FIG. 7.29b. Stability relations among some nickel compounds in water at 25 °C and 1 atmosphere total pressure. Total dissolved sulfur species = 10^{-1}. [Courtesy J. Anthony.]

Unfortunately, data are not available for NiS_2, the pyrite analog, but it would, if stable, sandwich in between NiS and $Ni(OH)_2$.

Figure 7.29b shows the effect of increasing total sulfur, with the appearance of a field of native sulfur and an enlargement of the NiS-occupied area.

Figure 7.29c shows the effect of CO_2 on the nickel oxides. The value of P_{CO_2} chosen is just sufficient to wipe out the field of $Ni(OH)_2$, demon-

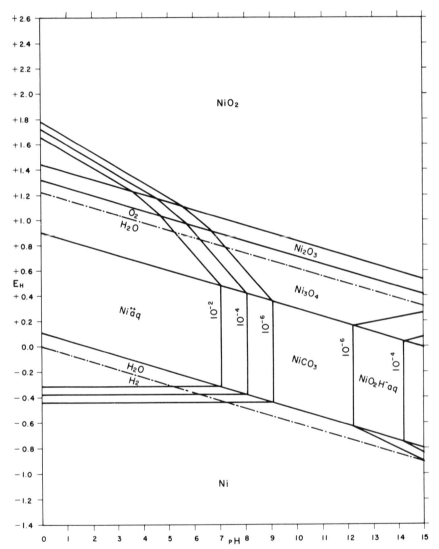

FIG. 7.29c. Stability relations among some nickel compounds in water at 25 °C and 1 atmosphere total pressure. $P_{CO_2} = 10^{-0.8}$. [Courtesy J. Anthony.]

strating that nickel carbonate is a good CO_2 indicator, requiring considerably greater P_{CO_2} than that of the atmosphere for its stable existence. Contours are labeled in terms of the activities of the ions shown. Note that native nickel, like lead, but unlike iron, has a small field of stability in water. Also, for essentially all the metals considered, a mere suspicion of sulfur in solution removes the possibility of stable existence of the native metal.

These three diagrams were constructed by John Anthony, Department of Geology, Harvard University Graduate School, 1957.

Figures 7.30a, b, c

Stability relations of cobalt compounds in water at 25 °C and 1 atmosphere total pressure in the presence of CO_2 and sulfur are given here.

Figure 7.30a demonstrates that $CoCO_3$ becomes stable relative to $Co(OH)_2$ at a very low total carbonate content of the system, as opposed to nickel carbonate. This is the basis for the separation of nickel and cobalt in various analytical schemes; $CoCO_3$ can be selectively precipitated by controlled carbonatization of the solution. However, the general similarity of cobalt and nickel behavior is striking.

Figure 7.30b demonstrates the great enlargement of the $CoCO_3$ field with increase of $\Sigma\ CO_2$ to $10^{-1.5}$.

Figure 7.30c shows typical development of a sulfide field with addition of sulfur to the system. Here we see interrelations of carbonate, hydroxides, and the sulfide in carbonated sulfur-bearing waters. If the various sulfides shown in these diagrams are compared, it will be found that the oxidation potential at which they go to appreciable concentrations of metal ions and sulfate is nearly the same. This occurs because of the remarkably sharp division on an Eh basis between divalent and sexivalent sulfur. The change is so marked that differences in the stability of the sulfides are largely obscured.

These cobalt diagrams were prepared by Ivan Barnes, Department of Geology, Harvard University Graduate School, 1957.

Figures 7.31a, b

Relations among the tungsten oxides and sulfides are given here. The strongly amphoteric behavior of tungsten is well shown by the large field of importance of the WO_4^{--} ion and by the stability of WO_3 in acid solutions. In nature, the WO_4^{--} ion tends to be fixed by cations not shown here, and forms such insoluble compounds as $CaWO_4$ and

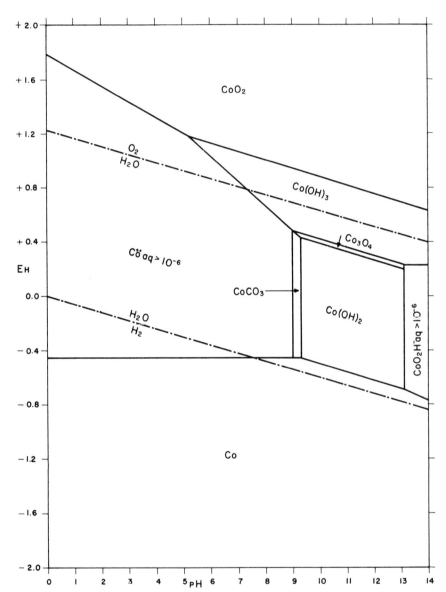

FIG. 7.30a. Stability relations among some cobalt compounds in water at 25 °C and 1 atmosphere total pressure. Total dissolved carbonate species = $10^{-4.9}$. [Courtesy I. Barnes.]

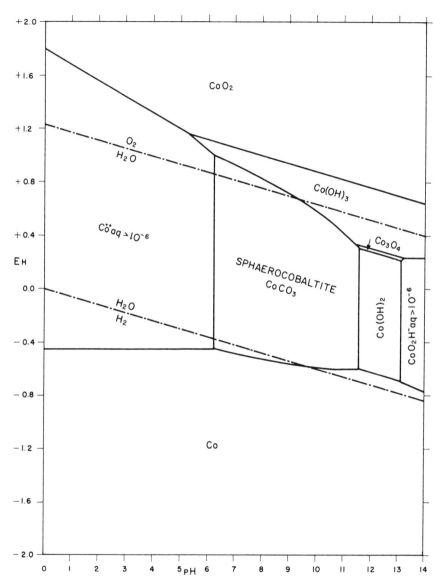

FIG. 7.30b. Stability relations among some cobalt compounds in water at 25 °C and 1 atmosphere total pressure. Total dissolved carbonate species = $10^{-1.5}$. [Courtesy I. Barnes.]

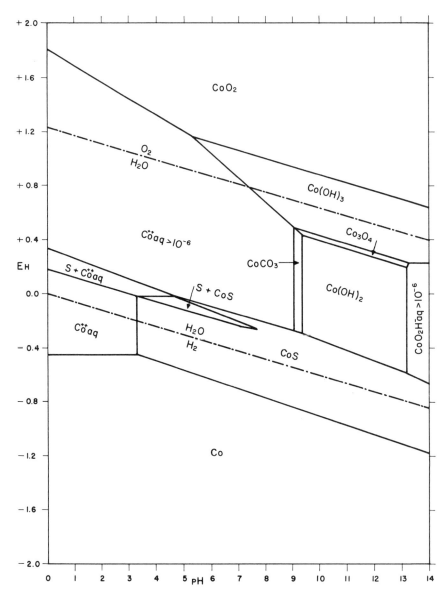

FIG. 7.30c. Stability relations among some cobalt compounds in water at 25 °C and 1 atmosphere total pressure. Total dissolved sulfur species = 10^{-1}, total dissolved carbonate species = $10^{-4.9}$. [Courtesy I. Barnes.]

FeWO$_4$. The sulfide is moderately stable, but also tends to be displaced by insoluble tungstates in the presence of various cations. Note that WO$_2$ is metastable with respect to water under the conditions shown

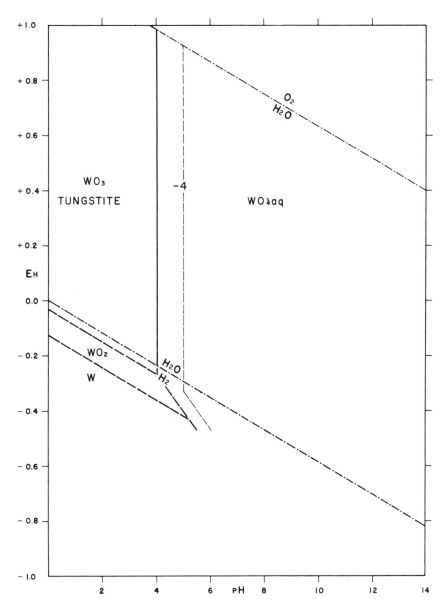

FIG. 7.31a. Stability relations among some tungsten compounds in water at 25 °C and 1 atmosphere total pressure. Boundaries of solids at activity of dissolved species of 10^{-6}; light dashed line at activity = 10^{-4}. [Courtesy K. Linn.]

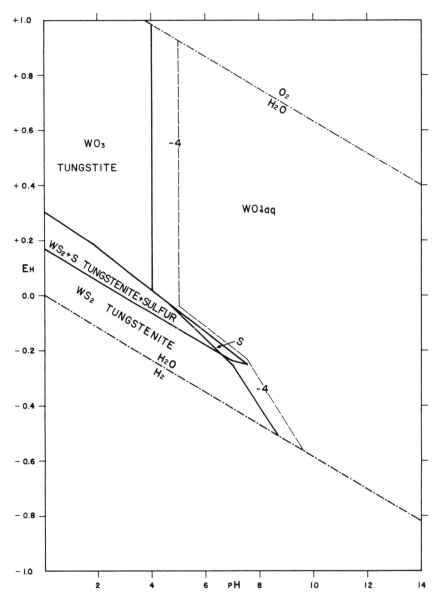

FIG. 7.31b. Stability relations among some tungsten compounds in water at 25 °C and 1 atmosphere total pressure. Total dissolved sulfur species = 10^{-1}. Boundaries of solids at activity of dissolved species of 10^{-6}; dashed line at activity = 10^{-4}. [Courtesy K. Linn.]

here, and that metallic tungsten is not to be expected as a stable phase, whatever its resistance to corrosion may be when once formed.

The diagrams were prepared by Kurt Linn, Department of Geology, Harvard Graduate School, 1958.

Figures 7.32a, b, c

These figures show some relations among uranium compounds and ions at 25 °C and 1 atmosphere total pressure.

Figures 7.32a and b compare the effect of CO_2 on uranium solubility in the open system (P_{CO_2}) and the closed system $(\Sigma\ CO_2)$. In both instances, sexivalent uranium is complexed strikingly as the uranyl dicarbonate and uranyl tricarbonate ionic species, so that with appreciable P_{CO_2} or $\Sigma\ CO_2$, the field of stability of the uranyl oxide hydrate (schoepite?) is wiped out. These complexes are so effective that they "eat" down into the field of stability of UO_2 (uraninite) when P_{CO_2} and $\Sigma\ CO_2$ are relatively high. It should be clear that carbonate-bearing solutions are excellent solvents for uranium.

In Figure 7.32c a technique that has been little developed, but which should have widespread application, is illustrated. A major geological problem in the oxidation of uranium ores is a deciphering of the conditions of migration and fixation of uranium. As shown in Figures 7.32a and b, carbonated water prevents precipitation of uranyl oxide hydrates, but it is well known that carnotite, the potassium uranyl vanadate, is a persistent mineral in the zone of oxidation. In Figure 7.32c a section of the multicomponent system $U-O_2-H_2O-K-V-CO_2$ has been drawn at fixed activities of K, V, and CO_2, and the "solubility" relations of uranium are examined for this particular set of conditions. The values chosen for $\Sigma\ K$, $\Sigma\ V$, and $\Sigma\ CO_2$ are reasonable ones for the waters of oxidizing uranium-vanadium ores. Although the fields of the solids should be examined through a whole range of these values, a single section serves to illustrate the gross relations. The uranium-carbonate complex ions contribute a "solubility trough" under reducing alkaline conditions, but the formidable field of carnotite is plainly shown. Increase of dissolved V or K would serve to enlarge the carnotite field, whereas increase of CO_2 would diminish its size.

The diagrams were prepared by P. B. Hostetler and R. M. Garrels, in collaboration with C. L. Christ and A. D. Weeks (R. M. Garrels, P. B. Hostetler, C. L. Christ, and A. D. Weeks. Stability of uranium, vanadium, copper, and molybdenum minerals in natural waters at low temperatures and pressures. Paper delivered at meeting of Geological Society of America, Atlantic City, November, 1957).

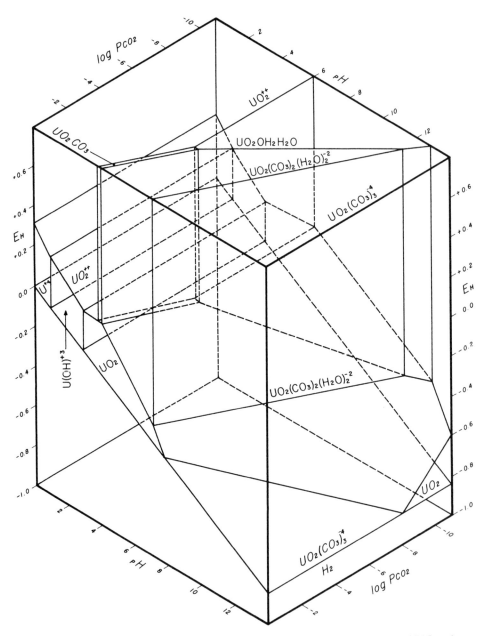

FIG. 7.32a. Stability relations among some uranium compounds in water at 25 °C and 1 atmosphere total pressure as a function of pH, Eh, and P_{CO_2}. Boundaries of solids at activity of total dissolved uranium-bearing species of 10^{-6}. [Courtesy R. Garrels, P. Hostetler, A. Weeks, C. Christ.]

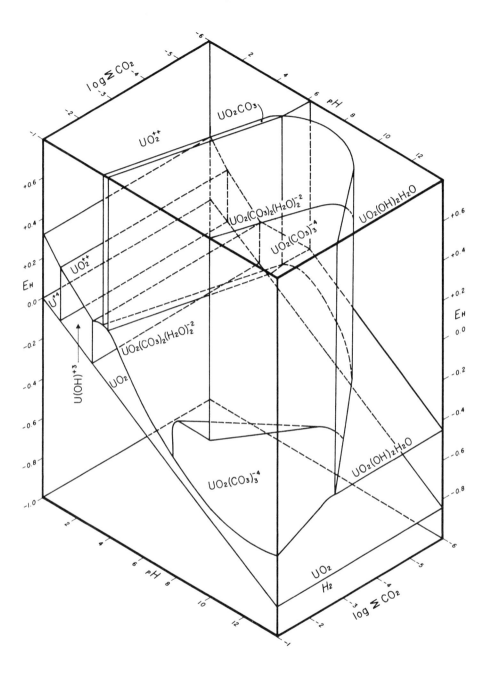

FIG. 7.32b. Stability relations among some uranium compounds in water at 25 °C and 1 atmosphere total pressure as a function of pH, Eh, and total dissolved carbonate species. Boundaries of solids at activity of total dissolved uranium-bearing species of 10^{-6}. [Courtesy R. Garrels, P. Hostetler, A. Weeks, and C. Christ.]

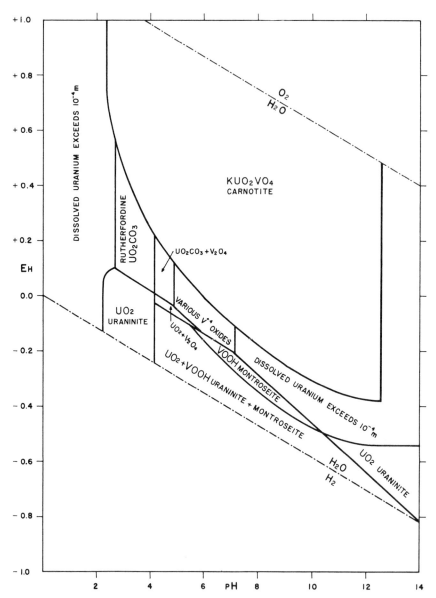

FIG. 7.32c. Stability relations among some uranium and vanadium compounds in water at 25 °C and 1 atmosphere total pressure. Total dissolved vanadium species = 10^{-3}; total dissolved carbonate species = 10^{-1}; total dissolved potassium species = 10^{-3}. [Courtesy R. Garrels, P. Hostetler, A. Weeks, and C. Christ.]

Eh-pH DIAGRAMS

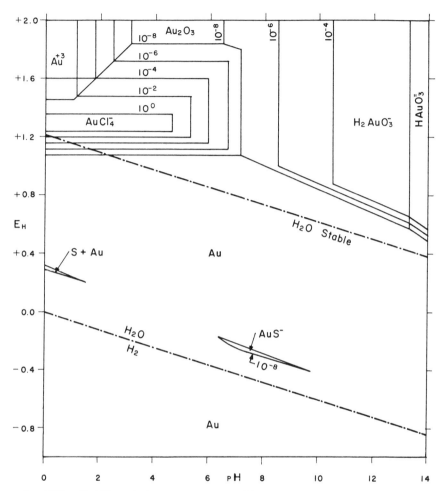

FIG. 7.33a. Stability relations among some gold compounds in water at 25 °C and 1 atmosphere total pressure. Total dissolved chloride species = 10^{-3}; total dissolved sulfur species = 10^{-6}. [Courtesy J. Phillips.]

Figures 7.33a, b

These figures show how such diagrams can be used to investigate the effects of complexes on metal solubility. John Phillips, as a graduate student in geology at Harvard in 1957, undertook to portray the effects of sulfide and chloride complexing of gold on an Eh-pH diagram. Many of the relations are based on Krauskopf's study of the solubility of gold [Konrad B. Krauskopf, The solubility of gold. *Econ. Geol.*, 46, 858–870 (1951)]. The figures show ionic activities at low and high

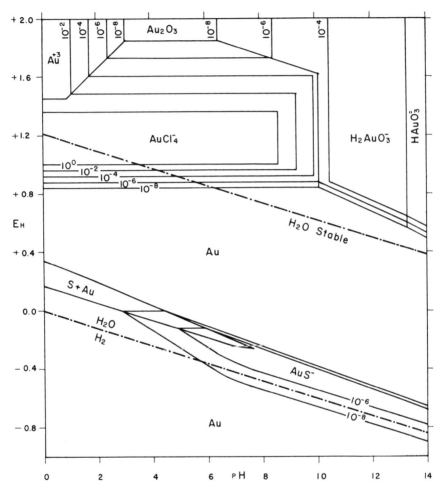

FIG. 7.33b. Stability relations among some gold compounds in water at 25 °C and 1 atmosphere total pressure. Total dissolved chloride species = 10^0; total dissolved sulfur species = 10^{-1}. [Courtesy J. Phillips.]

chloride and sulfur contents of the system. Eh values far above those in equilibrium with water are included so that the overall relations can be seen, even though the reactions above the water boundary are not truly valid. An Eh-pH diagram for gold in oxygenated water is simple indeed—only native gold appears, with no ions exceeding 10^{-6}. In the presence of high chloride, gold is somewhat soluble in acid oxidizing solutions as $AuCl_4^-$; with high sulfur, a little gold dissolves as the AuS^- complex over a wide range of strong reducing conditions. Both relations are of geological interest, the first because of the possibility of transport of gold during oxidation of sulfide ores, and the second because of the problem of transporting gold in vein-forming fluids.

SUMMARY OF THE USE AND LIMITATIONS OF Eh-pH DIAGRAMS

In the preceding text there are, from place to place, precautionary statements concerning the use of Eh-pH diagrams. Here an attempt is made to discuss various aspects of this problem, with the hope of showing the kinds of conclusions that can be unequivocally drawn from these diagrams, and the areas in which dangers await the unwary. There is no question that the best safeguard to the user is a thorough grounding in solution chemistry. It is not possible here to cover all the questions that may arise, or even a small fraction of them, but at least the major strengths and weaknesses can be pointed out.

Free-Energy Data

The standard free energy of formation data on which the diagrams are based are, unfortunately, very uneven in their accuracy. No easy guide to estimation of the accuracy of the value is available. It can be said, somewhat cynically, that the values listed are probably seldom as good as the significant figures given. If one lops off one digit from the right-hand side of the value listed, a more realistic picture is perhaps presented. Most of the uncertainty stems from difficulties in defining the precise nature of the substance listed in the tables. Many solids have several polymorphs, and not all workers have been careful to establish the exact crystalline nature of the substance with which they have worked. For example, in experiments in which a free energy is determined by precipitation of a solid, sufficient time may not have been allowed for complete aging of the precipitate, or it might not even be possible to produce a reasonably coarse-grained solid. Furthermore, there are many paths of experiment and calculation by which free-energy values can be obtained, and without critical examination of these steps the validity of the final result is questionable.

However, these difficulties are not restricted to the use of free-energy values in the synthesis of Eh-pH diagrams—they are inherent in any thermochemical calculation. The best guide, perhaps, is common sense; if a calculation shows that phase A is stable relative to phase B by a small free-energy margin, phase B should be eliminated with reservations. This is especially true if the calculation path is a devious one and involves the difference of several large free energy of formation values.

The ΔF_f° values available for solids are characteristically for materials that correspond analytically as closely as possible to the formulas given. Minerals are characteristically "impure" as compared to these standards. Thus, a diagram illustrating relations among "pure" chemical compounds must not be mistaken for one representing actual minerals. Generally speaking, the difference in free energy of formation caused by minor substitution of other elements in the structure of a given chemical species is not large (the effect is ordinarily approximately proportional to the mole fraction of the substituent). The most serious error likely to be made by the calculator is in forgetting how far minerals do deviate from pure end-member composition, or to what extent certain minerals differ structurally from the analog chemical compound. For example, some natural compounds of dolomite composition are disordered; what this does to stability we would like to know. Also, some sphalerites, which we might tend to regard as ZnS for purposes of calculation, contain in fact 10 to 14 percent iron.

Continuing in the same vein, we find ourselves in a geological difficulty when we attempt to construct a diagram simulating conditions of deposition of the chemically precipitated iron minerals. Should stability relations be calculated for crystalline hematite or goethite, or is the natural system controlled by the freshly precipitated and much less stable hydrous oxide of indefinite water content? The problem again is the choice of the right chemical compound to simulate the natural situation.

THE EFFECT OF TEMPERATURE AND PRESSURE

All calculations given have been for 25 °C and 1 atmosphere total pressure. Yet the implication has been clear throughout that the results from using data derived under these conditions would differ but slightly from those that might be obtained within the extremes of the surface and near-surface environments of the earth. Just what is the order of magnitude of the error expected if several atmospheres are impressed upon the system, or if the temperature is 15 °C or 35 °C?

The effects of changes in temperature and pressure on mineral equilibria are treated in some detail in Chapter 9. It will be sufficient here to consider a few results.

Specifically, if we consider the Eh-pH boundary between magnetite and hematite

$$2Fe_3O_{4\,c} + H_2O_l = 3Fe_2O_{3\,c} + 2H^+_{aq} + 2e,$$

the equation for the boundary at 25 °C is

$$Eh = 0.221 - 0.059\,pH;$$

at 35 °C the equation is

$$Eh = 0.227 - 0.061 \text{ pH}.$$

Thus, the difference can barely be plotted. This order of magnitude is about average for the reactions considered; a few boundaries would be affected appreciably more; a few, less. In fact, several diagrams have been constructed for 110 °C, and whereas the positions of the fields of various solids shift relative to Eh-pH values of the axes, the shapes and sizes of the fields tend to be maintained. In general, a temperature change of a few degrees does not alter the diagrams more than the width of the lines used to indicate the phase boundaries.

Now let us consider, at 25 °C, the following reaction

$$3\text{Fe}_c + 4\text{H}_2\text{O}_1 = \text{Fe}_3\text{O}_{4\,c} + 4\text{H}_{2\,g}.$$

At equilibrium

$$\Delta F_r^\circ = \Delta F_{f\,\text{Fe}_3\text{O}_4}^\circ + 4\Delta F_{f\,\text{H}_2}^\circ - 3\Delta F_{\text{Fe}}^\circ - 4\Delta F_{\text{H}_2\text{O}}^\circ$$

$$\Delta F_r^\circ = -242.6 + 4(0) - 3(0) - 4(-56.69)$$

$$\Delta F_r^\circ = -15.6 \text{ kcal}.$$

Then

$$-15.6 = -1.364 \log K = -1.364 \log f_{\text{H}_2}^4$$

$$f_{\text{H}_2} = 10^{2.85} \text{ atm}$$

$$f_{\text{H}_2} = 708 \text{ atm}.$$

Thus, the equilibrium fugacity of hydrogen at 25 °C is 708 atmospheres. This large a value of the fugacity cannot be equated to the partial pressure, and in fact, from Figure 2.3 we find that the corresponding equilibrium pressure is approximately 500 atmospheres. In this calculation we have neglected only the effects of the increased pressure on the activities of the solids, and the liquid water. Since the changes in activities tend to cancel, such corrections are small.

In summary, then, for those stability diagrams in which the partial pressures of the reacting gases are small, an increase in total pressure of some tens of atmospheres can be safely ignored. On the other hand, where the equilibrium fugacity for a reaction is large, it can still be calculated; the corresponding equilibrium partial pressure will be related to the fugacity through the gas activity coefficient [equation (2.4)].

DIAGRAMS VALID ONLY FOR SPECIES CONSIDERED

In using Eh-pH diagrams, one must remember that the only answers that come out of thermochemical calculations are in response to the questions asked. For instance, one can calculate that FeO and Fe_3O_4 are in equilibrium at a given pair of Eh and pH values, but if one were to assume from such a calculation that FeO and Fe_3O_4 are a stable pair in the iron-water-oxygen system, he would be making a gross error. A given calculation does not tell the results of others that have not been performed. If one considers various possible reactions, he will find that FeO_c will disproportionate at equilibrium into Fe_c and $Fe_3O_4{}_c$

$$4FeO_c = Fe_3O_{4c} + Fe_c; \quad \Delta F_r^\circ \text{ is negative.}$$

For the geologist this type of restriction is not always a serious difficulty. He knows the mineral phases that occur in a certain deposit, and he can ask his questions regarding these minerals. Many of the useful results are negative; one may receive an unequivocal answer that phases A and B *cannot* coexist at equilibrium. Thus, calculations give permissive or negative results. A and B *may* coexist at equilibrium in a given system, if neither is unstable with respect to some other unconsidered species, or if A and B do not interact: alternately, A and B *cannot* coexist at equilibrium.

ACTIVITIES VERSUS SOLUBILITIES

The calculation of the activities of dissolved species in equilibrium with solids is fraught with no more and no fewer dangers than the calculation of the fields of stability of solids. But there is apparently some insidious influence that forces an investigator to attempt to determine solubilities from activities. As pointed out in Chapter 2, the activity of an ion differs from its molality by a factor γ, which can sometimes be estimated fairly accurately from knowledge of the ionic strength. To calculate the solubility of a solid at a given pair of Eh-pH conditions, the activities of *all* contributing dissolved species must be known, as well as their activity coefficients. Ignorance of the presence of a given ion, or an error in estimation of an activity coefficient, leads to error in solubility. This point was made firmly in Chapter 2.

A more serious difficulty, from the geological point of view, stems from the reverse situation: given the composition of a natural water, its Eh and pH, and knowledge of the mineral species in contact with the water, can one deduce whether the minerals present are in equilibrium with their environment? If the stability can be expressed as a function of Eh and pH alone (e.g., Fe_2O_3), then no difficulty arises, because both these values are measurements of activity. But, as for calcite, if it is necessary to know the activity of calcium ion we are back to the problem of calculating the activity of a given ion from an analysis for total dissolved element. The situation is clearly the reverse of an attempt to calculate solubility.

Despite these difficulties, a useful generalization can be made as follows: if a given solid is in equilibrium with a solution, the sum of the activities of all known dissolved species containing a given element is almost always less than the concentration of that element as determined by chemical analysis.

Therefore, it is possible to make calculations of the sum of the activities of ions in equilibrium with a solid and to assign a *minimum solubility* value to this sum. Again we are faced with a positive or a permissive situation: if the sum of the activities of known ions is a large value, it can be said that the solid in question is soluble; if the sum of the activities of known ions is small, the solid in question *may not* be soluble.

In passing it must be emphasized that few data are currently available on the mineral associations in rocks, together with chemical analyses of the associated aqueous solutions. With such information we could make rapid strides in assessing our chemical ignorance. By assuming equilibrium between minerals and solution, one can calculate the solubility of a given element to the best of his knowledge. The discrepancy between this value and the analytical concentration of the element in the solution represents ignorance, and one can make tests to see if the discrepancy is due to failure to obtain equilibrium, to unsuspected complexes, to incorrect thermochemical values, or to still other factors.

SELECTED REFERENCES

Pourbaix, M. J. N., J. Van Muylder, and N. de Zhoubov, *Atlas d'Équilibres Électrochimiques* à *25 °C*. Paris, Gauthier-Villars, 1963.

Pourbaix, M. J. N., *Thermodynamics of Dilute Solutions*. London, E. Arnold, 1949.

Technical Reports of the Belgian Center for Study of Corrosion, 24 Rue des Chevaliers, Brussels.
    1. Pourbaix, M., Sur l'interprétation thermodynamique de courbes de polarisation (1952).
    2. Pourbaix, M., Applications de diagrammes tension-pH relatifs au fer et a l'eau oxygénée. Expériences de démonstration (1954).
    3. Deltombe, E., and M. Pourbaix, Comportement électrochimique du cadmium (1953).
    4. Schmets, J., and M. Pourbaix, Comportement électrochimique du titane (1953).
    5. Deltombe, E., and M. Pourbaix, Comportement électrochimique des cyanures (1953).
    6. Deltombe, E., and M. Pourbaix, Comportement électrochimique du cobalt (1954).

7. Deltombe, E., and M. Pourbaix, Comportement électrochimique du fer (1954).
8. Deltombe, E., and M. Pourbaix, Comportement électrochimique du fer en solution carbonique (1954).
9. Pourbaix, M., Sur la phosphation oxydante des aciers ordinaires (1953).
10. Van Muylder, J., La corrosion et la protection des gaines en plomb des câbles enterrés (1953).
11. Van Muylder, J., and M. Pourbaix, Corrosion et protection cathodique du fer. Expérience de démonstration (1953).
12. Van Muylder, J., Au sujet des hydrures de plomb et de la pulvérisation cathodique du plomb (1954).
13. Van Muylder, J., and M. Pourbaix, Sur le comportement électrochimique du plomb. Corrosion, protection cathodique, passivation (1953).
14. Van Muylder, J., and M. Pourbaix, Corrosion et protection cathodiques du plomb (1954).
15. Magee, G. M., Corrosion electrolytique de l'acier dans le béton (trad. d'un article de G. M. Magee paru dans *Corrosion*, 5, 11 novembre 1949, 378–382) (1954).
16. Wattecamps, P., Progrès récents de la technique américaine d'application des revêtements protecteurs (1953).
17. Bureau of Ships, U.S. Navy (Trad. par P. Wattecamps), Études sur des systèmes de revêtements anticorrosif et antisalissant pour carènes de navires.
18. Moussard, A. M., J. Brenet, F. Jolas, M. Pourbaix, and J. Van Muylder, Comportement électrochimique du manganèse (1954).
19. Abd El Wahed, A. M., and M. Pourbaix, Utilisation des corbes de polarisation pour l'étude des circonstances de corrosion et de protection du fer en présence de chlorures. Phosphatation et phosphatation oxydante (1954).
20. Pourbaix, M., Applications de l'électrochimique à des études de corrosion (1954).
21. Pourbaix, M., Vue d'ensemble sur le comportement électrochimique des métaux (1ère partie) (1953).
22. Comptes-rendus des Journées d'Études du Cebelcor, 13 et 14 avril 1955.
23. Deltombe, E., N. de Zoubov, and M. Pourbaix, Comportement électrochimique du nickel (1955).
24. Laureys, J., J. Van Muylder, and M. Pourbaix, Note sur l'efficacité d'un appareil de traitement magnétique des eaux (1955).
25. Deltombe, E., N. de Zoubov, and M. Pourbaix, Comportement électrochimique de l'étain (1955).
26. de Zoubov, N., and E. Deltombe, Enthalpies libres de formation standard de l'hydrure d'étain gazeux.
27. de Zoubov, N., E. Deltombe, and M. Pourbaix, Comportement électrochimique du germanium (1955).

28. Enthalpies libres de formation standards, à 25 °C, 1955.
29. Deltombe, E., N. de Zoubov, and M. Pourbaix, Comportement électrochimique du vanadium (1956).
30. Pourbaix, M., Leçons sur la corrosion électrochimique, 2me fascicule (1956).
31. Deltombe, E., N. de Zoubov, and M. Pourbaix, Comportement électrochimique de l'uranium (1956).
32. Deltombe, E., N. de Zoubov, and M. Pourbaix, Comportement électrochimique du tungstène (1956).
33. Deltombe, E., N. de Zoubov, and M. Pourbaix, Comportement électrochimique du tellure (1956).
34. Van Muylder, J., and M. Pourbaix, Comportement des anodes réactives en magnésium et en zinc (1956).
35. Deltombe, E., N. de Zoubov, and M. Pourbaix, Comportement électrochimique du molybdène (1956).
36. Van Eijnsbergen, J. F. H., La protection contre la corrosion de l'acier par der couches métalliques, spécialement par la galvanisation à chaud (1956).
37. Vandervelden, F., and M. Pourbaix, Protection contre la corrosion dans l'emballage et au cours du stockage (1956).
38. Pourbaix, M., Services que peuvent rendre à l'industrie pétrollière les centres de recherche contre la corrosion (1956).
39. Van Muylder, J., and M. Pourbaix, Comportement électrochimique du magnésium (1956).
40. Pourbaix, M., Sur la corrosion du fer et des aciers par les eaux. Influence du pH, des oxydants, des réducteurs, des chlorures, des phosphates et de la température (1956).
41. Deltombe, E., N. de Zoubov, and M. Pourbaix, Comportement électrochimique du chrome (1956).
42. Deltombe, E., and M. Pourbaix, Comportement électrochimique de l'aluminium (1956).
43. Pourbaix, M., and N. de Zoubov, Sur les conditions des passivation du fer par les chromates, molybdates, tungstates et vanadates (1957).
44. Valensi, G., E. Deltombe, N. de Zoubov, and M. Pourbaix, Comportement électrochimique du chlore (1957).
45. Maraghini, M., P. Van Rysselberghe, E. Deltombe, N. de Zoubov, and M. Pourbaix, Comportement électrochimique du zirconium (1957).
46. Van Muylder, J., and M. Pourbaix, Comportement électrochimique de l'arsenic (1957).
47. Deltombe, E., N. de Zoubov, and M. Pourbaix, Comportement électrochimique du bore (1957).
48. Van Muylder, J., and M. Pourbaix, Comportement électrochimique du bismuth (1957).
49. Pourbaix, M., Leçons sur la corrosion électrochimique, 3me fascicule (1957).
50. de Zoubov, N., and M. Pourbaix, Comportement électrochimique du technétium (1957).

51. de Zoubov, N., and M. Pourbaix, Comportement électrochimique du rhénium (1957).
52. Van Muylder, J., and M. Pourbaix, Comportement électrochimique du tantale (1957).
53. Van Muylder, J., N. de Zoubov, and M. Pourbaix, Comportement électrochimique du niobium (1957).
54. Pitman, A. L., M. Pourbaix, N. de Zoubov, Comportement électrochimique de l'antimoine (1957).

Blumer, Max, Die Existenzgrenzen anorganischer Ionen bei der Bildung von Sedimentgesteinen: *Helv. Chim. Acta, 33*, fasc. VI, No. 206 (1950).
Delahay, Paul, Marcel Pourbaix, and Pierre Van Rysselberghe, Potential-pH diagrams. *J. Chem. Educ.*, 27, 683–688 (1950).
Charlot, G., *Théorie et méthode nouvelle d'Analyse qualitative*, 3rd edition. Paris, Masson, 1949.
Schmitt, H. H. (ed.), *Equilibrium Diagrams for Minerals*. Cambridge, Mass., Geological Club of Harvard University, 1962.
Sillén, Lars G., Redox diagrams. *J. Chem. Educ.*, 29, 600–608 (1952).

PROBLEMS

7.1. a. Write the reaction for equilibrium between SnO_c and $SnO_{2\,c}$ in terms of water, hydrogen ions, and electrons.
b. Calculate the standard free energy change of the reaction at 25 °C.
    Ans. $\Delta F_r^\circ = -5.0$ kcal.
c. Calculate E° for the reaction at 25 °C.   *Ans.* $E^\circ = -0.108$ volt.
d. Write the equation for the phase boundary between SnO_c and $SnO_{2\,c}$ in terms of Eh and pH.   *Ans.* Eh $= -0.108 - 0.0592$ pH.
e. What is the activity of Sn^{4+} in equilibrium with $SnO_{2\,c}$ at pH $= 8$?
    Ans. $[Sn^{4+}] = 10^{-39.7}$.
f. What is the activity of Sn^{++} in equilibrium with $SnO_{2\,c}$ at pH $= 8$ and Eh $= 0.3$ volt?   *Ans.* $[Sn^{++}] = 10^{-44.8}$.

7.2. Construct a diagram showing the stability of SnO_c and $SnO_{2\,c}$ as a function of Eh and pH at 25 °C. Draw contours showing activities of Sn^{4+}_{aq}, Sn^{++}_{aq}, and $HSnO_{2\,aq}^-$ where they exceed 10^{-6}. What conclusion can you draw concerning the stability of SnO_c in water? Is SnO_c stable with respect to $Sn(OH)_{2\,c}$?

    Ans. SnO_c reacts with water to release hydrogen.
    Yes, SnO_c is stable with respect to $Sn(OH)_{2\,c}$.

CHAPTER 8

Ion Exchange and Ion-Sensitive Electrodes

INTRODUCTION

Many solid substances, when placed in contact with an aqueous solution, show a marked tendency to lose certain components while retaining others. When this occurs, the components that are lost are replaced with similar species from solution, but the basic structure of the original material is preserved.

When the species lost or gained are ions, the phenomenon is called *ion exchange*, and this process can be further divided into *cation exchange* and *anion exchange*. Cation exchange has received far more attention experimentally than anion exchange, and will be considered almost exclusively here. However, the processes involved are essentially the same, so that the detailed treatment of cation exchange will provide the principles required for analogous treatment of anion exchange.

Substances of geological interest that exhibit marked cation-exchange behavior usually consist of a negatively charged repetitive structural framework, which has well-defined negatively charged sites occupied by singly or doubly charged cations. The classic examples are the clay minerals, especially the montmorillonites and illites (hydrous micas). In this case the negative framework consists essentially of aluminum silicate sheet structures, and the exchangeable cations are located in interlayer positions or adjacent to the particle surfaces. When particles of these aluminum silicates are placed in aqueous solution, it is found that part of the interlayer and surface cations can be displaced and replaced quickly by cations from solution, and the mineral grains subsequently recovered without appreciable gross structural damage.

The same kind of behavior as exhibited by the clay minerals, although perhaps not so striking or so well known, occurs with a great variety of materials generally characterized by durable negatively charged frameworks studded with cations of low charge. Among these are included most silicate minerals, silicate glasses, mineral and synthetic arsenates, vanadates, molybdates, and related species. In addition, there has been a vast development of synthetic materials, chiefly organic, that have the required characteristic of exchanging cations while retaining overall structural integrity.

The importance of cation exchange in influencing the characteristics of natural solutions that pass over exchangers is well known; there is another aspect of exchange materials that is potentially of comparable importance, but which is much less studied. If a material with cation-exchange properties is placed as a membrane between two solutions of different composition, there is a tendency for the two solutions to equilibrate by cation migration through the ion exchanger. The exchanger, under these conditions, behaves as a semipermeable membrane. Electric potentials are developed, and the two solutions change their compositions through time in a manner dictated by the exchange membrane. Geologically, the essentials of this situation develop wherever a shale lies between two sandstones carrying waters of different compositions.

The behavior of ion exchangers as semipermeable membranes has led to the development of their use as cation-sensitive electrodes, some of which are selective for certain cations. The glass electrode for measurement of hydrogen ion activity is the best known example of such an application of an exchange membrane. Recently, glasses selectively sensitive to alkali metal and alkaline earth cations have been developed, and permit measurement, for example, of $pNa = -\log[Na^+]$, analogous to the familiar pH. The use of stearates and allied compounds has resulted in electrodes selective for alkaline earth cations. The clay minerals have been used, although they are in general not highly selective between alkali metal and alkaline earth cations. Synthetic exchange materials also have been used as electrodes.

In the following pages the cation exchange processes involving particles immersed in solution will be treated first, and then the behavior of exchangers that can be used as electrodes when placed as membranes between solutions of different compositions.

CATION EXCHANGE

Many of the materials that perform obviously as cation-exchangers have been studied in some detail. It is quite remarkable that their behavior with regard to a pair of cations is almost invariably described fairly simply. If the exchanger, with its exchangeable sites occupied by a given cation, is placed in a solution of a different cation, exchange

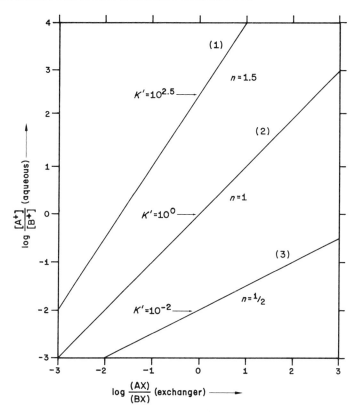

FIG. 8.1. Relation of ion activity in solution and ion concentration in exchanger for several ion-exchangers. From equation (8.1) $\log \frac{[A^+]}{[B^+]} = \log K'_{AB} + n \log \frac{(AX)}{(BX)}$; as shown, if $\log \frac{[A^+]}{[B^+]}$ is plotted against $\log \frac{(AX)}{(BX)}$, a straight line results with slope n and intercept equal to $\log K'_{AB}$. Examples are given for: $K' = 10^{2.5}, n = 1.5$; $K' = 10^0$, $n = 1$; $K' = 10^{-2}, n = 0.5$.

takes place. The ratio of the concentrations of ions in the exchanger can be determined as a function of the activities of the ions in the solution by varying the solution composition. If the logarithm of the ratio of the *activities* of the ions in the aqueous solution is plotted against the logarithm of the ratio of the *concentration* of the ions on the exchanger, a straight-line relation is observed. Figure 8.1 shows typical plots of the logarithm of the ratio of the activities of the monovalent cations in solution versus the logarithm of their concentration ratios.

Walton[1] lists the results of a whole series of experiments on exchange

---

[1] H. F. Walton, Ion exchange equilibria, in *Ion Exchange Theory and Practice*, F. C. Nachod (ed.). New York, Academic, 1949.

relations for various exchangers, and finds the kind of relation illustrated in Figure 8.1. These relations can be summarized by the equation

$$\frac{[A^+]}{[B^+]} = K'_{AB}\left(\frac{(AX)}{(BX)}\right)^n. \qquad (8.1)$$

$[A^+]$ and $[B^+]$ are the activities of cations in solution; (AX) and (BX) are the concentrations of ions in the exchanger, n is an exponent, and K'_{AB} is the exchange constant.

Table 8.1 is adapted from Walton (*op. cit.*), and lists values for K' and for n for a variety of substances.

TABLE 8.1. Exchange Constants for Various Exchangers: Values of K' and n for the Exchange Equations

$$\frac{[A^+]}{[B^+]} = K'_{AB}\left(\frac{(AX)}{(BX)}\right)^n$$

$$\frac{[A^+]^2}{[B^{++}]} = K'_{AB}\left(\frac{(A_2X_2)}{(BX_2)}\right)^n$$

Exchanger		n	K'
NH$_4$ synthetic aluminosilicate (fused)	+ Na$^+$	1.50	0.22
NH$_4$ synthetic aluminosilicate (fused)	+ K$^+$	1.39	1.00
Ag synthetic aluminosilicate (fused)	+ Li$^+$	1.45	0.00065
Ag synthetic aluminosilicate (fused)	+ K$^+$	2.17	0.01
Ag synthetic aluminosilicate (fused)	+ Tl$^+$	2.00	0.35
Li bentonite	+ NH$_4^+$	1.25	1.52
K bentonite	+ NH$_4^+$	0.91	0.54
Na synthetic aluminosilicate (gel)	+ Ca^{++}	1.58	0.55
Na synthetic aluminosilicate (gel)	+ Ba^{++}	1.82	3.2
NH$_4$ synthetic aluminosilicate (fused)	+ Ca^{++}	3.84	0.81
NH$_4$ synthetic aluminosilicate (fused)	+ Mg^{++}	5.88	0.0029
Ca synthetic aluminosilicate (fused)	+ Na$^+$	1.11	0.028
Ca synthetic aluminosilicate (gel)	+ Na$^+$	1.39	0.115
Ca casein	+ Ba^{++}	1.33	1.28

The ionic activities in solution are obtained from solution composition and from activity coefficients obtained by the methods described in Chapter 2. The concentrations of ions in the exchanger can be determined in a number of ways. Usually, the characteristics of an exchanger are examined by first placing it in a series of solutions containing high concentrations of a given cation, so that the exchange sites are presumably occupied by a single cation, such as sodium ion. Then the exchanger is placed in water, and another cation added in the form of a soluble salt, such as KCl. By recording the amount of Na$^+$ released to solution during addition of KCl, and by eventually displacing all Na$^+$ from the exchanger by KCl addition, data are obtained on the *exchange capacity* of the exchanger, and on the required values of concentration ratios of ions in the exchanger and in the solution.

For example, if a 100-gram sample of montmorillonite is treated with a series of NaCl solutions so that all exchange sites are occupied by sodium ions, and then is placed in a series of strong KCl solutions, it is found that sodium ions appear in moderate concentration in the first treating solution, and in lesser and lesser concentrations in succeeding KCl solutions. When the Na^+ released becomes analytically insignificant, it is assumed that the exchange sites are occupied by K^+ and that the total amount of Na^+ drawn off the exchanger represents its capacity to hold Na^+. Of course, the exchange takes place in aqueous solution, so that H^+ is always available as a competitor for exchange sites. It is common practice to define the exchange capacity as that observed at pH 7. In general, H^+ occupation of negative sites is small for solutions of pH 7 or higher, but serious H^+ interference may occur in some systems. For most exchangers, convenient units are *milliequivalents of exchangeable cation per 100 grams of exchanger*. Table 8.2 lists the exchange capacity of some common clay minerals as determined in this way.

TABLE 8.2. Cation-Exchange Capacity of Clay Minerals

Mineral	Structural Control	Exchange Capacity (meq/100 g at pH 7)
Kaolinite	Unsatisfied valences on edges of structural units	3–15
Halloysite (2H$_2$O)	Unsatisfied valences on edges of structural units	5–10
Halloysite (4H$_2$O)	Unsatisfied valences on edges of structural units and on internal surface between the layers	40–50
Montmorillonite group	Substitutions in the octahedral and tetrahedral units giving excess negative charge; unsatisfied valences on edges of units	70–100[a]
"Illites" (hydrous micas)	As in montmorillonite, plus deficiency of K^+ between the layers	10–40
Vermiculite	Replacement of interlayer cations, substitution within the units, and unsatisfied valences on edges of units	100–150
Chlorite	No data. Possible deficiency of charge due to substitution in the brucite layer	10–40?
Glauconite	As in "illites"	11–20+
Palygorskite group	Substitution of Al^{3+} for Si^{4+} in structural units, unsatisfied exchange sites within channels in the structure	20–30
Allophane	Porous amorphous structure with unsatisfied valences	~70

[a] Certain members of the montmorillonite group have a much lower exchange capacity because there is no substitution in the tetrahedral or octahedral units—for example, stevenite with 36 meq/100 g.
SOURCE: Dorothy Carroll, Ion exchange in clays and other minerals, *Bull. Geol. Soc. Am.*, 70, 754 (1959).

The use of *equivalents* of exchangeable cations, as opposed to *moles*, is convenient because a given sample of an exchanger has a given number of negatively charged sites, so that it tends to be occupied by a fixed number of positive ions. Ideally, it should require only half the number of moles of calcium ion to occupy completely a given sample of exchanger as it does sodium ion. In general this relation holds fairly well, although one of the complications in dealing with an exchanger is that its exchange capacity, conceived in the ideal case as a fixed number of negative sites, differs experimentally depending upon the occupying cation.

Because of the great variation in exchange capacity of various exchangers, or *exchange substrates*, it is convenient for many purposes to use the fraction of the total exchange capacity occupied by a given cation, rather than its absolute concentration. For this purpose the *mole fraction* of a given ion in the substrate seems to be the most useful quantity. If the total exchange capacity of a montmorillonite is 10 milliequivalents for a 10-gram sample, and if the amount of Na^+ ion is 5 milliequivalents and that of K^+ is 5 milliequivalents, then the number of moles of each is 0.005. The mole fraction of Na^+ is

$$N_{NaX} = \frac{n_{Na^+}}{n_{Na^+} + n_{K^+}} = \frac{0.005}{0.005 + 0.005} = 0.5,$$

and that of K^+ (N_{KX}) is 0.5.

INTERPRETATION OF EXCHANGE ON THE BASIS OF THE LAW OF MASS ACTION

The phenomena of cation exchange have been interpreted in a variety of ways, all of which permit derivation of the observed behavior. For details of the various derivations, the reader is referred to the extended treatments of ion exchange listed at the end of this chapter. In the following discussion the interpretation is made in terms of the Law of Mass Action, a plan that is in harmony with the extensive use of this principle throughout this book.

Two-Cation Exchange

When an exchanger, with its negative sites occupied by a monovalent cation, is placed in a solution containing only one different monovalent cation, exchange takes place, and the empirical relations, as indicated previously, are described by the equation

$$\frac{[A^+]}{[B^+]} = K'_{AB}\left(\frac{(AX)}{(BX)}\right)^n, \tag{8.1}$$

where (AX) and (BX) refer to the concentrations of A^+ and B^+ in the exchanger in moles per unit weight of exchanger. The concentrations can also be expressed as mole fractions. If the numerator and denominator of equation (8.1) are divided by $\{(AX) + (BX)\}^n$, we have

$$\frac{[A^+]}{[B^+]} = K'_{AB}\left(\frac{\frac{(AX)}{(AX)+(BX)}}{\frac{(BX)}{(AX)+(BX)}}\right)^n = K'_{AB}\left(\frac{N_{AX}}{N_{BX}}\right)^n. \quad (8.2)$$

The empirical equation (8.1) can be derived using the Law of Mass Action, as follows: the exchange reaction is

$$AX + B^+ = BX + A^+, \quad (8.3)$$

and assuming chemical equilibrium, we have

$$\frac{[A^+][BX]}{[B^+][AX]} = K_{AB}. \quad (8.4)$$

Rearranging

$$\frac{[A^+]}{[B^+]} = K_{AB}\frac{[AX]}{[BX]}. \quad (8.5)$$

The expression given by (8.5) is similar in form to that given by (8.1) except that activities instead of concentrations appear on the right-hand side of the equation, and the exponent n does not appear. If we treat the exchanger as a binary solid solution with components AX and BX as the pure end-members, then by equation (2.28) we can write

$$[AX] = \lambda_{AX} N_{AX} \quad (8.6)$$

$$[BX] = \lambda_{BX} N_{BX} \quad (8.7)$$

where λ_{AX} and λ_{BX} are the rational activity coefficients. Substituting (8.6) and (8.7) in (8.5)

$$\frac{[A^+]}{[B^+]} = K_{AB}\frac{\lambda_{AX} N_{AX}}{\lambda_{BX} N_{BX}}. \quad (8.8)$$

If the exchange substrate forms a regular solution, then by equation (2.42)

$$\frac{\lambda_{AX}}{\lambda_{BX}} = \frac{\exp\left(\frac{B}{RT} N_{BX}^2\right)}{\exp\left(\frac{B}{RT} N_{AX}^2\right)} = \exp\left[-\frac{B}{RT}(N_{AX}^2 - N_{BX}^2)\right]$$

$$= \exp\left[-\frac{B}{RT}(N_{AX} - N_{BX})\right].$$

Substituting this relation into equation (8.8)

$$\frac{[A^+]}{[B^+]} = K_{AB}\left(\frac{N_{AX}}{N_{BX}}\right) \exp\left[-\frac{B}{RT}(N_{AX} - N_{BX})\right]. \quad (8.9)$$

Taking logarithms of both sides

$$\ln\frac{[A^+]}{[B^+]} = \ln K_{AB} + \ln\frac{N_{AX}}{N_{BX}} - \frac{B}{RT}(N_{AX} - N_{BX}). \quad (8.10)$$

The quantity $\ln(N_{AX}/N_{BX})$ can be expanded as the series

$$\ln\frac{N_{AX}}{N_{BX}} = 2\left\{\left(\frac{\frac{N_{AX}}{N_{BX}} - 1}{\frac{N_{AX}}{N_{BX}} + 1}\right) + \frac{1}{3}\left(\frac{\frac{N_{AX}}{N_{BX}} - 1}{\frac{N_{AX}}{N_{BX}} + 1}\right)^3 + \frac{1}{5}\left(\frac{\frac{N_{AX}}{N_{BX}} - 1}{\frac{N_{AX}}{N_{BX}} + 1}\right)^5 + \cdots\right\}. \quad (8.11)$$

This series converges rapidly for values of N_{AX} and N_{BX} commonly obtained from experimental work on exchangers, i.e. for values of N between about 0.1 and 0.9. As an approximation, then, only the first term in the series of (8.11) need be considered, and we have

$$\ln\frac{N_{AX}}{N_{BX}} \simeq 2(N_{AX} - N_{BX}). \quad (8.12)$$

The substitution of this value into equation (8.10) yields

$$\ln\frac{[A^+]}{[B^+]} \simeq \ln K_{AB} + \ln\frac{N_{AX}}{N_{BX}} - \frac{B}{2RT}\ln\frac{N_{AX}}{N_{BX}}. \quad (8.13)$$

Taking antilogarithms of both sides of (8.13)

$$\frac{[A^+]}{[B^+]} \simeq K_{AB}\left(\frac{N_{AX}}{N_{BX}}\right)^{1-(B/2RT)}. \quad (8.14)$$

Thus, equation (8.14) is identical with the empirical exchange equation (8.2), with

$$n = 1 - \frac{B}{2RT}, \quad (8.15)$$

and

$$K_{AB} = K'_{AB}. \quad (8.16)$$

The approximation that was made in obtaining equation (8.12) from equation (8.11) is a very good one when $N_{AX} \simeq N_{BX}$, and is poor only when N_{AX} is much larger or much smaller than N_{BX}. This suggests that the empirical relation of Walton [equation (8.1)] is an approximation that is valid only for values of N_{AX} or N_{BX} significantly different from

ION EXCHANGE AND ION-SENSITIVE ELECTRODES

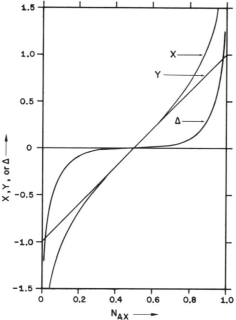

FIG. 8.2. Plots of $X = \frac{1}{2} \ln \frac{N_{AX}}{N_{BX}}$, $Y = (N_{AX} - N_{BX})$, and $\Delta = (X - Y)$ vs. N_{AX}. In equations (8.10) and (8.13) (see text) all the terms are the same except X and Y. Note in the figure that Δ becomes large only when N_{AX} departs appreciably from 0.5.

zero or unity, and that the direct application of regular solution theory, as expressed in equation (8.9) (without series approximation) may lead to more accurate results. The differences obtained in using equations (8.9) and (8.14), as a function of the ratio of N_{AX} to N_{BX}, are shown in Figure 8.2.

For a monovalent-divalent ion exchange, the empirical equation is

$$\frac{[A^+]^2}{[B^{++}]} = K'_{AB}\left(\frac{(A_2X_2)}{(BX_2)}\right)^n. \tag{8.17}$$

The reaction for this exchange can be written

$$A_2X_2 + B^{++} = BX_2 + 2A^+ \tag{8.18}$$

whence

$$\frac{[A^+]^2}{[B^{++}]} = K_{AB}\frac{[A_2X_2]}{[BX_2]}, \tag{8.19}$$

or

$$\frac{[A^+]^2}{[B^{++}]} = K_{AB}\frac{\lambda_{A_2X_2}N_{A_2X_2}}{\lambda_{BX_2}N_{BX_2}}. \tag{8.20}$$

Again assuming that the substrate forms a regular solution, by analogy with the steps used in obtaining equation (8.14) from equation (8.8), we proceed from (8.20) to

$$\frac{[A^+]^2}{[B^{++}]} \cong K_{AB}\left(\frac{N_{A_2X_2}}{N_{BX_2}}\right)^n = K_{AB}\left(\frac{(A_2X_2)}{(BX_2)}\right)^n. \qquad (8.20a)$$

Exchange Involving More than Two Cations

Very little experimental work seems to have been done for exchanges involving more than two different cations. This is surprising, inasmuch as natural systems involve many different cations. The relations for three monovalent cations can be derived from the Law of Mass Action on the basis of the previous assumption concerning the relation between cation activity in solution and concentration in the exchanger.

For three ions A^+, B^+, and C^+, the equations for the equilibria can be written

$$\frac{[A^+]}{[B^+]} = K_{AB}\left(\frac{N_{AX}}{N_{BX}}\right)^n \qquad (8.21)$$

$$\frac{[A^+]}{[C^+]} = K_{AC}\left(\frac{N_{AX}}{N_{CX}}\right)^m. \qquad (8.22)$$

Multiplying equation (8.15) by equation (8.16), we have

$$\frac{[A^+]^2}{[B^+][C^+]} = K_{AB}K_{AC}\frac{(N_{AX})^{n+m}}{(N_{BX})^n(N_{CX})^m}. \qquad (8.23)$$

It is known that if the exponents n and m are unity, equation (8.23) is at least approximately obeyed.

SOME LIMITATIONS ON THE BEHAVIOR OF SUBSTANCES AS CATION EXCHANGERS

When the nature of the reactions involved in the process of cation exchange is considered, it is remarkable that the fairly simple equations used describe the process so well. The bonding sites on the negatively charged membrane must behave nearly identically; cations must go on and off the membrane without appreciable breakup of the membrane by solution or structural change. The total number of available sites must remain essentially constant. The activities of the cations on the negative framework must be related to their concentrations by exponential equations of identical form.

ION EXCHANGE AND ION-SENSITIVE ELECTRODES

Any real material fails to fulfill each of these requirements in some degree. Some of the kinds of failure that cause important deviations from the equations given are as follows.

Nonequivalent Bonding Sites

In substances that have particles of irregular shapes and of small size, the surface of the solid consists of a considerable proportion of edges and corners, in addition to more or less plane surfaces. The energy required to exchange a cation on a corner is markedly different from that on an edge, and that on an edge is different from that on a surface. No single exchange constant can be used to describe the exchange behavior of the material; instead, each small increment of an exchange titration curve requires a separate constant. If hydrogen ions are placed in exchange positions, for example, and if they are displaced by sodium ions, the first addition of sodium ions frees hydrogen ions relatively easily, but successive additions must displace hydrogen ions from more and more strongly held positions.

Finely-ground hydrous aluminum silicates show behavior of this type. The negatively charged sites available on such solids result only from broken bonds at the surfaces of the particles, and the sites differ markedly in their character, both because of surface geometry and because of the nature of the bonds broken during grinding.

A second example of nonequivalent sites is exhibited by numerous clay minerals when they are placed in acid solution and then titrated with an alkali-metal hydroxide. The hydrogenated clays behave as if there were two distinct kinds of hydrogen sites: one easily replaced by alkali metal cations, the other much more recalcitrant. A titration curve showing such behavior is illustrated in Figure 8.3. On the assumption that the exchange reaction is

$$HX + K^+ = KX + H^+$$

the exchange constant for the first site is $10^{-2.5}$; that for the second, $10^{-6.7}$. In this example both constants hold well through their range of application; the interpretation is that the bonding energy difference for K^+ and H^+ is uniform for each substrate, but that the two substrates are nonequivalent. A satisfactory physical picture of the reason for two such distinct types of negative sites is not available.

Disruption of the Negative Framework

It should always be remembered that the negatively charged framework is itself not impervious to the action of the aqueous solution while cations are being added and removed. For example, a magnesian

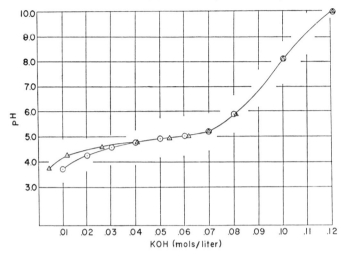

FIG. 8.3. Calculated (triangles) vs. observed (circles) titration curves for Putnam beidellite. Calculation assumes two exchange reactions represented by equations:

$$\frac{[H^+]}{[K^+]} = 10^{-2.5} \frac{(HC)}{(KC)} \quad \text{and} \quad \frac{[H^+]}{[K^+]} = 10^{-6.7} \frac{(HE)}{(KE)}.$$

[From R. M. Garrels and C. L. Christ, Am. J. Sci., 254, 372 (1956).]

montmorillonite, if placed in a sodium-rich solution, so that essentially all its exchange sites are occupied by sodium, is not in equilibrium with respect to all possible chemical reactions. There is a tendency for some of the magnesium in octahedral coordination to emerge into the solution, and a tendency for some of the sodium ions to enter the framework structure. There is little information concerning the rate at which such changes occur, but over a period of time the framework must alter. Furthermore, all the components of the negative framework must dissolve into the surrounding solution if complete equilibrium is to be attained. This means that there is a certain amount of dissolution of the framework, perhaps with surface reconstitution when it is allowed to remain in water. Such an effect is best observed in the reaction of acid solutions on aluminosilicates: high H^+ concentration tends to destroy their structures fairly rapidly, and immersion in a strong acid usually causes a change in exchange values determined before and after immersion. In some cases, as the structure is attacked, aluminum or other ions are released and occupy the exchange sites ordinarily reserved for less highly charged cations.

SELECTIVITY OF CATION-EXCHANGERS

For pairs of monovalent cations, the exchange constant K' [equation (8.1)] is the most convenient measure of the *selectivity* of the exchanger, that is, its tendency to bond one ion more strongly than the other. For example, if the ratio of ionic activities of A^+ and B^+ in solution is unity when their mole fractions in the exchanger are equal, the exchanger has no selectivity: A^+ and B^+ are bonded to it with equal strength. But if K'_{AB} is 10, then equal occupation of the exchanger takes place only when the ratio of the activities of the ions in solution is 10; conversely, when the ratio of the activities of the ions in solution is unity, the exchanger will be largely occupied by one of the ions.

The exchange constants in Table 8.1 show that some exchangers are not strikingly selective. Silicate glasses are marked exceptions to this generalization. These glasses are described in detail in the following section on membrane electrodes. K' values of the order of 10^9 have been observed for some of these.

An elegant theory to explain relative bonding strengths of cations in various types of exchangers has been worked out by Eisenman.[2] This investigator derives the free energy change (and thus the exchange constant) for a given cation exchange from a consideration of the energy required to remove a cation from its exchange site to its hydrated state in solution, and the free energy change involved in simultaneously removing the second cation from solution and placing it on the exchange site. The energy calculations are based entirely on considerations of coulombic forces. From analysis of the charge distribution within an exchanger, Eisenman has developed the concept of an "equivalent anion." In other words, each unit negative charge on the exchanger is treated as if it were associated with a spherical atom. It is then possible, knowing the cation size and charge, to calculate the energy required to remove the cation to an infinite distance in a vacuum. Similarly, it is possible to calculate the energy change in taking the cation from a vacuum to a hydrated state in solution, basing the calculation, as before, on the cation size and charge and the size and charge distribution of the polar water molecules.

From the interplay between cation size and charge, "equivalent anion" size, and the hydration energies of the cations, eleven sequences of bonding energies of the monovalent cations Na^+, K^+, Li^+, Cs^+, and Rb^+ can be predicted. These sequences, as well as the order of magnitude of the differential bonding energies, have been shown to correspond with the behavior of real exchangers.

Figure 8.4 (adapted from Eisenman, *op. cit.*) shows relative bonding energies plotted against "equivalent anion" size (negative electrostatic field strength). The positions of a few real exchangers are noted along

---

[2] George Eisenman, On the elementary origin of equilibrium ionic specificity, in *Symposium on Membrane Transport and Metabolism*, A. Kleinzeller and A. Kotyk (eds.). New York, Academic, 1962.

the ordinate. The abscissa shows differential bonding energies as well as exchange constants for cations relative to K^+. For example, the sodium-aluminum glass has a high negative field strength, or a small equivalent anionic size. The strength of bonding of cations to this glass is in the order $H^+ > Na^+ > Li^+ > K^+ > Rb^+ > Cs^+$. The differential bonding energy for H^+ over K^+ is about 6.7 kilocalories, i.e.,

$$KX + H^+ = HX + K^+$$

$$\Delta F_r^\circ = -6.7 \text{ kcal.}$$

The exchange constant (equilibrium constant) for the reaction (at 25 °C) is

$$\frac{(HX)[K^+]}{(KX)[H^+]} = 10^{4.9}.$$

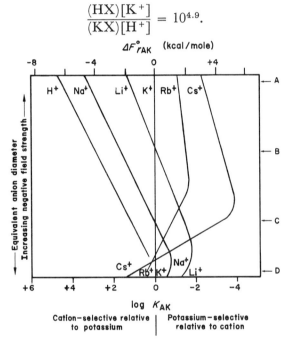

FIG. 8.4. Plot showing the exchange constants, K_{AK}, for various monovalent ions $A = H^+$, Na^+, etc., relative to K^+, as a function of the "equivalent anion diameter" of Eisenman. Noted on the ordinate are the real exchangers: (A) Na-Al glass (11.6% Na_2O, 17.7% Al_2O_3), (B) Na-sodalite, (C) Ag-permutite, (D) clay minerals. [Adapted from Eisenman, op. cit.]

Consequently, this particular glass is highly *selective* for H^+ over K^+, inasmuch as the glass would be equally occupied by H^+ and K^+ only when the ratio of K^+ to H^+ activity in solution is $10^{4.9}$. Also, this particular glass is an excellent sodium electrode. Although it prefers

ION EXCHANGE AND ION-SENSITIVE ELECTRODES 281

H^+ to Na^+, the ratio of Na^+ to H^+ in many solutions is so high that H^+ response is negligible. Sodium is preferred strongly over all other alkali cations.

Clay minerals have low field strength, a characteristic correlated with their low selectivity. Also, they occur in a region of numerous crossovers of selectivity lines, which explains the difficulties that have heretofore been encountered in listing a selectivity sequence for them.

CATION ELECTRODES

Introduction

Electrodes selectively sensitive to several cations in natural waters have recently been developed. These electrodes are typical materials that behave as cation exchange substrates in aqueous solution. When such exchangers are placed as membranes between solutions of different compositions, EMF measurements in the solution can be used to obtain activities of the ions in the solution. Consequently, the EMF values provide numbers directly comparable to those yielded by calculations from thermodynamic data. The difficulties of attempting to obtain activities of ions from knowledge of the total concentrations of various elements, as obtained in the usual chemical analysis, are bypassed. Furthermore, our knowledge of the extent of complexing of ions in natural waters will be rapidly enhanced as comparisons of activities and concentrations are made.

By sampling waters in contact with known mineral assemblages, activities obtained from electrode measurements can be compared with those calculated for equilibrium with the minerals, and the extent to which equilibrium is achieved can be assessed.

Cation Exchangers as Cation Electrodes

Cation-sensitive electrodes all operate on the same principle. An exchange substrate can be looked upon as a negatively charged, rigid, porous framework through which cations tend to migrate, but which forbid anion movement. Migration of cations in a membrane possibly should be looked upon as a displacive process, in which entry of a cation on one side of an exchanger causes emergence of a cation on the other side.

Figure 8.5 shows schematically a cation exchange membrane. Because of the higher activity of A^+ in solution A, there is a tendency for transfer of A^+ to solution B. This tendency can be considered quantitatively through the use of the free energy relation. For the process of transferring a cation A^+ in a solution at an activity of unity to a solution where the A^+ activity is 0.01, we write

$$A^+(a = 1) = A^+(a = 0.01).$$

By virtue of equation (1.13)

$$\Delta F_r = \Delta F_r^\circ + RT \ln Q.$$

For the process under consideration, ΔF_r per mole of cation transferred is

$$\Delta F_r = RT \ln \frac{0.01}{1}.$$

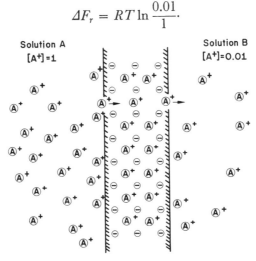

FIG. 8.5. Schematic diagram showing negatively charged rigid framework, with cations diffusing through membrane under activity gradient.

ΔF_r is negative, meaning that the reaction will proceed spontaneously to the right. ΔF_r° is zero for this reaction, since it is the free-energy change involved for the process

$$A^+(a = 1) = A^+(a = 1).$$

The free-energy change may be expressed as an electrical potential, and with an appropriate experimental arrangement this potential can be measured. Such an experimental system is illustrated in Figure 8.6, where a sodium glass performs the membrane function. When sodium ions leave solution A, there is a tendency for electrons to leave the solution through the inert electrode, in order to maintain electrical neutrality. Similarly, as sodium ions appear in solution B, electrons tend to move into that solution, to neutralize the excess positive charge. The movement of Na^+ from left to right in the cell through the membrane corresponds *formally* to a movement of negative sites in the membrane from right to left. Thus, the reaction in the right-hand portion of the cell corresponds to the dissociation

$$\text{NaX}_B = \text{Na}_B^+(a = 0.01) + \text{X}_B^- \quad \text{(right-hand portion of cell)} \quad (8.24)$$

i.e., the release of Na^+, at an activity of 0.01, with the negative sites X^- moving to the left. In the left-hand portion of the cell the excess negative sites combine with Na^+, at activity of 1, according to the reaction

$$\text{Na}_A^+(a = 1) + \text{X}_A^- = \text{NaX}_A \quad \text{(left-hand portion of cell)} \quad (8.25)$$

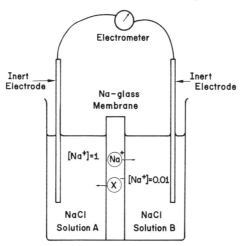

FIG. 8.6. Glass-membrane cell illustrating the development of a potential between two different solutions because of the tendency of cation migration.

These two chemical reactions are exactly analogous to half-cell reactions involving the release and gain of electrons, and can be treated by the methods given in Chapter 1 for such half-cell reactions. Thus, we write for the oxidation potential corresponding to (8.25), for 25 °C,[3]

$$\text{Eh}_A = E_A^\circ + \frac{0.0592}{n_{\text{X}^-}} \log \frac{[\text{Na}_A^+]}{[\text{NaX}_A]}, \quad (8.26)$$

and to (8.24)

$$\text{Eh}_B = E_B^\circ + \frac{0.0592}{n_{\text{X}^-}} \log \frac{[\text{Na}_B^+]}{[\text{NaX}_B]}. \quad (8.27)$$

[3] To write (8.26) and subsequent equations for any temperature T, replace the factor 0.0592 by $(2.303RT)/\mathscr{F}$.

Subtracting equation (8.27) from equation (8.26), we have

$$E_{obs} = Eh_A - Eh_B$$
$$= (E_A^\circ - E_B^\circ) + \frac{0.0592}{n_{X^-}} \left(\log \frac{[Na_A^+]}{[NaX_A]} - \log \frac{[Na_B^+]}{[NaX_B]} \right). \quad (8.28)$$

Simplifying, and taking into account that $n_{X^-} = 1$, for this process,

$$E_{obs} = \Delta E_{AB}^\circ + 0.0592 \log \frac{[NaX_B]}{[NaX_A]} + 0.0592 \log \frac{[Na_A^+]}{[Na_B^+]}. \quad (8.29)$$

If the two surfaces of the membrane behave in exactly the same way toward the solutions, we shall have $\Delta E_{AB}^\circ = 0$, and $[NaX_B]/[NaX_A] = 1$. In general, this will not be found to be true; it is actually found when solutions of the same Na^+ activity are put on both sides of the membrane that an EMF is developed. However, for a given membrane ΔE_{AB}° and $[NaX_B]/[NaX_A]$ are constant during the period of measurement, so that equation (8.29) becomes

$$E = C + 0.0592 \log \frac{[Na_A^+]}{[Na_B^+]}. \quad (8.30)$$

This included empirical potential, C, is known as the *asymmetry* potential.

By maintaining a constant activity of Na^+ on the right-hand side of the cell (solution B), and using it for measurement of unknown Na^+ activity in solution A, the observed EMF becomes

$$E = C' + 0.0592 \log [Na_A^+] \quad (8.31)$$

where

$$C' = C - 0.0592 \log [Na_B^+].$$

Thus, the EMF of the cell changes by 0.0592 volt for each tenfold change in $[Na^+]$ in solution A.

Membrane electrodes function accurately only as long as the framework remains "rigid" and permits movement only of cations. If, for example, the openings in the membrane are sufficiently large to permit migration of negative ions as a countercurrent to migration of cations, the membrane tends to act as a simple conductor, and the EMF between the solutions becomes small, even if the cationic activity difference is large. In fact, the residual values, when there is free communication of cations and anions between the solutions, depends only upon the relative ease of migration of positive and negative ions—a difference depending upon their transference numbers. Such residual potentials are often known as liquid junction potentials, and range from zero to about 100 millivolts.

At any rate, a 0.0592-volt change in EMF per tenfold change in activity of a cation in the test solution is a good indication that the membrane is performing as if it were permitting only cationic displacement.

Types of Membrane Electrodes

There are many substances that can be used as membrane electrodes. The obvious suitability of silicates, which consist of negatively charged repetitive structural frameworks of high chemical stability, with cations studded through the framework, explains their extensive utilization to date.

The silicate glasses are the most extensively used membrane electrodes. They have the advantage of relatively low electrical resistivity, because of the ease with which the cations can move through the loose glass structure. Even so, the resistance of these glasses is high enough so that electrometers with high input impedance had to be developed to make measurements in circuits involving these as membranes.

Clay minerals, with their loosely bound interlayer cations, have been used in making electrodes. The aluminum silicate sheet structures perform as the negatively charged membrane framework, and the interlayer spacing is of the right order of magnitude to permit cation migration while preventing an anionic countercurrent.

Synthetic cation-exchange materials have also been used as electrodes. As in the clay minerals, they consist of an inert negatively charged complex, to which cations can be added or subtracted without affecting the negative complex.

Well-crystallized substances, such as the three-dimensional silicates, which have strong cation bonding, have rarely been used as membranes because of their directional properties, which cause problems of membrane orientation. Also, electrodes of such materials have such high resistance, with their relatively immobile cations, that the problem of potential measurement is still essentially unsolved. Some progress has been made by grinding such substances and mounting them in an inert substrate. If the mounting is done properly, the openings between the surfaces of the ground material (such as feldspar) and the inert medium (such as paraffin) is of the right order of magnitude to permit displacive migration of cations along the surfaces of the feldspar particles, but not large enough to allow anion countercurrents. Construction of efficient electrodes of this type is still largely an art, and most attempts to synthesize them result in potential measurements that indicate that part of the membrane is behaving ideally, that is, by cation-displacive migration, whereas the rest of the electrode has openings sufficiently large to permit both cations and anions to move with little restriction.

An interesting recent development is in the use of thin films of soaps, which have been carefully oriented by deposition of a series of molecular monolayers, as cation sensitive membranes. They give great promise of high selectivity for certain cations, and will be discussed in detail later in this chapter.

Cation Electrode Equations

INTRODUCTION. From the similarity of the materials used as ion exchangers and as ion-sensitive electrodes, it is evident that the ion exchange and the electrode functions of these materials are closely related phenomena. In the preceding discussion of the behavior of a negatively charged membrane with cations relatively loosely bound to the negative sites, it has been assumed that the surfaces of the membrane are occupied by the kinds of cations found in the interior of the membrane. For most membranes it is found that the EMF values observed are functions of more than one cation if several cations are present in the test solution. In other words, *the electrode measures the change of activity of a given ion only if the electrode surface is essentially completely occupied by that ion.* Moreover, for a given substance, it is the ion that occupies the electrode surface that determines the EMF; the cations in the interior of the glass membrane merely transmit the charge by displacing each other.

Three situations are sketched in Figure 8.7 to show that the sodium glass electrode, with a reference solution of fixed Na^+ activity inside, can behave as an Na^+ electrode, or as a K^+ electrode, or as an electrode with a mixed Na^+-K^+ function. For most glasses the external surfaces exposed to the test solution rarely will be occupied by a single cation, and therefore will be responsive to more than a single cation.

Thus, many membrane electrodes have been developed that permit measurement of the activity of a given cation when the electrode is immersed in a solution that contains only that cation. But if an electrode is to be selectively sensitive to a given cation in a solution containing several cations, the desired cation must bond so strongly to the membrane that all others are excluded from its surface.

A few electrodes sufficiently selective for a given cation so that they can be used with assurance to measure that cation in the presence of equal or greater activities of other cations are available. There are many more substances that can be used as membranes that exhibit selection for two or three cations, while essentially ignoring all others. These can be utilized if the nature of the electrode response, when the membrane surface in the test solution is partially occupied by each of these selected cations, can be expressed in terms of the EMF.

EMF FOR A SINGLE CATION. As was previously shown, in developing equations relating electrode potentials to cation activities, it is convenient

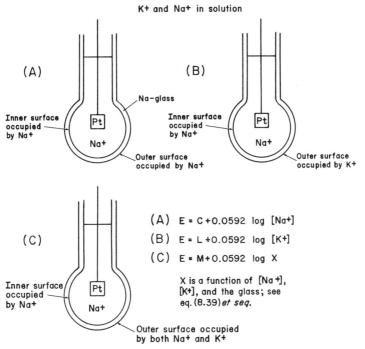

FIG. 8.7. Three situations are given to show how a Na-glass electrode, containing an internal reference solution of fixed Na$^+$ activity, can behave as a Na$^+$ electrode (A), a K$^+$ electrode (B), or an electrode having a mixed Na$^+$-K$^+$ function (C). The glass electrode is to be considered as dipping into a solution containing both Na$^+$ and K$^+$, and is connected in the measuring circuit by means of the wire leading from the internal Pt electrode.

to consider the membrane electrode in terms of the half-cells involved. A sodium glass bulb electrode, with a solution of fixed Na$^+$ activity inside the bulb, such as is illustrated in Figure 8.7, constitutes a half-cell of constant potential. Following equation (8.26), we write for the potential of the electrode

$$E_{\text{inner}} = \text{constant} = E^\circ_{\text{inner}} + 0.0592 \log \frac{[\text{Na}^+]_{\text{inner}}}{[\text{NaX}]_{\text{inner}}}. \tag{8.32}$$

If the electrode is placed in a solution containing Na$^+$ and the outer surface is occupied only by Na$^+$, we may write, corresponding to equation (8.27)

$$E_{\text{outer}} = E^\circ_{\text{outer}} + 0.0592 \log \frac{[\text{Na}^+]_{\text{outer}}}{[\text{NaX}]_{\text{outer}}}. \tag{8.33}$$

Subtracting equation (8.32) from equation (8.33)

$$E = E_{outer} - E_{inner}$$
$$= (E°_{outer} - \text{constant}) + 0.0592 \log \frac{[Na^+]_{outer}}{[NaX]_{outer}}. \quad (8.34)$$

If the outer surface is occupied only by Na^+, then the activity of NaX is fixed—$[NaX]_{outer}$ is a constant; call it W. Then

$$E = (E°_{outer} - \text{constant} - \log W) + 0.0592 \log [Na^+]_{outer} \quad (8.35)$$

and

$$E = C' + 0.0592 \log [Na^+]_{outer} \quad (8.36)$$

which is the same as equation (8.31), previously derived.

If the electrode is placed in a test solution containing only K^+, so that the outer surface of the electrode is occupied entirely by K^+, the potential of the inner cell remains constant; that of the outer cell is given by

$$E_{outer} = E°_{outer} + 0.0592 \log \frac{[K^+]}{[KX]}, \quad (8.37)$$

where [KX] is constant. Combining the two half-cells, as before, we have

$$E = L' + 0.0592 \log [K^+]. \quad (8.38)$$

Under these conditions, with the external electrode surface occupied entirely by potassium ions, the electrode functions as a K^+ electrode. The constant L' is different from that when the electrode is functioning as an Na^+ electrode, but the EMF changes 0.0592 volt for each tenfold change of K^+ activity in the external solution.

In summary, a membrane electrode with an inner solution of fixed composition responds to the cation in the outer or test solution that occupies the entire surface of the electrode. By calibration of the electrode in a solution of known cation activity, the 0.0592-volt change with tenfold change of cation activity can be used to determine cation activity in unknown solutions.

EMF FOR TWO MONOVALENT CATIONS. Equation (8.33) can be used in deriving a relation for an electrode whose surface is occupied by two cations. If both K^+ and Na^+ are present at the electrode surface, the activity of NaX or KX is no longer constant, and the term [NaX] in equation (8.33), for example, cannot be contained in the constant C' in equation (8.36). Instead [NaX] will range from a maximum, corresponding to complete occupation of the electrode surface by Na^+, to zero, corresponding to complete occupation of the surface by K^+.

ION EXCHANGE AND ION-SENSITIVE ELECTRODES

Starting with equation (8.34), we write for the case under consideration

$$E = L + 0.0592 \log \frac{[\text{Na}^+]}{[\text{NaX}]}. \qquad (8.39)$$

We have previously shown that if the substrate, which in this application is the electrode surface, obeys regular solution theory, then by equation (8.14)

$$\frac{[\text{Na}^+]}{[\text{K}^+]} \cong K_{\text{NaK}} \left(\frac{N_{\text{NaX}}}{N_{\text{KX}}}\right)^{1-(B/2RT)}.$$

Replacing the approximately equal sign by an equal sign for convenience in manipulation, and writing

$$n = 1 - \frac{B}{2RT},$$

we have

$$\frac{[\text{Na}^+]}{[\text{K}^+]} = K_{\text{NaK}} \left(\frac{N_{\text{NaX}}}{N_{\text{KX}}}\right)^n. \qquad (8.40)$$

Since, by the Law of Mass Action

$$\frac{[\text{Na}^+]}{[\text{K}^+]} = K_{\text{NaK}} \frac{[\text{NaX}]}{[\text{KX}]},$$

by virtue of equation (8.40)

$$\frac{[\text{NaX}]}{[\text{KX}]} = \left(\frac{N_{\text{NaX}}}{N_{\text{KX}}}\right)^n. \qquad (8.41)$$

Equation (8.41) may be rewritten as the pair of simultaneous equations

$$[\text{NaX}] = k N_{\text{NaX}}^n \qquad (8.42a)$$

$$[\text{KX}] = k N_{\text{KX}}^n. \qquad (8.42b)$$

If the standard state of NaX is defined in the usual way, so that $[\text{NaX}] = 1$ when $N_{\text{AX}} = 1$ ($[\text{KX}] = 1$ when $N_{\text{KX}} = 1$), then $k = 1$, and

$$[\text{NaX}] = N_{\text{NaX}}^n \qquad (8.43a)$$

$$[\text{KX}] = N_{\text{KX}}^n. \qquad (8.43b)$$

Substituting equation (8.43a) in equation (8.39)

$$E = L + 0.0592 \log \frac{[\text{Na}^+]}{N_{\text{NaX}}^n}$$

or

$$E = L + 0.0592 \log \left(\frac{[\text{Na}^+]^{1/n}}{\frac{(\text{NaX})}{(\text{NaX}) + (\text{KX})}}\right)^n. \qquad (8.44)$$

Rewriting (8.44)

$$E = L + 0.0592 \log \left(\frac{[\text{Na}^+]^{1/n}((\text{NaX}) + (\text{KX}))}{(\text{NaX})} \right)^n \quad (8.45)$$

or

$$E = L + 0.0592 \log \left([\text{Na}^+]^{1/n} + \frac{[\text{Na}^+]^{1/n}(\text{KX})}{(\text{NaX})} \right)^n. \quad (8.46)$$

Rewriting equation (8.40) as

$$\frac{[\text{Na}^+]^{1/n}}{[\text{K}^+]^{1/n}} = K_{\text{NaK}}^{1/n} \left(\frac{(\text{NaX})}{(\text{KX})} \right) \quad (8.47)$$

and substituting (8.47) into (8.46)

$$E = L + 0.0592 \log \left([\text{Na}^+]^{1/n} + [\text{Na}^+]^{1/n} K_{\text{NaK}}^{1/n} \frac{[\text{K}^+]^{1/n}}{[\text{Na}^+]^{1/n}} \right)^n; \quad (8.48)$$

after cancellation of the common term

$$E = L_{\text{NaK}} + 0.0592 \log \left([\text{Na}^+]^{1/n} + K_{\text{NaK}}^{1/n} [\text{K}^+]^{1/n} \right)^n. \quad (8.49)$$

(where L is rewritten as L_{NaK} to give explicit recognition to the particular electrode process being considered).

The constant L_{NaK} is commonly written as $E°$ in the literature, but as shown here is actually an empirical constant, specific for the particular electrode being considered, and is obtained by calibration of that electrode. Equation (8.49) has been found to furnish an accurate description of the behavior of glass electrodes.[4]

In the special case of $n = 1$, there is ideal solution behavior between NaX and KX, and

$$E = L_{\text{NaK}} + 0.0592 \log ([\text{Na}^+] + K_{\text{NaK}}[\text{K}^+]); \quad (8.50)$$

K_{NaK}, the exchange constant for the reaction, is called the *selectivity factor* of the electrode. If K_{NaK} has the numerical value of 10, for example, and the activities of NaX and KX are equal, the ratio of $[\text{Na}^+]$ to $[\text{K}^+]$ in solution is 10 [equation (8.1)]. If for a given electrode sensitive to two monovalent cations the exchange constant K_{AB} has some high value, such as 10^6 for example, equation (8.50) becomes

$$E = L_{AB} + 0.0592 \log ([A^+] + 10^6[B^+]). \quad (8.51)$$

From this result it is seen that most of the measured potential will result from the value of $[B^+]$. Such an electrode would be a B^+-sensitive electrode.

---

[4] George Eisenman, Donald O. Rudin, and James U. Casby, *Science*, 126, 831 (1957).

ION EXCHANGE AND ION-SENSITIVE ELECTRODES

Thus, exchange constants can be used to describe electrode behavior, and potential measurements can be used to determine exchange constants.

EMF FOR MORE THAN TWO CATIONS. Equations can be developed to relate the response of an electrode to more than two cations, but these relations are untested except for the special case in which the surface of the electrode behaves as an ideal solution for all of the cations involved ($n = 1$ for all exchanges).

The appropriate equation for this *ideal* case, involving the three ions A^+, B^+, and C^+, can be derived by starting with equation (8.39), remembering that $n = 1$, and therefore $[AX] = N_{AX}$

$$E = L + 0.0592 \log \left(\frac{[A^+]}{\frac{(AX)}{(AX) + (BX) + (CX)}} \right) \qquad (8.52)$$

or, by rearrangement

$$E = L + 0.0592 \log \left([A^+] + [A^+]\frac{(BX)}{(AX)} + [A^+]\frac{(CX)}{(AX)} \right). \qquad (8.53)$$

Since, for the ideal case

$$\frac{[A^+]}{[B^+]} = K_{AB}\frac{[AX]}{[BX]} = K_{AB}\frac{(AX)}{(BX)}$$

$$\frac{[A^+]}{[C^+]} = K_{AC}\frac{[AX]}{[CX]} = K_{AC}\frac{(AX)}{(CX)},$$

equation (8.53) can be rewritten as

$$E = L_{ABC} + 0.0592 \log \left([A^+] + K_{AB}[B^+] + K_{AC}[C^+] \right). \qquad (8.54)$$

Thus, the potential given by equation (8.54) is that obtained for a system of these cations behaving ideally. It is seen that equation (8.54) has the same form as equation (8.50) for two cations. In fact, the equation may be extended to cover any number of cations

$$E = L_{A \ldots X} + 0.0592 \log \left([A^+] + K_{AB}[B^+] + K_{AC}[C^+] + \cdots + K_{AX}[X^+] \right). \qquad (8.55)$$

Equation (8.55) tells us that the EMF of a glass electrode that behaves ideally toward the ions in solution is a logarithmic function of the sum of the activities of the ions, with the effect of each cation modified by the exchange constant. Physically, the exchange constants are related to the differential bonding energies of the ions.

From Table 8.1 it is seen that the number of exchangers for which n is nearly unity is few; consequently, equation (8.55) has limited application.

The difficulty in deriving a relation for the EMF for three cations lies in the fact that it cannot be safely assumed that the exchange constant K'_{AB} will be independent of $[C^+]$, or K'_{AC} independent of $[B^+]$, as was assumed in deriving equation (8.55). Nor is information available concerning changes in n in equation (8.1), for a given exchange, in the presence of a third ion.

EMF FOR A SYSTEM CONTAINING A MONOVALENT AND A DIVALENT CATION OR TWO DIVALENT CATIONS. For a membrane electrode that is sensitive to a divalent cation as well as a monovalent cation, we can write the exchange reaction by equation (8.18)

$$A_2X_2 + B^{++} = BX_2 + 2A^+,$$

and the expression for the equilibrium constant by equation (8.19)

$$\frac{[A^+]^2}{[B^{++}]} = K_{AB}\frac{[A_2X_2]}{[BX_2]}.$$

The appropriate electrode equation for this exchange is derived from equation (8.39), and has the form

$$E = L + \frac{0.0592}{2} \log \frac{[B^{++}]}{[BX_2]}. \tag{8.56}$$

For this reaction two electrons are involved; hence the factor 2 giving the multiplier $0.0592/2$. Assuming regular solution for the electrode surface, by equation (8.43)

$$[BX_2] = N_{BX_2}^n = \left(\frac{(BX_2)}{(BX_2) + (A_2X_2)}\right)^n. \tag{8.57}$$

Substituting for $[BX_2]$ in (8.56), we have

$$E = L + \frac{0.0592}{2} \log \left(\frac{[B^{++}]}{\left(\frac{(BX_2)}{(BX_2) + (A_2X_2)}\right)^n}\right), \tag{8.58}$$

which on simplification yields

$$E = L + \frac{0.0592}{2} \log \left([B^{++}]^{1/n} + [B^{++}]^{1/n}\frac{(A_2X_2)}{(BX_2)}\right)^n. \tag{8.59}$$

For regular solution, by equation (8.41)

$$\frac{[A_2X_2]}{[BX_2]} = \left(\frac{N_{A_2X_2}}{N_{BX_2}}\right)^n = \left(\frac{(A_2X_2)}{(BX_2)}\right)^n. \tag{8.60}$$

Substituting this result in equation (8.19), we have

$$\frac{[A^+]^2}{[B^{++}]} = K_{AB}\left(\frac{(A_2X_2)}{(BX_2)}\right)^n, \tag{8.61}$$

or

$$\frac{1}{K_{AB}^{1/n}} \times \frac{[A^+]^{2/n}}{[B^{++}]^{1/n}} = \frac{(A_2X_2)}{(BX_2)}. \tag{8.62}$$

Replacing $(A_2X_2)/(BX_2)$ in (8.59) by its equivalent in (8.62)

$$E = L + \frac{0.0592}{2} \log \left([B^{++}]^{1/n} + \frac{1}{K_{AB}^{1/n}} [A^+]^{2/n}\right)^n. \tag{8.63}$$

Combining the constant $-0.0296 \log K$ with L, we have finally

$$E = L_{AB} + 0.0296 \log ([A^+]^{2/n} + K_{AB}^{1/n}[B^{++}]^{1/n})^n. \tag{8.64}$$

This equation has been checked experimentally with glass membranes as electrodes and has been found to be accurate.[5]

For an electrode sensitive to two divalent cations, the electrode equation is analogous to that given by (8.64), and is

$$E = L_{AB} + 0.0296 \log ([A^{++}]^{1/n} + K_{AB}^{1/n}[B^{++}]^{1/n})^n. \tag{8.65}$$

GLASS ELECTRODES

The pH Electrode

The glass pH electrode, which has had such wide use over the last 20 years, is now seen as a specific type of membrane electrode. The usual pH electrode is made of sodium-calcium-silicate glass, and has a relatively low electrical resistance. The pH electrode contains inside a bulb of the glass, a solution of fixed H^+ activity with an electrode (usually Ag-AgCl) dipping into this solution; this inner electrode conducts electrons reversibly into and out of the contained solution. The external, or test solution, is connected to the circuit with a reference electrode, such as the saturated calomel electrode. Details have previously been given in Chapter 5.

One might expect that such an electrode would be sensitive to sodium ions in the external solution, but experiment shows that for pH values up to about 9, the EMF depends only upon the activity of hydrogen ion. The explanation of this H^+ sensitivity is that the Na^+ is so loosely bound to the glass that in most aqueous solutions the sodium ions in the external surface of the bulb are replaced by hydrogen ions. Therefore, the

[5] R. M. Garrels, M. Sato, M. E. Thompson, and A. H. Truesdell, *Science*, *135*, 1045 (1962).

tendency for displacive transfer of cations through the glass depends not upon the Na⁺ activity of the test solution, but upon the H⁺ activity.

Such a model accounts for the necessity for a "sodium ion correction" to pH values measured in alkaline solutions with high activities of alkali metals. When H⁺ activity falls low enough, and alkali metal ion activity is high enough, part of the glass surface becomes occupied by alkali metal ions, and the EMF of the electrode is a function of both H⁺ and alkali metal cations. This effect is sufficiently regular to permit empirical correction to pH measurements in alkaline media.

It has been found that by varying glass composition, the relative strength of H⁺ bonding over that of other cations can be so enhanced that the pH electrodes of the proper glass can be used without correction even in strong solutions of alkali metal hydroxides.

When the glass surface is occupied by hydrogen ions, the electrode reaction is described by an equation of the same type as equation (8.36)

$$E = C_H + 0.0592 \log [H^+], \quad (8.66)$$

at 25 °C. The more general equation appropriate to any temperature is

$$E = C_H + \frac{2.303 RT}{\mathscr{F}} \log [H^+]. \quad (8.66a)$$

Since pH = $-\log [H^+]$

$$E = C_H - \frac{2.303 RT}{\mathscr{F}} pH \quad (8.67)$$

or

$$E = C_H - 0.0592 \, pH, \quad (8.67a)$$

at 25 °C.

The value of C_H differs from electrode to electrode, and depends upon the specific composition of glass used, the standard solution inside the bulb, and the particular reference electrode used in the circuit. By calibration against test solutions of known pH, the electrode can be used to measure hydrogen ion activity in unknown solutions, as discussed in Chapter 5.

Alkali Metal Ion-Sensitive Glass Electrodes

The sensitivity of the glass pH electrode to alkali metal cations at high pH leads to the suggestion that a glass of appropriate composition might be sufficiently sensitive even at low pH values to permit general use as an alkali metal ion electrode. This possibility has been investigated by a number of researchers; the most systematic study of the influence of glass composition on cation sensitivity is perhaps that of Eisenman and his coworkers.[6] They investigated glasses in the system Na_2O-Al_2O_3-SiO_2 in terms of their sensitivity to both monovalent and divalent cations. Figure 8.8 shows some of the results of these investigators, in the form of

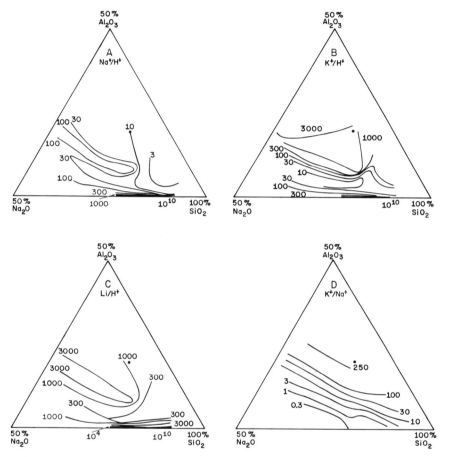

FIG. 8.8. Isosensitivity contours of electrode response for glasses in the system Na_2O-Al_2O_3-SiO_2; plots are on mole percentage basis. Contours are of the ratio of the activity of cation activity of H^+ in solution (K^+/Na^+ in D) at which the electrode surface is equally occupied by cation and H^+ (or K^+ and Na^+ in D), i.e., the selectivity constant. The solid dot indicates glass of composition 11% Na_2O, 18% Al_2O_3, 71% SiO_2. [Adapted from George Eisenman, D. O. Rudin, and J. U. Casby, op. cit.]

isosensitivity contour diagrams for various pairs of ions plotted on the glass composition diagram.

The effect of small amounts of Al_2O_3 in the glass on the sensitivity ratios of various pairs of cations is remarkable. The exchange constant

[6] George Eisenman, D. O. Rudin, and J. U. Casby, op. cit. See additional references at end of this chapter.

K_{NaH}, for example, changes from 10^{10} to 10 on addition of only a few percent Al_2O_3. This means that the electrode response to Na^+ changes by a factor of 10^9 in favor of increased response to Na^+.

CALIBRATION AND USE OF CATION-SENSITIVE GLASS ELECTRODES

Calibration of cation-sensitive glass electrodes is accomplished in much the same way as that for the pH electrode. However, because few of the electrodes are entirely selective for a given ion, except under carefully controlled conditions, there are a number of extra calibration steps that must be performed, and numerous precautions that have to be taken, to assure that the observed EMF values can be converted properly into ionic activities.

Determination of Electrode Equation for Two Cations

Figure 8.8 shows that a glass containing 11 percent Na_2O and 18 percent Al_2O_3 has a high selectivity factor for Na^+ relative to all other monovalent cations except H^+. The first step in using such an electrode is to establish the selectivity factor for H^+. This is ordinarily done by placing the electrode in a solution of a sodium salt of known concentration at 25 °C, and then titrating with a standard acid. The original concentration of the sodium salt should be so low (about 0.01 m) that the individual ion activity coefficient can be obtained with confidence from the Debye-Hückel equation

$$-\log \gamma_{\text{Na}^+} = \frac{Az_{\text{Na}^+}^2 \sqrt{I}}{1 + \mathring{a}_{\text{Na}^+} B\sqrt{I}}. \qquad (2.76)$$

In practice it is convenient to use a 0.0112 m NaCl solution, which has a γ_{Na^+} of 0.895; consequently, $[\text{Na}^+] = 0.0112 \times 0.895 = 0.01$. The acid used for the titration is usually HCl solution; a strength is chosen so that the volume of acid required for titration does not affect the volume of the solution appreciably. The EMF of the Na electrode-calomel electrode pair is measured, a simultaneous measurement is made of pH, and a plot constructed of the relations of these two variables. The results of a typical titration are shown in Figure 8.9. The titration curve has three distinct regions. In the first region (A in Figure 8.9) additions of HCl solution have no measurable effect on the EMF of the electrode. In theory, this is the range of $[\text{Na}^+]/[\text{H}^+]$ through which the electrode surface is almost entirely occupied by Na^+, and does not "see" the hydrogen ions in the solution. In the second region (B in Figure 8.9)

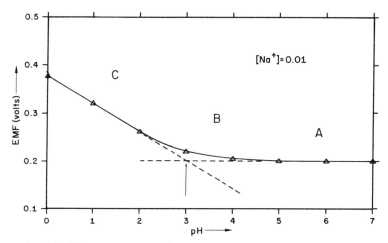

FIG. 8.9. Titration curve used to establish equation for glass electrode. Solid line is best fit to measured EMF vs. pH values obtained upon adding HCl solution with [Na⁺] = 0.01. Triangles indicate values calculated for E = 0.318 + 0.0592 log([Na⁺] + 10[H⁺]).

the EMF changes as HCl is added, but by a different amount for each change in pH unit. In this range the electrode surface presumably is occupied by both Na^+ and H^+, and is responding to both cations in the test solution. In the third region (C in Figure 8.9) the titration curve is linear and has a slope of 0.0592 volt per pH unit. In this range the electrode is behaving like an ordinary pH electrode; its surface is presumably entirely occupied by H^+. At pH 3, the point of intersection of the lines represented by region A and region C, the electrode is occupied equally by Na^+ and H^+. Therefore, because

$$\frac{[Na^+]}{[H^+]} = K_{NaH}\left(\frac{(NaX)}{(HX)}\right)^n$$

the term on the right becomes unity when $(NaX) = (HX)$, and the ratio of $[Na^+]/[H^+]$ in the solution is equal to K_{NaH}, the exchange constant or selectivity factor. For the titration shown, equal occupation is at $[Na^+] = 0.01$ and $[H^+] = 0.001$ (pH = 3); therefore the selectivity factor is 10.

The electrode equation can be determined from the numerical data. Since

$$E = L_{NaH} + 0.0592 \log ([Na^+]^{1/n} + K_{NaH}^{1/n}[H^+]^{1/n})^n, \quad (8.68)$$

the initial EMF (region A in Figure 8.9) in the practical absence of H^+ is

$$E = L_{NaH} + 0.0592 \log [Na^+]. \quad (8.69)$$

Substituting 0.200 for E, and 0.01 for [Na$^+$]

$$0.200 = L_{\text{NaH}} + 0.0592 \log 0.01$$

$$L_{\text{NaH}} = 0.318 \text{ volt}.$$

Using this value for L_{NaH}, and $K_{\text{NaH}} = 10$ in equation (8.68) we have

$$E = 0.318 + 0.0592 \log ([\text{Na}^+]^{1/n} + 10^{1/n}[\text{H}^+]^{1/n})^n. \quad (8.70)$$

Finally, values of E can be calculated for various pH values, assuming that n is unity. Such calculated points are shown as triangles on Figure 8.9. Agreement with the observed values shows that the electrode behaves as an ideal exchange substrate for which $n = 1$. Thus, the behavior of the electrode can be described by a final equation

$$E = 0.318 + 0.0592 \log ([\text{Na}^+] + 10[\text{H}^+]). \quad (8.71)$$

Fortunately, the cation-sensitive glass electrodes perform similarly for most cation pairs, and n is close to unity for these electrodes. The simplest method of determining n from a calibration titration is to compare the observed relations to a series of curves calculated for various n values.[7]

Determination of Activities of Two Cations by Simultaneous Use of Two Electrodes

The equation for the sodium electrode [(8.71)] shows that the EMF is a function of both Na$^+$ and H$^+$ over a considerable range of solution composition. In a solution that has [Na$^+$] = 0.01, the electrode functions as a pH electrode at pHs below 2, and as a Na$^+$ electrode above pH 4 (Figure 8.9). Between pH 2 and pH 4 it has a mixed function. However, the usual pH electrode can be used to measure pH in this range without Na$^+$ interference; and the value for [H$^+$] thus obtained can be substituted in equation (8.71) to obtain [Na$^+$]. Similarly, for solutions containing Na$^+$ and K$^+$, at pH values too high for H$^+$ interference, an electrode is available that measures Na$^+$ without "seeing" K$^+$ ($K_{\text{NaK}} \cong 0.004$), and can be used in conjunction with an electrode that is somewhat selective for K$^+$ (e.g., $K_{\text{NaK}} \cong 4$) to obtain [K$^+$].

If electrodes must be used in solutions of two cations in which both electrodes have a mixed function, the activities of the cations still can be obtained by solving the electrode equations simultaneously, provided, of course, that there is a sufficient difference in their selectivities.

As a specific example of the determination of the activities of Na$^+$ and K$^+$ in a solution of KCl and NaCl, let us assume that two electrodes of different glass compositions are available. Neither shows H$^+$ sensitivity

---

[7] We thank Dr. Eisenman for telling us about this method.

above pH 5; the test solution has a pH of 7. Electrode A is calibrated by placing it in an NaCl solution of known activity of Na^+ and then titrating with KCl to get its selectivity factor; electrode B is similarly treated. The resultant equations are ($n = 1$)

$$E_A = 0.300 + 0.0592 \log ([Na^+] + 5.0[K^+]) \quad (8.72a)$$

$$E_B = 0.250 + 0.0592 \log ([Na^+] + 0.1[K^+]). \quad (8.72b)$$

Then a measurement is made of the unknown solution. E_A is 0.283 volt; E_B is 0.150 volt. Substituting and transforming

$$[Na^+] + 5.0[K^+] = 10^{(0.282-0.300)/(0.0592)} \quad (8.73a)$$

$$[Na^+] + 0.1[K^+] = 10^{(0.150-0.250)/(0.0592)}. \quad (8.73b)$$

Eliminating $[Na^+]$ and solving for $[K^+]$

$$4.9[K^+] = 0.50 - 0.02 = 0.48$$

$$[K^+] = 0.1.$$

Substituting in equation (8.73b)

$$[Na^+] = 0.01.$$

ELECTRODES SENSITIVE TO ALKALINE EARTH METAL CATIONS. Recently, glass compositions have been found suitable for making glass membrane electrodes that are sensitive to divalent cations.[8] These glasses are quaternary systems of composition varying from 50 to 90 mole percent SiO_2, with the remaining 50 to 10 mole percent consisting of the oxides of a monovalent, a divalent, and a trivalent cation, e.g., K_2O, BaO, and Al_2O_3.

It has been found that such electrodes obey equation (8.65)

$$E = L_{AB} + 0.0296 \log ([A^{++}]^{1/n} + K_{AB}^{1/n}[B^{++}]^{1/n})^n,$$

quite accurately for pairs of divalent cations, and that for these pairs $n \cong 1$ and K_{AB} varies from approximately 1.5 to 1. The electrodes are selective toward the alkaline-earth cations in the sequence $Ba^{++} > Sr^{++} > Ca^{++} > Mg^{++}$. This means that $K_{AB} > 1$ for any pair of ions in the sequence, where A stands to the right of B in the sequence. For example, for a typical electrode $K_{MgCa} = 1.5$, so that

$$E = L_{MgCa} + 0.0296 \log ([Mg^{++}] + 1.5[Ca^{++}]).$$

These divalent cation sensitive electrodes will also respond to monovalent cations contained in the solution. However, the EMF observed does not always follow that predicted by equation (8.64)

$$E = L_{AB} + 0.0296 \log ([A^+]^{2/n} + K_{AB}^{1/n}[B^{++}]^{1/n})^n,$$

[8] R. M. Garrels et al., op. cit.

but in some cases obeys an empirical equation

$$E = L_{AB} + 0.0296 \log{([A^+]^{2m/n} + K_{AB}^{1/n}[B^{++}]^{1/n})^n}. \quad (8.74)$$

It is found that m differs from unity for thick electrodes. However, if the glass membrane is made thin enough, and the electrode is properly conditioned by soaking in the electrolyte solution of interest, m will become equal to unity.

Another approach to the problem of making electrodes suitable for measuring low concentrations of divalent metal ions in the presence of moderate to high concentrations of monovalent ions has been made by Gregor.[9] He noted that the stearates bond calcium, strontium, and barium strongly, whereas they hold alkali metal ions weakly. However, the stearates are hydrated compounds at room temperature, and it is not possible to make a stearate glass, so as to provide a continuous, rigid structural framework. Gregor solved the problem of a continuous framework by plating a succession of monolayers of various divalent- and trivalent-cation stearates on a glass support. These electrodes function well when used for measuring solutions containing only the ion used in the stearate comprising the membrane. For example, a calcium stearate shows the response predicted by the electrode equation [see equation (8.31) with $n_{X^-} = 2$] where used in calcium chloride solutions, even in those of high $CaCl_2$ content.

Unfortunately, the specific volume of stearates increases as the bonding energy of the cation decreases; Gregor found, for example, that calcium stearate electrodes, when immersed in strong Na^+ solutions, tend to swell and thus to be destroyed. The sequences of bonding energies and of specific volumes of the stearates studied are

bonding energy: $Fe^{3+} > Ba^{++} > Sr^{++} > Ca^{++} > Mg^{++} > Na^+$;

specific volume: $Fe^{3+} < Ba^{++} < Sr^{++} < Ca^{++} < Mg^{++} < Na^+$;

Thus, a ferric stearate electrode is selective for ferric ion in the presence of any of the other ions in the sequence given, but the presence of any of the other ions tends to cause the ferric stearate to swell and thus to destroy the rigid framework. Similarly, a calcium stearate electrode is selective for Ca^{++} over Mg^{++} and Na^+, but will respond to Sr^{++}, Ba^{++}, and Fe^{3+}. Also, a calcium stearate electrode, in the presence of Sr^{++}, Ba^{++}, or Fe^{3+}, tends to shrink. Shrinkage ruins the electrode just as effectively as does swelling. Gregor has made calcium stearate electrodes that function selectively as calcium electrodes in the presence of fairly high ratios of Na^+/Ca^{++} by using a vise to prevent swelling, but the

[9] H. P. Gregor and Harold Schonhorn, Multilayer membrane electrodes, *J. Am. Chem. Soc.*, *709*, 1507 (1957).

ION EXCHANGE AND ION-SENSITIVE ELECTRODES

mechanical difficulties have so far prevented widespread production and use of the stearate electrodes.

SUMMARY

Cation exchangers obey the relations

$$\frac{[A^+]}{[B^+]} = K_{AB}\left(\frac{(AX)}{(BX)}\right)^n \tag{8.1}$$

$$\frac{[A^+]^2}{[B^{++}]} = K_{AB}\left(\frac{(A_2X_2)}{(BX_2)}\right)^n \tag{8.17}$$

where $[A^+]$ and $[B^+]$, or $[B^{++}]$ are activities of ions in solution, and (AX), (BX), (BX_2), or (A_2X_2) are concentrations of ions in the exchanger; K_{AB} is the exchange or selectivity constant, and $n = 1 - B/2RT$ is an exponent usually close to unity but ranging in value from 0.25 to 5.0. These equations are derivable on the basis of the Law of Mass Action and the assumption that the exchange substrate forms a regular solution.

Monovalent cation-sensitive glass electrodes obey the empirical equation (at 25 °C) for the voltage E

$$E = L_{AB} + 0.0592 \log\left([A^+]^{1/n} + K_{AB}^{1/n}[B^+]^{1/n}\right)^n. \tag{8.49}$$

For monovalent-divalent systems the equation is

$$E = L_{AB} + 0.0296 \log\left([A^+]^{2/n} + K_{AB}^{1/n}[B^{++}]^{1/n}\right)^n. \tag{8.64}$$

For divalent-divalent systems the equation is

$$E = L_{AB} + 0.0296 \log\left([A^{++}]^{1/n} + K_{AB}^{1/n}[B^{++}]^{1/n}\right)^n. \tag{8.65}$$

L_{AB} is an empirical constant, different for each electrode. These EMF equations are derivable on the bases of the usual half-cell equation and the empirical exchange equations given above.

Glass electrodes can be used to measure the activities of individual monovalent cations or divalent cations in solutions containing several cations, by using glass electrodes of different compositions and hence different selectivities toward the several cations involved.

Electrodes fabricated from stearates are available for measurement of alkaline-earth cations in homoionic solutions; mechanical difficulties have so far prevented their use in polyionic solutions.

The phenomena of ion exchange and of membrane electrodes are shown to be aspects of the behavior predicted for rigid, negatively charged frameworks that permit gain, loss, or migration of cations without disturbance of the framework.

SELECTED REFERENCES

ION EXCHANGE

Kelly, Walter P., *Cation Exchange in Soils*. (ACA Monograph 109). New York, Reinhold, 1948.

Marshall, C. E., *The Colloid Chemistry of the Silicate Minerals*. New York, Academic, 1949.

Salmon, J. E., and D. K. Hale, *Ion Exchange, A Laboratory Manual*. New York, Academic, 1959.

    See especially Chapter 3. Properties of ion exchange resins.

Schachtschabel, P., Untersuchungen über die sorption der tonmineralien und Bodenkolloide, und die Bestimmung des Anteils dieser Kolloide und der sorption im boden: *Kolloid Beih.*, *51*, 199 (1940).

Walton, H. F., Ion exchange equilibria, *in Ion Exchange Theory and Practice*, F. C. Nachod (ed.). New York, Academic, 1949.

Wiklander, Lambert, Cation and anion exchange phenomena, *in Chemistry of the Soil*, Firman E. Bear (ed.). (ACA Monograph 126), New York, Reinhold, 1955.

GLASS ELECTRODES

Bower, C. A., Determination of sodium in saline solutions with a glass electrode: *Proc. Soil Sci. Am.*, *23*, 29 (1959).

Eisenman, George, Cation selective glass electrodes and their mode of operation: *Biophysical J.*, *2*, pt. 2 (supplement), 259 (1962).

Eisenman, George, On the elementary origin of equilibrium ionic specificity, *in Symposium on Membrane Transport and Metabolism*, A. Kleinzeller and A. Kotyk (eds.). New York, Academic, 1962.

Eisenman, George, D. O. Rudin, and J. U. Casby, Glass electrode for measuring sodium ion: *Science*, *126*, 831 (1957).

Fedotov, N. A., The electrode properties of lithium glass: *Zhur. Fiz. Khim.*, *32*, 1951 (1958).

Friedman, S. M., J. D. Jamieson, M. Nakashima, and C. L. Friedman, Sodium- and potassium-sensitive glass electrodes for biological use: *Science*, *130*, 1252 (1959).

Garrels, R. M., M. Sato, M. E. Thompson, and A. H. Truesdell, Glass electrodes sensitive to divalent cations: *Science*, *135*, 1045 (1962).

Goremykin, V. E., and P. A. Kryukov, Potentiometric method of determining sodium ion by means of a glass electrode with a sodium function: *Izvest. Akad. Nauk S.S.S.R., Otdel. Khim Nauk No. 11*, 1387 (1957).

Isard, J. O., Alkali ion determination by means of glass electrodes: *Nature*, *184*, 1616 (1959).

Kryukov, P. A., M. M. Shul'ts, and V. E. Goremykin, Use of glass electrodes having a sodium function for water analysis: *Gidrokhim. Materialy*, *24*, 23 (1955).

Nikolskii, B. P., M. M. Shul'ts, and N. V. Peshekonova, Theory of glass electrodes VII. Effect of foreign ions on the sodium and potassium functions of the glass electrodes: *Zhur. Fiz. Khim*, *32*, 19 (1958).

    The preceding papers in this series are of much interest.

Nikolskii, B. P., M. M. Shul'ts, and N. V. Peshekonova, Theory of glass electrodes VIII. The transition of the glass electrode from one metallic function to another: *Zhur. Fiz. Khim.*, *32*, 262 (1958).

Shul'ts, M. M., Investigation of the sodium function of the glass electrode: *Uchenie Zapiski Leningrad. Gosudarst. Univ. No. 169, Ser. Khim. Nauk No. 13*, 80 (1954).

MEMBRANE ELECTRODES

Belinskaya, F. A., and E. A. Materova, Electrode properties of ion-exchange membranes: *Vestnik Leningrad. Univ.*, *12*, No. 16, Ser. Fiz. Khim. No. 3, 85 (1957).

Fischer, R. B., and R. F. Babcock, Electrodes consisting of membranes of precipitates: *Anal. Chem.*, *30*, 1732 (1958).

Gregor, H. P., and H. Schonhorn, Multilayer membrane electrodes: *J. Am. Chem. Soc.*, *79*, 1507 (1957).

Gregor, H. P., and H. Schonhorn, Multilayer membrane electrodes II. Preparation and use in double concentration cells: *J. Am. Chem. Soc.*, *81*, 3911 (1959).

Marshall, C. E., The electrochemical properties of mineral membranes: *J. Am. Chem. Soc.*, *63*, 1911 (1941).

Sollner, K., Recent advances in the electrochemistry of membranes of high ionic selectivity, *J. Electrochem. Soc.*, *97*, 139C (1950).

PROBLEMS

A. CATION EXCHANGE

8.1. If $\dfrac{[A^+]}{[B^+]} = 10\left(\dfrac{N_{AX}}{N_{BX}}\right)^{1/3}$, draw a graph plotting $\log \dfrac{[A^+]}{[B^+]}$ as ordinate against $\log \dfrac{N_{AX}}{N_{BX}}$. (a) What is the slope of the line obtained? (b) What is the intercept of the line on the ordinate?

Ans. (a) Slope = 1/3; (b) intercept = log 10.

8.2. In the preceding problem, what is the ratio of $[A^+]$ to $[B^+]$ when $N_{AX} = N_{BX}$? *Ans.* 10.

8.3. If an exchanger behaves as an ideal binary solution, what is the activity of BX when $N_{AX} = 0.3$? *Ans.* $a_{BX} = 0.7$.

8.4. If an exchanger behaves as a regular solution for which B is 0.70 kcal/mole at 25 °C, what is the activity of AX when $N_{AX} = 0.30$?

Ans. $a_{AX} = 0.54$.

8.5. If the exchange equilibrium in problem 8.4 is expressed in the form $\dfrac{[A^+]}{[B^+]} = K_{AB}\left(\dfrac{N_{AX}}{N_{BX}}\right)^n$, what is the approximate value of n at 25 °C?

Ans. $n = 0.41$.

8.6. A certain montmorillonite has $K_{NaK} = 2.5$ at 25 °C and behaves as an ideal solution in its exchange behavior. The montmorillonite has equilibrated with a ground water that contains only Na^+ and K^+ as cations in significant concentrations. The ionic strength of the water is approximately 0.01, and it contains 230 ppm Na^+ and 39 ppm K^+. What is the ratio of Na^+ to K^+ in the exchange positions of the montmorillonite at 25 °C?

Ans. $(NaX)/(KX) = 4$.

8.7. A sample of montmorillonite is reacted with solutions containing Ca^{++} and Na^+. Its behavior is described by $\dfrac{[Ca^{++}]}{[Na^+]^2} = 0.01 \left(\dfrac{N_{CaX}}{N_{Na_2X}}\right)^{1.5}$, at 25 °C. At equilibrium, what is $[Ca^{++}]/[Na^+]$ in solution, when $N_{CaX} = N_{Na_2X}$?

Ans. Indeterminate. Either $[Ca^{++}]$ or $[Na^+]$ must be specified.

8.8. For the montmorillonite of problem 8.7, what is $\dfrac{N_{CaX}}{N_{Na_2X}}$

a. When $[Ca^{++}] = 0.01$ and $[Na^+] = 0.01$?
b. When $[Ca^{++}] = 0.01$ and $[Na^+] = 0.10$?
c. When $[Ca^{++}] = 0.01$ and $[Na^+] = 10$?

Ans. (a) $10^{2.67}$; (b) $10^{1.33}$; (c) $10^{-1.33}$.

8.9. For the same montmorillonite, which ions would you expect to occupy most of the exchange positions if the montmorillonite were in the suspended load of an average river—alkali metal cations or alkaline earth cations?

Ans. Alkaline earth cations.

B. CATION ELECTRODES

8.10. In the measurement of pH, what is the change of EMF of the glass electrode-calomel electrode pair at 25 °C when the pH is changed by two units?

Ans. 0.118 volt.

8.11. If the inner solution of a glass electrode (a 0.1N HCl solution) begins to evaporate because of a leak, in which direction would the pH of a solution of constant composition appear to drift—up or down?

Ans. pH will appear to drift up.

8.12. A glass electrode sensitive to sodium ions is used in conjunction with a saturated calomel electrode in two NaCl solutions. The first solution has a known $[Na^+]$ of 10^{-3} and an observed EMF of 0.2231 volt at 25 °C. The observed EMF of the second solution is 0.3242 volt.

a. What is $[Na^+]$ in the second solution?
b. What is m_{Na^+}, approximately?

Ans. (a) $[Na^+] = 10^{-1.29}$; (b) $\gamma_{Na^+} \simeq 0.83$, $m_{Na^+} \simeq 0.06$.

8.13. The behavior of an electrode sensitive to both Na^+ and K^+ is described at 25 °C by the equation

$$E_{NaK} = 0.5002 + 0.0592 \log([Na^+] + 4.0[K^+]).$$

A second electrode sensitive only to Na^+ follows the relation

$$E_{Na} = 0.4921 + 0.0592 \log[Na^+].$$

If, in a given solution at 25 °C, $E_{Na} = 0.3737$ volt, and $E_{NaK} = 0.4231$ volt, what are the activities of Na^+ and K^+?

Ans. $[Na^+] = 0.01$; $[K^+] = 0.01$.

8.14. An electrode sensitive to both Na^+ and K^+ is placed in a solution containing Na^+ at an activity of 0.001, at 25 °C. The solution is then

titrated by adding a potassium salt. The following table gives E values as a function of the activity of the added K^+ (at constant $[Na^+]$).

$[Na^+]$	$[K^+]$	E_{NaK} (volts)
0.001	0.0	0.3226
0.001	0.0001	0.3293
0.001	0.001	0.3582
0.001	0.01	0.4109
0.001	0.1	0.4701

Draw a graph with E_{NaK} as ordinate versus $\log [K^+]$. From the graph determine
    a. The selectivity constant, K_{NaK}, for the electrode.
    b. Whether the electrode performs as an ideal or a regular solution.
                                    Ans. (a) $K_{NaK} \simeq 3.6$; (b) ideal solution.

8.15. A glass electrode sensitive to Na^+, K^+, and Li^+ performs at 25 °C in NaCl solutions according to the equation

$$E_{Na} = 0.2500 + 0.0592 \log [Na^+].$$

If K_{NaK} is 0.1 and K_{NaLi} is 2.0, what is E_{NaKLi} in a solution at 25 °C, with $[Na^+] = 0.01$, $[K^+] = 0.1$, and $[Li^+] = 0.02$? The electrode behaves as an ideal solution for these cations.   *Ans.* $E_{NaKLi} = 0.1777$ volt.

8.16. For the electrode described in problem 8.15, what is ΔF_r° at 25 °C for the exchange equilibrium

$$LiX_{gls} + Na^+_{aq} = NaX_{gls} + Li^+_{aq}.$$

                                                          Ans. $\Delta F_r^\circ = 0.411$ kcal.

8.17. In sea water $[Na^+]$ is about 0.35, and $[Ca^{++}]$ is about 0.003. Two electrodes are available, one of which is entirely selective for Na^+ and the other sensitive only to Na^+ and Ca^{++}. The latter electrode has a K_{CaNa} of 20 at 25 °C. Can these electrodes be used in conjunction to measure $[Ca^{++}]$ in sea water at 25 °C, if used with a voltmeter having a precision of 0.1 millivolt? Explain answer.
                                    Ans. No; much greater sensitivity is needed.

8.18. A sample of water from a public water supply contains 9 ppm Na^+, 1 ppm K^+, 20 ppm Ca^{++}, and 5 ppm Mg^{++}, and has an ionic strength of 0.003. Can an electrode that has K_{CaMg} of 1, K_{CaNa} of 20, and K_{CaK} of 10, be used to measure total hardness (Ca + Mg) to ± 2 percent, if no corrections are made for the presence of Na and K ions?
        Ans. Yes. Corrections for sodium and potassium are unimportant. A 2 percent change in concentration of Ca + Mg will change the measured EMF by approximately 1 millivolt.

CHAPTER 9

Effects of Temperature and Pressure Variations on Equilibria

INTRODUCTION

The procedures that have been illustrated for the calculation of stability relations at 25 °C (298.15 °K) and 1 atmosphere total pressure can be applied to systems at other temperatures and pressures if thermodynamic data are available for the conditions of interest. Whereas required data are directly available for the construction of many diagrams of fair to excellent validity at low temperatures and pressures, relatively few data are available in the same form for elevated temperatures and for elevated pressures. Therefore, construction of diagrams usually requires calculation of free-energy values by methods of widely varying accuracy.

In this chapter various methods of obtaining free energy values at elevated temperatures will be discussed and illustrated, and a few diagrams will be constructed at elevated temperatures and 1 atmosphere pressure. Then the effect of increase in total pressure will be assessed.

FREE-ENERGY VALUES AS A FUNCTION OF TEMPERATURE

To calculate stability relations among compounds at some elevated temperature and 1 atmosphere total pressure, one must be able to obtain the standard free energies of the various reactions at the temperature of interest. Ideally, one would simply have tables of standard free energies of formation of compounds, dissolved species, and elements (if not in their reference states) at all temperatures of interest, or tables with values at sufficiently small temperature intervals so that interpolation would be easy and accurate. With such tables, the pattern of calculation would be identical to that at 298.15 °K. An alternative would be to have $\Delta F_f°$

values for all species of interest, expressed as algebraic functions of temperature, or possibly in graphical form.

Because of the fragmentary nature and lack of availability of the data, no such comprehensive compilation can be made; instead it is still necessary to use data ranging from the excellent and accurate analytical functions or tables with small temperature increments, to data obtained by calculations of a validity ranging from good to poor.

In the following, we consider briefly a number of ways in which ΔF_f° values at various temperatures are presented in the literature, either directly, or in terms of other functions.

Third Law Calculation

For any chemical reaction, the following relation holds

$$\Delta F_r^\circ = \Delta H_r^\circ - T \Delta S_r^\circ. \tag{9.1}$$

Here ΔF_r° is the usual standard free-energy change, ΔH_r° is the change in *enthalpy*, or *heat content*, and ΔS_r° is the change in *entropy*. The superscript symbol signifies, as always, that the value of each quantity, ΔF_r°, ΔH_r°, and ΔS_r°, entering the equation is that appropriate to *standard conditions*, i.e., the value at unit fugacity and the temperature T. The temperature may have any value of interest, but is the same for each of the three thermodynamic variables.

Obviously, if values of ΔH_r° and ΔS_r° were available for various temperatures, for all reactions of interest, it would be quite easy to calculate the corresponding ΔF_r°, using equation (9.1). Before going further into this question, we first discuss the nature of the quantities ΔH_r° and ΔS_r°. In this text, we have no detailed interest in the thermodynamic significance of these quantities, but are concerned with the use of their numerical values for calculating free energy changes.[1]

THE ENTHALPY. The *enthalpy change* incurred in an isothermal reaction, at constant total pressure,[2] is equal to the *total heat absorbed* in the reaction. Thus, the enthalpy change is called the heat of reaction. For example, for the reaction

$$H_{2\,g} + \tfrac{1}{2}O_{2\,g} = H_2O_1 \qquad \Delta H_{r\,298.15}^\circ = -68.3174 \text{ kcal}. \tag{9.2}$$

---

[1] For a discussion of the thermodynamic basis of the matters treated in Chapter 9, see, for example, I. Klotz, *Chemical Thermodynamics*. Englewood Cliffs, N.J., Prentice-Hall, 1950.

[2] For all processes or reactions, the thermodynamic functions ΔF, ΔH, and ΔS depend only upon the initial and final states of the substances involved, and not upon how the process is carried out. Hence, specification of constant temperature or constant pressure for a reaction means only that the temperature or pressure be the same for the initial and final states.

That is, at 298.15 °K (25 °C) and 1 atmosphere total pressure, the heat of reaction, $\Delta H_r^\circ = -68.3174$ kcal/mole. Since the sign of ΔH° is negative, heat has been evolved.

The heat of formation of a compound (or element not in its standard state) is defined as the heat absorbed in the reaction involved in the formation of the compound from its elements in their standard states[3] at any specified temperature, and is written ΔH_f°. The heat of formation of an element in its standard state is zero, *by convention*. Reaction (9.2) may be rewritten in more detail to illustrate these conventions

$$\underset{\substack{\text{(gas at 298.15 °K} \\ \text{zero pressure)}}}{H_2} + \underset{\substack{\text{(gas at 298.15 °K} \\ \text{zero pressure)}}}{\tfrac{1}{2}O_2} = \underset{\substack{\text{(liquid at 298.15 °K} \\ \text{1 atmos pressure)}}}{H_2O}$$

$$\Delta H_r^\circ = \Delta H_{f\,H_2O}^\circ - \Delta H_{f\,H_2}^\circ - \tfrac{1}{2}\Delta H_{f\,O_2}^\circ = -68.3174 - 0 - 0$$
$$\Delta H_{f\,H_2O}^\circ = -68.3174 \text{ kcal/mole}.$$

From tabulated values of heats of formation ΔH_f°, heats of reaction ΔH_r° can be calculated by the same procedure as was previously used for obtaining ΔF_r° from ΔF_f° (Chapter 1). For any reaction, at temperature T,

$$\Delta H_r^\circ = \Sigma \Delta H_{f\,\text{products}}^\circ - \Sigma \Delta H_{f\,\text{reactants}}^\circ. \tag{9.2a}$$

For example, let us calculate the heat of reaction at 298.15 °K for the hydration of corundum to form gibbsite. We write

$$\underset{\text{corundum}}{Al_2O_3{}_c} + 3H_2O_1 = \underset{\text{gibbsite}}{Al_2O_3 \cdot 3H_2O}{}_c$$

and

$$\Delta H_r^\circ = \Delta H_{f\,\text{gibbsite}}^\circ - \Delta H_{f\,\text{corundum}}^\circ - 3\Delta H_{f\,H_2O}^\circ. \tag{9.3}$$

Values for $\Delta H_{f\,298}^\circ$ for a number of substances are listed in Appendix 2; selecting the appropriate values for (9.3), we have

$$\Delta H_r^\circ = -613.7 - (-399.09) - 3(-68.317)$$
$$\Delta H_r^\circ = -9.66 \text{ kcal/mole}. \tag{9.3a}$$

The change in enthalpy with temperature of a substance, at constant pressure, is equal to the heat capacity,[4] C_P, of the substance. That is

$$\left(\frac{\partial H}{\partial T}\right)_P = C_P. \tag{9.4}$$

---

[3] For enthalpy and entropy, as for free energy, the standard state of a pure solid or liquid is the most stable form at 1 atmosphere and the specified temperature (unless otherwise specified). For a pure gas, the entropy is referred to the hypothetical ideal gas at 1 atmosphere, in which state the enthalpy has the same value as that of the real gas at zero pressure. Under these conditions, ΔF_f°, ΔH_f°, and ΔS_f° will have values all referred to the same standard gas at unit fugacity. For solutions, definitions become more complicated; see e.g., I. Klotz, *op. cit.*, chap. 19.

[4] The heat capacity of a substance is defined to be the number of calories required to raise the temperature of the substance 1 °C.

On partial integration (9.4) becomes

$$H_{T_2} - H_{T_1} = \int_{T_1}^{T_2} C_P \, dT. \tag{9.5}$$

The right-hand member of equation (9.5) may be evaluated in various ways. Over a small range of temperature, or in general, when C_P is constant, (9.5) yields

$$H_{T_2} - H_{T_1} = C_P(T_2 - T_1). \tag{9.6}$$

The general method for evaluating the integral in (9.5) is to plot the experimentally measured values of C_P against T; the measured area, from T_1 to T_2, under the resulting curve, is equal to $H_{T_2} - H_{T_1}$.

Alternatively, C_P can be represented by an empirical equation, which when substituted into equation (9.4) permits the solution of that equation. If C_P has the form

$$C_P = a + 2bT - cT^{-2}, \tag{9.7}$$

then the corresponding enthalpy equation has the general form

$$H_T - H_{T'} = aT + bT^2 + cT^{-1} + d. \tag{9.8}$$

This four-constant equation is recommended by Maier and Kelley[5] as one that will fit a large variety of substances above the base temperature $T' = 298.15\ °K$. H and C_P as written hold for any pressure; when the substance is measured at unit fugacity, at each value of T, H is replaced everywhere in equations (9.4) to (9.8) by $H°$, and C_P by $C_P°$.

Table 9.1 for sillimanite illustrates the tabular presentation of $H_T° - H_{298}°$ data for various temperatures above 298.15 °K, corresponding entropy data (discussed in a subsequent section), the heat content equation corresponding to (9.8), and the heat capacity equation corresponding to (9.7). This table was taken from the work of Kelley,[6] in which like data are listed for 893 elements and inorganic compounds, many of which are of geochemical interest.

The quantity $H_{T_2}° - H_{T_1}°$ is to be contrasted with the heat of formation $\Delta H_f°$. The first of these quantities is the change in heat content of a substance in passing from temperature T_1 to T_2; $\Delta H_f°$ is the change in heat content in the formation of a compound from its elements at any *constant* temperature T. In particular, $\Delta H_f°$ for an element in its standard state is zero at every specified temperature; on the other hand, $H_{T_2} - H_{T_1} \neq 0$ for an element.

[5] C. G. Maier and K. K. Kelley, *J. Am. Chem. Soc.*, **54**, 3243 (1932).
[6] K. K. Kelley, High-temperature heat-content, heat-capacity, and entropy data for the elements and inorganic compounds: *U.S. Bureau of Mines Bulletin 584* (1960).

TABLE 9.1. Heat Content and Entropy of Sillimanite, Al_2SiO_5
[Base, crystals at 298.15 °K; mol wt, 162.05]

T (°K)	$H_T^\circ - H_{298.15}^\circ$ (cal/mole)	$S_T^\circ - S_{298.15}^\circ$ (cal/deg mole)
400	3,300	9.49
500	6,940	17.60
600	10,900	24.82
700	15,300	31.60
800	19,900	37.74
900	24,400	43.03
1,000	28,900	47.77
1,100	33,400	52.06
1,200	37,900	55.98
1,300	42,500	59.66
1,400	47,000	62.99
1,500	51,600	66.17
1,600	56,300	69.20

Al_2SiO_5 (sillimanite):
$H^\circ_T - H^\circ_{298.15} = 40.09T + 2.93 \times 10^{-3}T^2 + 10.13 \times 10^5 T^{-1} - 15,611$ (1.5%; 298° – 1600 °K);
$C^\circ_P = 40.09 + 5.86 \times 10^{-3}T - 10.13 \times 10^5 T^{-2}$.
SOURCE: From K. K. Kelley, *Bulletin 584* (op. cit.), p. 13.

By the rules of differential calculus, we can expand the previous equations in H to parallel equations in ΔH_r. Proceeding from equation (9.4), the change with temperature of the heat of reaction, at constant pressure, becomes

$$\left(\frac{\partial \Delta H_r}{\partial T}\right)_P = \Delta C_P, \tag{9.9}$$

where

$$\Delta C_P = \Sigma\, C_{P\text{ products}} - \Sigma\, C_{P\text{ reactants}}. \tag{9.9a}$$

The subscript P indicates only that the pressure on each substance entering the reaction be the same at each temperature considered. Thus, if we know ΔH_r for the conversion of 1 mole of liquid water at 10 atmospheres into water vapor at 1 atmosphere at temperature T_1, then equation (9.9) tells us how to calculate ΔH_r for the same reaction at some other temperature T_2.

Partial integration of (9.9) leads to

$$\Delta H_{r\, T_2} - \Delta H_{r\, T_1} = \int_{T_1}^{T_2} \Delta C_P\, dT. \tag{9.10}$$

Over a small range of temperature, or in general, where ΔC_P can be regarded as constant

$$\Delta H_{r\, T_2} - \Delta H_{r\, T_1} = \Delta C_P(T_2 - T_1). \tag{9.11}$$

EFFECTS OF TEMPERATURE AND PRESSURE VARIATIONS

In general, if ΔC_P for a given reaction is plotted against temperature, the area under the resulting curve from T_1 to T_2 is equal to $\Delta H_{r\ T_2} - \Delta H_{r\ T_1}$, according to equation (9.10).

If the C_P of each substance taking part in a reaction can be represented by an empirical equation of the type given by (9.7), then from (9.9a)

$$\Delta C_P = \Delta a + 2\Delta bT - \Delta cT^{-2}. \tag{9.12}$$

Substitution of equation (9.12) into equation (9.9) leads, upon integration, to

$$\Delta H_r = \Delta H_I + \Delta aT + \Delta bT^2 + \Delta cT^{-1}, \tag{9.13}$$

where ΔH_I is a constant of integration. Equation (9.13) is parallel to equation (9.8).

If all substances concerned are measured at unit fugacities, ΔC_P is replaced everywhere by ΔC_P°, and ΔH by ΔH°.

As an illustration of the use of equations (9.12) and (9.13), let us calculate the heat of reaction at 1000 °K for the reaction

$$\mathrm{H_{2\ g}} \underset{\text{(ideal state, 1000°)}}{} + \tfrac{1}{2}\mathrm{S_{2\ g}} \underset{\text{(ideal state, 1000°)}}{} = \mathrm{H_2S_g}, \underset{\text{(ideal state, 1000°)}}{} \tag{9.14}$$

from data given by Kelley in *Bulletin 584* (*op. cit.*). The C_P° equations for H_2S, H_2, and S_2 are found on pages 82, 79, and 181, respectively, of *Bulletin 584*. These are

H_2S_g: $C_P^\circ = 7.81 + 2.96 \times 10^{-3}T - 0.46 \times 10^5 T^{-2}$
$H_{2\ g}$: $C_P^\circ = 6.52 + 0.78 \times 10^{-3}T + 0.12 \times 10^5 T^{-2}$
$S_{2\ g}$: $C_P^\circ = 8.72 + 0.16 \times 10^{-3}T - 0.90 \times 10^5 T^{-2}$.

From these relations, the coefficients of equation (9.13) are calculated as follows:

$\Delta a = 7.81 - 6.52 - \tfrac{1}{2}(8.72) = -3.07$
$2\Delta b = [2.96 - 0.78 - \tfrac{1}{2}(0.16)]10^{-3} = 2.10 \times 10^{-3}$
$\Delta b = 1.05 \times 10^{-3}$
$-\Delta c = [-0.46 - 0.12 - \tfrac{1}{2}(-0.90)]10^5 = -0.13 \times 10^5$
$\Delta c = 0.13 \times 10^5$.

Substituting these numerical values into (9.13), we have

$$\Delta H_r^\circ = \Delta H_I^\circ - 3.07T + 1.05 \times 10^{-3}T^2 + 0.13 \times 10^5 T^{-1}. \tag{9.14a}$$

The integration constant ΔH_I° can be readily evaluated for this reaction. Consider the same reaction at 298.15 °K, i.e.,

$$\mathrm{H_{2\ g}} \underset{\text{(ideal state, 298°)}}{} + \tfrac{1}{2}\mathrm{S_{2\ g}} \underset{\text{(ideal state, 298°)}}{} = \mathrm{H_2S_g} \underset{\text{(ideal state, 298°)}}{}$$

The heat of reaction is given by

$$\Delta H_r^\circ = \Delta H_{f\,H_2S_g}^\circ - \Delta H_{f\,H_2\,g}^\circ - \tfrac{1}{2}\Delta H_{f\,S_2\,g}^\circ;$$

the values for the heats of formation (at 298 °K) are given in Appendix 2. Then

$$\Delta H_r^\circ = -4815 - 0 - \tfrac{1}{2}(29{,}860) = -19{,}745 \text{ cal/mole}.$$

Substituting this value and $T = 298$ into equation (9.14a), we see that

$$-19{,}745 = \Delta H_I^\circ - 3.07(298) + 1.05 \times 10^{-3}(298)^2 + 0.13 \times 10^5 (298)^{-1},$$

from which $\Delta H_I^\circ = -18{,}964$ cal/mole.

For any temperature T, equation (9.14a) then becomes

$$\Delta H_r^\circ = -18{,}964 - 3.07T + 1.05 \times 10^{-3} T^2 + 0.13 \times 10^5 T^{-1}. \tag{9.15}$$

Substituting $T = 1000$ into (9.15) and solving, we find[7]

$$\Delta H_{r\,1000}^\circ = -20{,}981 \text{ cal/mole}. \tag{9.16}$$

The standard heat of this reaction at 1000 °K can also be calculated directly from the tabulated data for $H_{1000}^\circ - H_{298}^\circ$ and $\Delta H_{r\,298}^\circ$.

In order to see how this is done in the general case, consider a reaction between A and B to yield compound C. Suppose the reaction is carried out at temperature T_1, and again at T_2; then for the reaction

$$A + B = C \tag{9.17}$$

$$\Delta H_{r\,T_2}^\circ = H_{C\,T_2}^\circ - H_{A\,T_2}^\circ - H_{B\,T_2}^\circ \tag{9.18}$$

$$\Delta H_{r\,T_1}^\circ = H_{C\,T_1}^\circ - H_{A\,T_1}^\circ - H_{B\,T_1}^\circ. \tag{9.19}$$

Subtracting (9.19) from (9.18)

$$\Delta H_{r\,T_2}^\circ - \Delta H_{r\,T_1}^\circ = (H_{T_2}^\circ - H_{T_1}^\circ)_C - (H_{T_2}^\circ - H_{T_1}^\circ)_A - (H_{T_2}^\circ - H_{T_1}^\circ)_B. \tag{9.20}$$

For data tabulated with base temperature 298.15 °K, equation (9.20) becomes

$$\Delta H_{r\,T}^\circ = \Delta H_{r\,298}^\circ + (H_T^\circ - H_{298}^\circ)_C - (H_T^\circ - H_{298}^\circ)_A - (H_T^\circ - H_{298}^\circ)_B. \tag{9.21}$$

---

[7] It will be recalled that the heat of formation of a compound is defined as the heat absorbed in the reaction involved in the formation of the compound from its elements in their *standard states at any specified temperature*. At temperatures of 1000 °K and above, sulfur vapor exists largely as S_2 molecules, as we have written. Thus, if the investigator chooses, the standard state of sulfur at these elevated temperatures may be chosen as $S_{2\,g}$. Then the ΔH_r° that we have calculated is equivalent to $\Delta H_{f\,H_2S}^\circ$, based upon $H_{2\,g}$ and $S_{2\,g}$; in contrast, the value of $\Delta H_{f\,H_2S}^\circ$ at 298.15 °K listed in tables, is based upon $H_{2\,g}$ and $S_{rhombic}$.

To solve (9.21) for the reaction to form H_2S at 1000 °K, it will be recalled that we have calculated $\Delta H_r^\circ = -19{,}745$ cal/mole at 298.15 °K. On the previously cited page references in Kelley, *Bulletin 584*, we find $(H_{1000}^\circ - H_{298}^\circ)_{H_2S} = 6695$; $(H_{1000}^\circ - H_{298}^\circ)_{H_2} = 4940$; and $\frac{1}{2}(H_{1000}^\circ - H_{298}^\circ)_{S_2} = 2988$. Substituting these values in equation (9.21) we find

$$\Delta H_{r\,1000}^\circ = -20{,}978 \text{ cal/mole.} \tag{9.22}$$

The value given in (9.22) is in excellent agreement with that found in (9.16). However, this is to be expected, because, although the calculations leading to (9.22) and (9.16), have been done in different ways, they are based on the same experimental data. Nevertheless, the two different methods have been given in some detail in order to illustrate alternative approaches to the same problem.

Equation (9.20) can be generalized to any number of products and reactants. In the general form it will read

$$\Delta H_{r\,T_2}^\circ - \Delta H_{r\,T_1}^\circ = \Sigma\,(H_{T_2}^\circ - H_{T_1}^\circ)_\text{products} \\ - \Sigma\,(H_{T_2}^\circ - H_{T_1}^\circ)_\text{reactants}. \tag{9.23}$$

The corresponding equation for the heat of formation of a compound follows directly from (9.23), and is

$$\Delta H_{f\,\text{cmpd}\,T_2}^\circ - \Delta H_{f\,\text{cmpd}\,T_1}^\circ = (H_{T_2}^\circ - H_{T_1}^\circ)_\text{cmpd} \\ - \Sigma\,(H_{T_2}^\circ - H_{T_1}^\circ)_\text{elements}. \tag{9.23a}$$

THE ENTROPY. When a substance absorbs heat from its surroundings, in any *reversible* process at constant temperature T, its *increase in entropy is given by the amount of heat absorbed divided by the absolute temperature*

$$dS = \frac{\delta q}{T}. \tag{9.24}$$

Since, by definition,

$$C_P = \left(\frac{\delta q}{\partial T}\right)_P \tag{9.25}$$

equation (9.24) becomes

$$dS = (C_P\,dT)/T \quad (P \text{ constant}). \tag{9.26}$$

Upon partial integration, (9.26) becomes

$$S_{T_2} - S_{T_1} = \int_{T_1}^{T_2} C_P\,d\ln T \tag{9.27}$$

or

$$S_{T_2} - S_{T_1} = 2.303 \int_{T_1}^{T_2} C_P\,d\log T. \tag{9.28}$$

If C_P is constant over the temperature range of interest, (9.28) yields

$$S_{T_2} - S_{T_1} = 2.303\, C_P \log \frac{T_2}{T_1}. \qquad (9.29)$$

Where C_P can be represented by some simple empirical equation such as (9.7), this empirical equation can be substituted into (9.26) and the resulting expression integrated (as was done previously in the case of the enthalpy). More generally, (9.28) is evaluated by plotting experimental values of C_P against $\log T$ and measuring the area under the resulting curve. This area is equal to $(S_{T_2} - S_{T_1})/2.303$. If a value of the entropy at temperature T_1 can be assigned, then that at T_2 is fixed.

There exists a very large body of experimental evidence to show that *the entropy of every substance in complete internal equilibrium is zero at 0 °K*. This is the substance of the third law of thermodynamics, first put into satisfactory form by Lewis and Randall.[8]

Because every well-crystallized pure substance can be assigned zero entropy at 0 °K, it is possible to evaluate and tabulate *absolute entropies*; this is the usual procedure. Returning now to equation (9.28) we set $T_1 = 0$; then, by the third law, S_{T_1} is zero, and for $T_2 = 298$ (298.15, precisely),

$$S_{298} = 2.303 \int_0^{298} C_P\, d\log T; \qquad (9.30)$$

S_{298} can be evaluated by the methods already discussed. A list of S°_{298} values is given in Appendix 2.

In particular, all elements have standard entropies of formation, S°_f, equal to zero at 0 °K; however, these have positive values at all other temperatures. In contrast, as previously stated, ΔF°_f and ΔH°_f for elements in their standard states are taken to be zero by convention, at any specified temperature.

The standard entropy change of reaction at temperature T, $\Delta S^\circ_{r,T}$, is given by

$$\Delta S^\circ_{r,T} = \Sigma\, S^\circ_{T,\text{ products}} - \Sigma\, S^\circ_{T,\text{ reactants}}. \qquad (9.31)$$

To illustrate the use of equation (9.31), we calculate ΔS°_r at 298.15 °K for a reaction we have already considered, namely

$$\underset{\text{corundum}}{Al_2O_{3\,c}} + 3H_2O_1 = \underset{\text{gibbsite}}{Al_2O_3 \cdot 3H_2O_c}. \qquad (9.31a)$$

From (9.31), for this reaction

$$\Delta S^\circ_r = S^\circ_{\text{gibbsite}} - S^\circ_{\text{corundum}} - 3 S^\circ_{H_2O}. \qquad (9.32)$$

[8] G. N. Lewis and Merle Randall, *Thermodynamics*. New York, McGraw-Hill, 1923.

Looking up the numerical values of $S°$ in Appendix 2 and substituting these in (9.32), we have

$$\Delta S_r° = 33.51 - 12.19 - 3(16.72)$$
$$\Delta S_r° = -28.84 \text{ cal/mole}. \tag{9.33}$$

Finally, we are in a position to accomplish our initial goal, namely, to take advantage of entropies evaluated by third-law considerations to calculate $\Delta F_r°$, using equation (9.1)

$$\Delta F_r° = \Delta H_r° - T\Delta S_r°.$$

For the example just considered, i.e., the hydration of corundum to gibbsite, we previously found, [(9.3a)], $\Delta H_r° = -9660$ cal/mole; combining this result with the $\Delta S_r° = -28.84$ cal/mole value just found, we have

$$\Delta F_r° = -9660 - 298(-28.84) = -1066 \text{ cal/mole}$$
$$\Delta F_r° = -1.07 \text{ kcal/mole}. \tag{9.34}$$

As was previously mentioned, entropy data are often given in tabulated form as $S_T - S_{T'}$ values. Kelley, in *Bulletin 584* (*op. cit.*), lists $S_T° - S_{298}°$ data for all the substances and temperatures for which he gives $H_T° - H_{298}°$ data. In another very useful publication, Kelley and King[9] have assembled the available 298.15 °K entropies of the elements and their inorganic compounds.

To summarize, absolute entropies are tabulated as $S_{298}°$ values, or for various temperatures as $S_T°$, or $S_T° - S_{T'}°$ values. The base temperature T' is usually 298.15 °K. Entropy data may be represented in terms of an empirical temperature function of $C_P°$.

The Free-Energy Function

A useful and accurate method of listing thermodynamic data is to tabulate values of the *free-energy function*, $-(F_T° - H_{T'}°)/T$. This function has the virtue that it varies slowly enough with temperature to permit accurate interpolation. Values are usually given for the base temperature $T' = 0$ °K or 298.15 °K. Tables of the function are available[10,11] for a considerable number of elements and compounds.

---

[9] K. K. Kelley and E. G. King, Entropies of the elements and inorganic compounds: *U.S. Bureau of Mines Bulletin 592* (1961).

[10] R. A. Robie, Thermodynamic properties of selected minerals and oxides at high temperatures: *Trace Elements Investigations Report 609, U.S. Geological Survey*, September 1959.

[11] G. N. Lewis and Merle Randall, *Thermodynamics*, 2nd ed., revised by K. S. Pitzer and Leo Brewer. New York, McGraw-Hill, 1961, Appendix 7, pp. 669–686.

To illustrate the use of the free-energy function, consider the following reaction at 1000 °K,

$$2CuO_c + \tfrac{1}{2}S_{2\,g} = Cu_2S_c + O_{2\,g}. \qquad (9.35)$$

The free-energy functions for the species involved are given at several temperatures in Table 9.2.

TABLE 9.2. Free Energies Based on H°_{298} for Some Solids and Gases

	$-(F^\circ - H^\circ_{298})/T$, cal/deg mole				$\Delta H^\circ_{f\,298}$ kcal/mole
	298.15 °K	500 °K	1000 °K	1500 °K	
CuO_c	10.19	11.43	16.09	20.07	−37.6
$S_{2\,g}$	54.51	55.42	58.72	61.35	30.84
Cu_2S_c	28.9	31.85	41.51	51.66	−19.0
$O_{2\,g}$	49.01	49.83	52.78	55.19	0

SOURCE: Data from Lewis and Randall, 2nd. ed. (op. cit.), Appendix 7.

By definition, the free-energy change in a reaction is equal to the sum of the free energies of the products minus the sum of the free energies of the reactants; the enthalpy of reaction is similarly defined. Hence, for any reaction, the following relation (based on H°_{298}) holds

$$\frac{\Delta F^\circ_{r\,T} - \Delta H^\circ_{r\,298}}{T} = \sum \left(\frac{F^\circ_T - H^\circ_{298}}{T}\right)_{\text{products}} - \sum \left(\frac{F^\circ_T - H^\circ_{298}}{T}\right)_{\text{reactants}}. \qquad (9.36)$$

In particular, for reaction (9.35)

$$\frac{\Delta F^\circ_r - \Delta H^\circ_{r\,298}}{T} = \left(\frac{F^\circ_T - H^\circ_{298}}{T}\right)_{Cu_2S} + \left(\frac{F^\circ_T - H^\circ_{298}}{T}\right)_{O_2}$$
$$- \tfrac{1}{2}\left(\frac{F^\circ_T - H^\circ_{298}}{T}\right)_{S_2} - 2\left(\frac{F^\circ_T - H^\circ_{298}}{T}\right)_{CuO}. \qquad (9.37)$$

Taking the appropriate values for 1000° from Table 9.2 and substituting these in (9.37), we have

$$\frac{\Delta F^\circ_r - \Delta H^\circ_{r\,298}}{1000} = -41.51 - 52.78 - \tfrac{1}{2}(-58.72) - 2(-16.09),$$

and therefore,

$$\Delta F^\circ_r - \Delta H^\circ_{r\,298} = -32.75 \text{ kcal/mole}. \qquad (9.38)$$

The quantity $\Delta H^\circ_{r\,298}$ can be readily evaluated from the heats of formation, $\Delta H^\circ_{f\,298}$, also listed in Table 9.2. It is found that $\Delta H^\circ_{r\,298} = 40.8$ kcal/mole. Substituting this result in (9.38) we have

$$\Delta F^\circ_{r\,1000} = 8.0 \text{ kcal/mole}. \qquad (9.39)$$

EFFECTS OF TEMPERATURE AND PRESSURE VARIATIONS

Also, since

$$-\Delta F_r^\circ = 2.303 RT \log K = 2.303 RT \log \frac{[Cu_2S][O_2]}{[CuO]^2[S_2]^{1/2}};$$

$$\log K_{1000} = -\frac{8.0}{2.303 \times 0.001987 \times 1000} = -1.75.$$

The activities of the solids are unity at 1 atmosphere total pressure, so we can write

$$\log \frac{P_{O_2}}{P_{S_2}^{1/2}} = -1.75$$

or

$$\log P_{O_2} = -1.75 + \tfrac{1}{2} \log P_{S_2} \qquad (T = 1000\ °K). \qquad (9.40)$$

A series of such calculations could be used to calculate partial pressure diagrams at 1000 °K (or other elevated temperatures) of the type constructed in Chapter 6 for 298.15 °K (25 °C).

As previously pointed out, Kelley has tabulated data for a great many compounds and elements. In *Bulletin 584* (*op. cit.*) data are given for $H_T^\circ - H_{298}^\circ$ and $S_T^\circ - S_{298}^\circ$, and in *Bulletin 592* (*op. cit.*) for S_{298}°. The free-energy function may be calculated readily from such tabulations as shown in the following.

By definition

$$F_T^\circ = H_T^\circ - TS_T^\circ. \qquad (9.41)$$

Subtracting H_{298}° from both sides of (9.41), and dividing by T,

$$\frac{F_T^\circ - H_{298}^\circ}{T} = \frac{H_T^\circ - H_{298}^\circ}{T} - S_T^\circ. \qquad (9.42)$$

Adding both S_{298}° and $-S_{298}^\circ$ to the right-hand side of (9.42), we have

$$\frac{F_T^\circ - H_{298}^\circ}{T} = \frac{H_T^\circ - H_{298}^\circ}{T} - (S_T^\circ - S_{298}^\circ) - S_{298}^\circ, \qquad (9.43)$$

which is in the desired form.

The free energy of formation of a compound, by definition, is the free-energy change that occurs in the reaction when the compound is formed from its component elements in their standard states. Hence, we can calculate the free energy of formation of a compound, at any desired temperature, using free-energy functions of the compound and the elements. The equation for $\Delta F_{f\,T}^\circ$ can be obtained directly from (9.36), and is

$$\frac{\Delta F_{f\,T}^\circ - \Delta H_{f\,298}^\circ}{T} = \left(\frac{F_T^\circ - H_{298}^\circ}{T}\right)_{\text{compound}} - \sum \left(\frac{F_T^\circ - H_{298}^\circ}{T}\right)_{\text{elements}}.$$

$$(9.44)$$

For example, let us calculate ΔF_f° for corundum, Al_2O_3, at 500 °K. The reaction involved is

$$2Al_c + \tfrac{3}{2}O_{2\,g} = Al_2O_{3\,c}.$$

Equation (9.44) becomes

$$\frac{\Delta F_f^\circ - \Delta H_{f\,298}^\circ}{T} = \left(\frac{F_T^\circ - H_{298}^\circ}{T}\right)_{Al_2O_3\,c} - 2\left(\frac{F_T^\circ - H_{298}^\circ}{T}\right)_{Al_c} - \frac{3}{2}\left(\frac{F_T^\circ - H_{298}^\circ}{T}\right)_{O_2\,g}. \quad (9.45)$$

Taking the appropriate free-energy function values (at 500 °K) for the species involved from Lewis and Randall, 2nd rev. ed. (*op. cit.*), Appendix 7, we have

$$\frac{\Delta F_f^\circ - \Delta H_{f\,298}^\circ}{500} = -14.61 - 2(-7.45) - \tfrac{3}{2}(-49.83),$$

$$\Delta F_f^\circ = \Delta H_{f\,298}^\circ + 37{,}518 \text{ cal/mole}.$$

The $\Delta H_{f\,298}^\circ$ for corundum is given in Appendix 2; it is -399.09 kcal/mole. Therefore

$$\Delta F_{f\,\text{corundum 500}}^\circ = -361.57 \text{ kcal/mole}.$$

This result may be compared with the value of the free energy of formation at room temperature, i.e., at 298 °K; this is -376.77 kcal/mole. Thus, the result of raising the temperature from 25 °C to 227 °C is to increase the free energy of the compound by about 4 percent.

Slope of the Plot of ΔF_r° vs. T

For use in the subsequent discussion, it is convenient at this point to determine the slope of the curve that results when ΔF_r° is plotted against T. Previously, in equation (9.1), we stated that

$$\Delta F_r^\circ = \Delta H_r^\circ - T\Delta S_r^\circ.$$

This equation holds for standard conditions. However, the relation is generally true, and we can write for any pressure

$$\Delta F_r = \Delta H_r - T\Delta S_r. \quad (9.46)$$

Hence, differentiating with respect to temperature, for constant pressure,

$$\left(\frac{\partial \Delta F_r}{\partial T}\right)_P = \left(\frac{\partial \Delta H_r}{\partial T}\right)_P - T\left(\frac{\partial \Delta S_r}{\partial T}\right)_P - \Delta S_r. \quad (9.47)$$

However, since by equation (9.26)

$$\left(\frac{\partial S}{\partial T}\right)_P = \frac{C_P}{T},$$

then, for a reaction
$$\left(\frac{\partial \Delta S_r}{\partial T}\right)_P = \frac{\Delta C_P}{T}. \tag{9.48}$$
Also, by equation (9.9)
$$\left(\frac{\partial \Delta H_r}{\partial T}\right)_P = \Delta C_P.$$
Substituting these results into (9.47), we get
$$\left(\frac{\partial \Delta F_r}{\partial T}\right)_P = -\Delta S_r. \tag{9.49}$$
In particular, for standard conditions,
$$\left(\frac{\partial \Delta F_r^\circ}{\partial T}\right)_P = -\Delta S_r^\circ. \tag{9.50}$$

This is the result sought. When the standard free-energy change of reaction is plotted against the absolute temperature, the slope of the curve, at each temperature, is the negative of the standard entropy change of reaction.

Free-Energy Changes and Types of Reactions

In discussing free-energy changes of reactions as a function of temperature, it is useful to consider the several categories of chemical reactions encountered; we exclude from consideration at this time reactions involving phases of variable composition. Reactions may include: (1) only solids; (2) solids plus liquids, or only liquids (or a liquid-like supercritical state); (3) only gases, or any combination of solids, liquids, and gases. We shall consider first reactions of type 1, solid-solid reactions.

SOLID-SOLID REACTIONS. In addition to the usual heat capacity, C_P, which is defined on a mole basis, we introduce here the *gram-atomic* heat capacity c_P, which is defined as the molal heat capacity for the substance, divided by the number of atoms contained in the formula of the substance. Thus, for example, for sillimanite, Al_2SiO_5, $c_P = C_P/8$.

At higher temperatures, the gram-atomic heat capacities of nearly all crystalline solids, measured at constant volume, approach the theoretical value $3R = 5.96$ cal/deg. For solids, this constant-volume heat capacity, C_V, and that at constant pressure, C_P, are nearly the same. Thompson[12] has estimated that the gram-atomic heat capacities of most rock-forming minerals approach $3R$ in the temperature range from 600 to 1000 °C.

[12] J. B. Thompson, Jr., The thermodynamic basis for the mineral facies concept: *Am. J. Sci.*, 253, 65 (1955).

The virtual constancy of the gram-atomic heat capacities implies that ΔC_P for a reaction involving only solids will be nearly zero. To show this, consider, for example, the reaction involving the formation of sillimanite from its component oxides at high temperatures,

$$\underset{\substack{c_P = 3R \\ C_P = 5(3R)}}{Al_2O_3} + \underset{\substack{3R \\ 3(3R)}}{SiO_2} = \underset{\substack{3R \\ 8(3R)}}{Al_2SiO_5}$$

$$\Delta C_P \cong 24R - 15R - 9R \cong 0.$$

We recall that for any reaction, by equation (9.48)

$$\left(\frac{\partial \Delta S_r}{\partial T}\right)_P = \frac{\Delta C_P}{T}.$$

Also, equation (9.9) states that

$$\left(\frac{\partial \Delta H_r}{\partial T}\right)_P = \Delta C_P.$$

Equations (9.9) and (9.48) tell us that if ΔC_P is very small, the change of ΔH_r and ΔS_r with temperature will likewise be very small. Again, Thompson (*op. cit.*) has estimated that for most solid-solid reactions of geological interest $(\Delta C_P)/T$ is less than 1.2×10^{-3} cal/mole deg². Presumably this numerical estimate is for temperatures of 1000 °K or higher.

In any event, as a first approximation, for temperatures above 25 °C it can be assumed that ΔS_r° and ΔH_r° are constant for solid-solid reactions. If this be true, then from equation (9.1) we can write

$$\Delta F_{r\,T}^\circ = \Delta H_{r\,298}^\circ - T\,\Delta S_{r\,298}^\circ. \tag{9.51}$$

Equation (9.51) enables us to calculate an approximate ΔF_r° value at some higher temperature from the usual listed 298 °K heat content of formation and entropy of formation values.

In Table 9.3 are listed the 25 °C molal entropies, the gram-atomic entropies, and heat of formation values for a number of minerals of geological interest. It is seen that the gram-atomic entropies at 25 °C and 1 atmosphere pressure vary only from about 2.4 to 4.8 cal/deg (excluding the liquid, water). This small spread in gram-atomic values means that the molal entropy change in any reaction involving these compounds will be small, since the total number of atoms in the reaction is constant. For example, in the formation of clinoenstatite from periclase and quartz at 25 °C, we have

$$\underset{\substack{S_{298}^\circ = 2(3.25) \\ S_{298}^\circ = 6.50}}{MgO} + \underset{\substack{3(3.293) \\ 9.879}}{SiO_2} = \underset{\substack{5(3.24) \\ 16.20}}{MgSiO_3} \tag{9.52}$$

$$\Delta S_r^\circ = 16.20 - 6.50 - 9.88$$

$$\Delta S_r^\circ = -0.18 \pm 0.1 \text{ cal/deg}$$

TABLE 9.3. Standard Heats of Formation and Entropies at 25 °C of Some Rock-forming Minerals

Mineral	Formula	ΔH_f° (kcal/mole)	Molal Entropy (cal/deg)	Gram-atomic Entropy (cal/deg)
Gibbsite	Al(OH)$_3$	$-306.8 \pm ?^a$	16.8 ± 0.1^b	2.400 ± 0.014
Corundum	Al$_2$O$_3$	-400.4 ± 0.3	12.17 ± 0.02	2.434 ± 0.004
Kyanite	Al$_2$(SiO$_4$)O	—	20.02 ± 0.08	2.502 ± 0.010
Andalusite	AlAl(SiO$_4$)O	$-642.2 \pm ?^a$	22.28 ± 0.10	2.785 ± 0.012
Sillimanite	Al$_2$SiO$_5$	$-648.9 \pm ?^a$	22.97 ± 0.10	2.871 ± 0.012
Brucite	Mg(OH)$_2$	-221.9 ± 0.5	15.09 ± 0.05	3.02 ± 0.01
Jadeite	NaAl(Si$_2$O$_6$)	—	31.9 ± 0.3	3.19 ± 0.03
Clinoenstatite	Mg(SiO$_3$)	-362.7 ± 0.7	16.22 ± 0.10	3.24 ± 0.02
Forsterite	Mg$_2$(SiO$_4$)	-512.9 ± 1.0	22.75 ± 0.20	3.25 ± 0.03
Periclase	MgO	-143.80 ± 0.09	6.50 ± 0.15	3.25 ± 0.075
Quartz (low)	SiO$_2$	-210.2 ± 0.4	9.88 ± 0.01	3.293 ± 0.003
Cristobalite (low)	SiO$_2$	-209.45 ± 0.25	10.38 ± 0.01	3.460 ± 0.003
Tridymite (low)	SiO$_2$	-209.42 ± 0.4	10.50 ± 0.10	3.500 ± 0.033
Portlandite	Ca(OH)$_2$	$-235.8 \pm ?^a$	17.4 ± 1.0^b	3.5 ± 0.3
Albite	Na(AlSi$_3$O$_8$)	—	50.2 ± 0.4	3.86 ± 0.03
Wollastonite	CaSiO$_3$	-383.2 ± 0.6	19.6 ± 0.2	3.92 ± 0.04
Nepheline	NaAlSiO$_4$	—	29.1 ± 0.2^b	4.16 ± 0.03
Aragonite	CaCO$_3$	$-288.5 \pm ?^a$	21.2 ± 0.3^b	4.24 ± 0.06
Calcite	CaCO$_3$	-288.3 ± 0.35	22.2 ± 0.2	4.44 ± 0.04
Lime	CaO	-151.8 ± 0.3	9.5 ± 0.2	4.75 ± 0.10
Water (liq.)	H$_2$O	-68.32 ± 0.01	16.75 ± 0.03	5.58 ± 0.01

SOURCE: Data are from R. A. Robie, op. cit., except as noted. Values in Table 9.3 are not in complete agreement with those given in Appendix 2; however, see footnote to table for Silicon in Appendix 2.
[a] From F. D. Rossini et al., National Bureau of Standards Circular 500, U.S. Dept. of Commerce, 1952.
[b] From J. B. Thompson, Jr., op. cit.; the form of the present table was suggested by a similar one in Thompson's paper.

There are two things to be noted from this result: first, that ΔS_r° is very small, and secondly, that the error associated with ΔS_r° is about 50 percent of its value. Calculating $\Delta H_{r\,298}^\circ$ for reaction (9.52) from the appropriate ΔH_f° values given in Table 9.3, we find $\Delta H_{r\,298}^\circ = 8700 \pm 1190$ cal/mole. Putting these values (together with their associated experimental errors) into equation (9.51), we get for $T = 700\ °K$,

$$\Delta F_{r\,700}^\circ = \Delta H_{r\,298}^\circ - 700 \Delta S_{r\,298}^\circ \qquad (9.53)$$

$$\Delta F_{r\,700}^\circ = -8700 \pm 1190 - 700(-0.18 \pm 0.1) \qquad (9.53a)$$

$$\Delta F_{r\,700}^\circ = -8700 \pm 1190 + 126 \pm 70 \qquad (9.53b)$$

$$\Delta F_{r\,700}^\circ = -8574 \pm 1260 \text{ cal/mole.} \qquad (9.53c)$$

The value of the contribution to $\Delta F_{r\,700}^\circ$ of the $T\Delta S_r^\circ$ term is just about one-tenth the value of the error associated with the ΔH_r° term. Thus, for all practical purposes in this particular reaction, we are just about as well off assuming $\Delta F_{r\,T}^\circ = \Delta H_{r\,298}^\circ$ as to use equation (9.53).

To carry consideration of this example one step further, we calculate the value of $\Delta F_{r\,700}^\circ$ without making any approximations at all. Starting with equation (9.43), we can write for a reaction for the formation of a compound from its component oxides

$$\Delta F^\circ_{rT} = \Delta H^\circ_{r\,298} - T\Delta S^\circ_{298}$$
$$+ [(H^\circ_T - H^\circ_{298})_{\text{cmpd}} - \Sigma\,(H^\circ_T - H^\circ_{298})_{\text{oxides}}]$$
$$- T[(S^\circ_T - S^\circ_{298})_{\text{cmpd}} - \Sigma\,(S^\circ_T - S^\circ_{298})_{\text{oxides}}]. \quad (9.54)$$

We may think of equation (9.54) as consisting of equation (9.51) plus correction terms. The values for these latter terms can be found in Robie (*op. cit.*); from these values and the result in (9.53c) we get

$$\Delta F^\circ_{r\,700} = -8574 + [9600 - 5630 - 4100]$$
$$- 700[19.95 - 11.62 - 8.60]$$
$$\Delta F_{r\,700} = -8574 - 130 + 189 = -8574 + 59$$
$$\Delta F_{r\,700} = -8515 \pm 1260 \text{ cal/mole} \quad (9.55)$$

(assuming that no new errors are introduced in the $H^\circ_{700} - H^\circ_{298}$ and $S^\circ_{700} - S^\circ_{298}$ terms).

It is seen that the correction terms amount to only 59 calories; this is considerably less than the error associated with the final result. In summary, then, for reaction (9.52) at 700 °K, we have obtained the following results

$$\Delta F^\circ_r = -8.7 \pm 1.3, \quad \text{assuming } \Delta S^\circ_r = 0;\ \Delta F^\circ_{r\,700} = \Delta H^\circ_{r\,298}$$
$$\Delta F^\circ_r = -8.6 \pm 1.3, \quad \text{assuming equation (9.51)}$$
$$\Delta F^\circ_r = -8.5 \pm 1.3, \quad \text{no approximations, equation (9.1)}$$

where the values, in kcal/mole, have been rounded to the nearest 0.1 kcal/mole.

Thus, for the particular reaction considered, namely, the formation of clinoenstatite from periclase and quartz, the experimental error associated with the numerical value of ΔF°_r is larger than the error made in approximate calculations. In general, the percentage experimental error associated with the ΔF°_r of a solid-solid reaction will be relatively large, as an examination of Table 9.3 shows; this fact must be kept in mind constantly by the investigator in attempting to apply the results of calculations to geologic situations.

REACTIONS INVOLVING LIQUIDS. Generally, the heat capacities of pure liquids do not differ greatly from those of the corresponding solids. As for solids, ΔC°_P for reactions involving solids and liquids, or only liquids, will be small and relatively constant with changing temperature. As a result, the change of ΔF°_r with temperature will be small, as for solid-solid reactions.

REACTIONS INVOLVING GASES. If a gas is involved in a reaction, ΔC°_P may be large, and it is not possible immediately to make the simplifying assumptions appropriate to solids and liquids. However, as we shall show in the succeeding section, here again it is possible to treat such reactions in a systematic manner.

Analytical and Graphical Representation of $\Delta F^\circ_{r\,T}$ vs. T

Equation (9.50) tells us that if $\Delta F^\circ_{r\,T}$ is plotted against T, the slope of the resulting curve is $-\Delta S^\circ_{r\,T}$. As we have shown for solid-solid reactions, and indicated for reactions involving liquids, equation (9.51) holds as a first approximation. On the basis of this latter equation the $\Delta F^\circ_{r\,T}$ vs. T plot would be a straight line of slope $-\Delta S^\circ_{r\,298}$.

As it turns out, a similar straight-line relationship holds, with good accuracy, for reactions of all categories. This result implies that each reaction can be expressed as an empirical equation of the form

$$\Delta F^\circ_{r\,T} = A - BT, \tag{9.56}$$

where A and B are constants for the particular reaction. To explain why equation (9.56) works, we compare it with equation (9.1), namely,

$$\Delta F^\circ_{r\,T} = \Delta H^\circ_{r\,T} - T\Delta S^\circ_{r\,T}.$$

There are two reasons involved. In the first place, ΔS°_r and ΔH°_r do not change greatly with temperature, except at phase changes (discussed subsequently). Secondly, changes in ΔH°_r and ΔS°_r, caused by changes in the heat capacities of the reactants and products, with changes in temperature, tend to counterbalance one another so that the change in ΔF°_r with temperature is a constant one.

In practice, where sufficiently accurate data are available, $\Delta F^\circ_{r\,T}$ values are calculated by methods equivalent to equation (9.1), for several temperatures, and the best straight line fitted to these points.

In a series of interesting papers, Richardson and coworkers[13] have given a number of these straight-line plots for several types of reactions, all of which are of interest to geochemists.

In the original papers of Richardson *et al.*, both the analytical expressions and the corresponding graphical plots for some 90-odd reactions are listed, together with much pertinent detail. We show here in Figures 9.1 and 9.2 a few of their results for reactions of various types.

Figure 9.1 represents, for the most part, solid-solid reactions. In general, the assigned accuracy of each reaction plot is based on the estimated accuracy with which the thermodynamic data are known,

---

[13] F. D. Richardson and J. H. E. Jeffes, The thermodynamics of substances of interest in iron and steel making from 0 °C to 2400 °C. I. Oxides.: *J. Iron Steel Institute*, *160*, 261 (1948). F. D. Richardson, J. H. E. Jeffes, and G. Withers, idem. II. Compounds between oxides: *J. Iron Steel Institute*, *166*, 213 (1950). F. D. Richardson and J. H. E. Jeffes, idem. III. Sulphides: *J. Iron Steel Institute*, *171*, 165 (1952).

rather than on any departure from linearity of plot. With the exception of the reaction involving the formation of a borate, it is seen that the slopes of the plots are very small indeed.

Various changes of phase can occur as the temperature increases. A

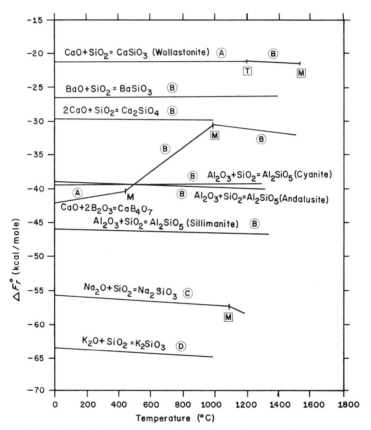

FIG. 9.1. Standard free energies of reaction as a function of temperature (in °C) for a number of oxide-oxide reactions. The symbols in this figure and in Figure 9.2 have the following meanings: probable accuracies (A) ±1 kcal, (B) ±3 kcal, (C) ±10 kcal, (D) > ±10 kcal. M, B, T, melting, boiling, or transition point of reactant; [M], etc., melting, etc., point of product. [Data from Richardson, Jeffes, and Withers, op. cit., II. Compounds between oxides.]

solid can undergo polymorphic transformation, it can melt to form a liquid, it can sublime to form a gas, or a liquid can boil. Such phase changes take place abruptly, i.e., over extremely small temperature intervals. All of these changes in state will be accompanied by the absorption of heat, except for polymorphic transformations where heat may be evolved or absorbed, depending upon the particular transformation.[14] The change in heat content of a substance undergoing (isothermal) transition, at constant pressure, is simply equal to this quantity of heat absorbed, by definition. Similarly, the change in entropy is equal to the quantity of heat absorbed (reversibly) divided by the absolute temperature. Hence, ΔH°_{fT} and S°_{fT}, after the transition,

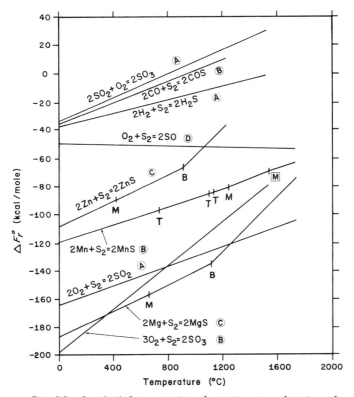

FIG. 9.2. Standard free energies of reaction as a function of temperature (in °C) for various reactions involving gaseous S_2. [Data from Richardson and Jeffes, op. cit., III. Sulphides.]

will, in general, have different values for the substance in question than they did before the transition. This in turn will result in new values for ΔH°_{rT} and ΔS°_{rT}, and the slope of the ΔF°_{rT} vs. T plot, will, in general, take on a new, constant value.

This change of slope is shown markedly in the reaction $CaO + 2B_2O_3 = CaB_4O_7$ (Figure 9.1). Up to 450 °C the reaction takes place between

[14] The quantity of heat absorbed on melting is called the (latent) heat of fusion; that absorbed on boiling, the (latent) heat of vaporization; that absorbed on sublimation, the (latent) heat of sublimation.

two crystalline solids. At that temperature, B_2O_3 melts, so that the reaction then involves a solid and a liquid. At 987 °C, the product CaB_4O_7 melts, and the slope again changes, becoming negative, but is not now so steep as before.

The reactions considered in Figure 9.1 are all concerned with pure solids and pure liquids at 1 atmosphere. The activities of these substances are all unity; hence, the fact that $\Delta F^\circ_{r\,T}$ is negative in each case means that the reacting oxides are unstable with respect to formation of the corresponding product, at all temperatures considered. This follows from equation (1.13), i.e.,

$$\Delta F_{r\,T} = \Delta F^\circ_{r\,T} + RT \ln Q.$$

Since $Q = 1$ for the type of reaction considered, then $\Delta F_{r\,T} = \Delta F^\circ_{r\,T}$. The reaction will be at equilibrium only when $\Delta F_{r\,T} = \Delta F^\circ_{r\,T} = 0$, i.e., if the slope of the reaction plot is steep enough so that the plot will cross the line where $\Delta F^\circ_{r\,T} = 0$; none of the plots shown do this.

In discussing equilibrium conditions for such reactions, we should remember that we have imposed two constraints on the conditions involved. In the first place, the pressure is defined to be that of 1 atmosphere, and secondly we have assumed that only pure phases are involved. However, in general, pressure can be varied, and solid and liquid solutions occur, so that the activities of the phases are variable.

The plots of the $\Delta F^\circ_{r\,T}$ reaction lines for the polymorphs cyanite and andalusite are instructive. In Figure 9.1 it is seen that these lines cross at approximately 500 °C. The point at which the lines do cross gives the temperature at which the two polymorphs have the same free energy content (at 1 atmosphere), and therefore the temperature at which they are in equilibrium. The accuracy with which this temperature can be fixed from this kind of plot is fairly low, as can be judged from the small slopes of the two lines and from the fact that the positions of the lines are accurate only to ± 3 kcal. In general, transition temperatures can be obtained far more accurately directly from pressure-temperature equilibrium studies.

In Figure 9.2 are shown some $\Delta F^\circ_{r\,T}$ vs. T plots for gas-gas and gas-solid reactions. In every case, sulfur is understood to be in the gaseous form as S_2 molecules. It will be noted immediately that the slopes of the reaction lines are much steeper than those of Figure 9.1, with the exception of the reaction involving the formation of SO. For the reactions involved here, the activities of any solids and liquids will be unity, but the activities of the gases can vary. These reactions correspond to equilibrium systems, with

$$\Delta F^\circ_{r\,T} = -RT \ln K.$$

For example, for the reaction $2Zn_c + S_{2g} = 2ZnS_c$, we have

$$\Delta F^\circ_{rT} = -RT \ln \frac{1}{P_{S_2}} = RT \ln P_{S_2}.$$

The data used in Figure 9.2 were taken from Richardson and Jeffes[15] paper on sulfides, in which they give plots for a large number of metal-sulfur gas reactions. In another paper, Richardson and Jeffes[16] have treated the similar problem of metal-oxygen gas reactions. In this latter paper, for each such reaction considered, an expression is obtained of the form

$$\Delta F^\circ_{rT} = RT \ln P_{O_2}.$$

Some Partial Pressure Diagrams at Elevated Temperatures

Making use of thermochemical data of the kind discussed in this chapter, Holland[17] has constructed a number of stability diagrams for ore metals in which the stability fields for the various compounds are delineated as functions of the fugacities of sulfur or oxygen, and temperature. It will be recalled that the fugacity of a gas is equal to its partial pressure at low to moderate pressures.

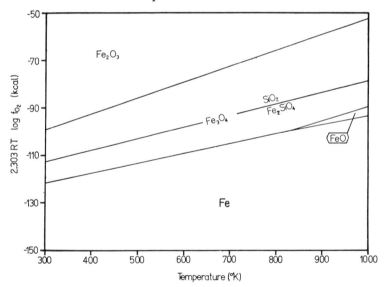

FIG. 9.3. Stability diagram for iron oxides plotted with temperature and $2.303 \, RT \log f_{O_2}$ as the defining variables. [From H. D. Holland, op. cit., p. 187; by permission.]

[15] op. cit. III. Sulfides.
[16] op. cit. I. Oxides.
[17] H. D. Holland, Some applications of thermochemical data to problems of ore deposits. I. Stability relations among the oxides, sulfides, sulfates, and carbonates of ore and gangue metals: Econ. Geol., 54, 184 (1959).

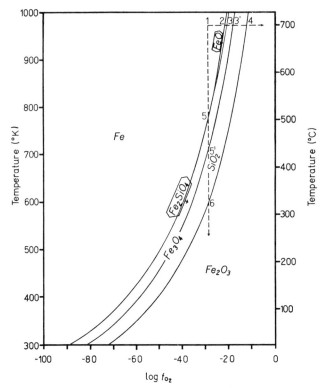

FIG. 9.4. Stability diagram for iron oxides plotted with temperature and log f_{O_2} as the defining variables. [From H. D. Holland, op. cit., p. 187; by permission.]

The construction of a partial pressure diagram as a function of temperature is carried out in the same way as was done in Chapter 6 for the single temperature, 25 °C, except that now the stability relationships are calculated for each temperature of interest. A diagram, taken from Holland's paper, showing the stability of the iron oxides, is shown in Figure 9.3. The Fe_2SiO_4-SiO_2 equilibrium has been superimposed on this diagram.

Because each reaction represented in Figure 9.3 is of a similar type, e.g.,

$$\tfrac{3}{2}Fe_c + O_{2\,g} = \tfrac{1}{2}Fe_3O_{4\,c},$$

the free energy of each of the reactions can be represented by the same form of equation

$$\Delta F^\circ_{r\,T} = RT \ln P_{O_2}.$$

EFFECTS OF TEMPERATURE AND PRESSURE VARIATIONS

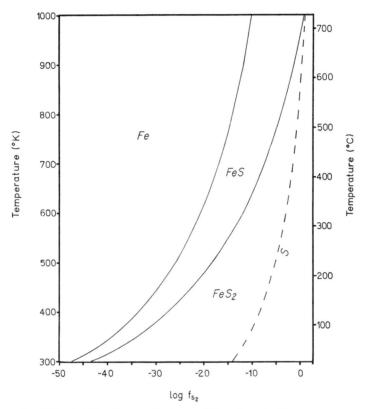

FIG. 9.5. Stability diagram for iron sulfides plotted with temperature and log f_{S_2} as the defining variables. [From H. D. Holland, *op. cit.*, p. 189; by permission.]

As we have shown (as in Figure 9.2 for reactions with $S_{2\,g}$), these $\Delta F°_{r\,T}$ vs. T plots are straight lines of form $A - BT$. Hence,

$$\Delta F°_{r\,T} = A - BT = 2.303 RT \log f_{O_2}.$$

Thus, in Figure 9.3, with the choice of ordinate shown, the boundaries of the stability fields are straight lines. A more conventional plot of the same system is shown in Figure 9.4.

In Figure 9.5 is shown the fugacity of sulfur vs. temperature diagram for the iron sulfides. In Figure 9.6, the results shown in Figures 9.4 and 9.5 have been combined in the form of four fugacity-fugacity isothermal sections for temperatures 400°, 600°, 800°, and 1000 °K.

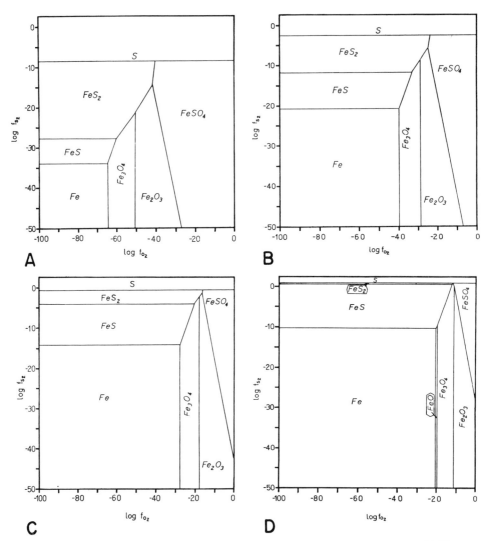

FIG. 9.6. Isothermal sections through the diagram log f_{O_2}-log f_{S_2}-T at: (A) 400 °K, (B) 600 °K, (C) 800 °K, (D) 1000 °K. [From H. D. Holland, op. cit., p. 191; by permission.]

Ionic Equilibria as a Function of Temperature

No really systematic treatment of the thermodynamic properties of electrolytes at elevated temperatures, of the kind presented in this chapter for phases of fixed composition, is yet available. Data for the free energies of formation, the enthalpies of formation, and the entropies of aqueous ions as a function of temperature are sparse and widely scattered throughout the literature.

In Chapter 4, we presented some results of the effect of temperature on the formation of complexes in aqueous solution; in Chapter 10 several silicate stability diagrams involving ionic species at elevated temperatures will be considered.

It may be that this lack of a satisfactory treatment of electrolyte solutions will soon be remedied. Recent work by Cobble and Criss[18] looks most promising. Other investigators are also attacking the general problem.

SUMMARY—FREE-ENERGY VALUES AS A FUNCTION OF TEMPERATURE

The problem of evaluating ΔF_r° or ΔF_r at all temperatures of interest is taken up. The thermodynamic functions, the enthalpy or heat content, and the entropy are defined, and the equations relating these to heat capacity and temperature are developed. The relationship (for standard conditions)

$$\Delta F_r^\circ = \Delta H_r^\circ - T \Delta S_r^\circ, \tag{9.1}$$

or alternatively (for any pressure),

$$\Delta F_r = \Delta H_r - T \Delta S_r, \tag{9.46}$$

provides the basis for the discussion of the nature of the change of ΔF_r with T for various categories of reactions. It is shown that for most reactions, a plot of ΔF_r° vs. T yields a straight line, and for condensed systems, this line usually has a small slope. Various methods of tabulating heat content, heat capacity, entropy, and free energy data are discussed, including the free-energy function $-(F_T^\circ - H_{T'}^\circ)/T$. A number of examples are given to show how the equations developed can

---

[18] J. W. Cobble, A method for predicting thermodynamic properties of electrolytes up to 350 °C: Paper presented at Symposium on High Temperature Solution Chemistry, 141 Meeting, Am. Chem. Soc., Wash., D.C., 1962; C. M. Criss, Thermodynamic properties of high-temperature aqueous solutions: Ph.D. dissertation, Purdue Univ., 1961 (Univ. Microfilms Inc., Ann Arbor, Mich., Mic. 61-2466).

be applied to practical problems. Several partial pressure diagrams at elevated temperatures illustrate the effect of changing temperature on mineral stability relationships.

FREE ENERGY AS A FUNCTION OF PRESSURE

Some General Considerations

The key relationship in assessing the effect of pressure variation on free energy changes is given by the equation

$$\left(\frac{\partial \Delta F_r}{\partial P}\right)_T = \Delta V_r = \Sigma V_{\text{products}} - \Sigma V_{\text{reactants}}. \quad (9.57)$$

Equation (9.57) states that at constant temperature, the rate of change of the free energy of reaction with total pressure (at any stated pressure and temperature) is equal to the sum of the molal volumes of the products of the reaction minus the sum of the molal volumes of the reactants.

For example, for the following reaction at 25 °C and 1 atmosphere

$$\underset{\substack{\alpha\text{ - quartz}\\ V(\text{cm}^3) = 22.69}}{SiO_2} + \underset{\substack{\text{corundum}\\ 25.57}}{Al_2O_3} = \underset{\substack{\text{sillimanite}\\ 49.92}}{Al_2SiO_5} \quad (9.58)$$

$$\Delta V_r = 49.92 - 22.69 - 25.57 = 1.66 \text{ cm}^3.$$

Therefore,

$$\left(\frac{\partial \Delta F_r}{\partial P}\right)_{25\,°C,\,1\,\text{atm}} = 1.66 \text{ cm}^3. \quad (9.59)$$

In equation (9.59), volume is expressed as cm^3, and if pressure is given in units of atmospheres, ΔF_r must be in units of cc-atmospheres. This latter unit is an energy unit: 1 cc-atmosphere is equal to 0.02422 calorie. From equation (9.59) we see that the increase in ΔF_r for each increase of 1 atmosphere pressure at 25 °C (near 1 atmosphere, or as long as ΔV_r is constant) is given by

$$\Delta(\Delta F_r) = 1.66 \times 1 \times 0.02422 = 0.0402 \text{ cal/mole atm.}$$

The quantity ΔV_r plays the same role with respect to the pressure change of ΔF_r as the quantity $-\Delta S_r$ does in connection with the temperature change of ΔF_r. Many of the considerations involving reaction types that apply to the magnitude and constancy of ΔS_r apply similarly to ΔV_r.

The volume of solids changes only slightly with temperature and pressure.[19] At 25 °C and 1 atmosphere, for most rock-forming minerals,

---

[19] The following discussion leading to the conclusion that ΔV_r is essentially constant for many reactions is based on a similar discussion in J. B. Thompson, *op. cit.*

the volume expansion, α, defined as

$$\alpha = \frac{1}{V}\left(\frac{\partial V}{\partial T}\right)_P \qquad (9.60)$$

lies between 1×10^{-5} per degree and 4×10^{-5} per degree. Similarly, the isothermal compressibility, β, defined as

$$\beta = -\frac{1}{V}\left(\frac{\partial V}{\partial P}\right)_T \qquad (9.61)$$

is between 0.5×10^{-6} per atmosphere and 3×10^{-6} per atmosphere. Both α and β generally rise slightly with increasing temperature and decrease slightly with increasing pressure.

The important point is, that since most solids have about the same thermal expansion and compressibility, ΔV_r will remain essentially constant over wide variations of temperature and pressure. That is, any increase or decrease in volume of the reactants will be offset by a corresponding increase or decrease in volume of the products. We illustrate the use of equation (9.57), for a reaction for which ΔV_r is constant, in the example given in the following discussion.[20]

The rhombic and monoclinic forms of sulfur are in equilibrium at 368.5 °K and 1 atmosphere pressure. That is, for the equilibrium

$$S_{\text{rhombic}} = S_{\text{monoclinic}}, \quad \Delta F^\circ_{r\ 368.5} = 0.$$

At temperatures above 100 °C and at high pressures

$$V_{\text{monoclinic}} - V_{\text{rhombic}} = \Delta V_r = 0.4 \text{ cc}.$$

We can safely assume that ΔV_r is constant, at least within the precision of the value given.

By virtue of equation (9.57), we have

$$\left(\frac{\partial \Delta F_r}{\partial P}\right)_T = 0.4.$$

Integrating, we get

$$\Delta F_{r\ P\ T} - \Delta F_{r\ P'\ T} = 0.4(P - P');$$

now let $P' = 1$; then

$$\Delta F_{r\ P\ T} - \Delta F^\circ_{r\ T} = 0.4(P - 1).$$

At equilibrium, at the new pressure P,

$$\Delta F_{r\ P\ T} = 0,$$

---

[20] The discussion is based on a similar one given in G. N. Lewis and Merle Randall, *Thermodynamics*, 2nd ed., revised by K. S. Pitzer and Leo Brewer. New York, McGraw-Hill, 1961, pp. 167, 169–170.

and therefore
$$-\Delta F^\circ_{r\,T} = 0.4(P - 1). \tag{9.61a}$$

Equation (9.61a) relates the equilibrium pressure P to the standard free energy of reaction at any temperature T (it is a condition of our choice of the value of ΔV_r that $T > 373$ °K). Let us then calculate P for $T = 380$ °K. In order to do this, we shall need to know the value of $\Delta F^\circ_{r\,380}$.

The values of S°_T for the two forms of sulfur are listed in Lewis and Randall, *rev. ed.* (*op. cit.*), p. 167. For 368.5 °K, these are $S^\circ_m = 9.07$, and $S^\circ_{rh} = 8.81$, and therefore $\Delta S^\circ_{r\,368.5} = 0.26$ cal/mole deg. Equation (9.50) states that

$$\left(\frac{\partial \Delta F^\circ_r}{\partial T}\right)_P = -\Delta S^\circ_r.$$

We assume that ΔS°_r is constant over the small temperature interval involved, and upon integration (and substitution of numerical values), we get

$$\Delta F^\circ_{r\,380} - \Delta F^\circ_{r\,368.5} = -0.26(380 - 368.5).$$

Since monoclinic and rhombic sulfur are in equilibrium at 368.5 °K and 1 atmosphere, $\Delta F^\circ_{r\,368.5} = 0$. Hence, $\Delta F^\circ_{r\,380} = -0.26 \times 11.5 = -3.0$ cal/mole.

We now have the data necessary to calculate the equilibrium pressure, using equation (9.61a). However, it is first necessary to convert the ΔF°_r value to units of cc-atm. We recall that 1 cc-atm = 0.02422 cal; hence, 1 cal = 41.29 cc-atm. Substituting in (9.61a), we get

$$-(-3.0) \times 41.29 = 0.4(P - 1)$$
$$P = 310 \text{ atm}.$$

To recapitulate, the two polymorphic forms of sulfur are in equilibrium at 368.5 °K and 1 atmosphere, and also at 380 °K and approximately 310 atmospheres. This latter pressure is only as accurate as the value of ΔV_r entering the calculation.

Molal volumes are readily determined, at or near room temperatures, from specific gravities, and at elevated temperatures, from x-ray diffraction measurements of unit cells. A recent compilation of precise molal volumes is given by Robie and Bethke.[21] Most of the values listed are for temperatures at or near 25 °C. Densities, values of α, and values of β are listed in Birch *et al.*[22]

[21] R. A. Robie and P. M. Bethke, Molar volumes and densities of minerals: *Trace Elements Report of Investigation 822*, U.S. Geological Survey, 1962.

[22] F. Birch, J. F. Schairer, and H. C. Spicer, Handbook of physical constants: *Geol. Soc. Am., Spec. Paper 36* (1942).

Molal Volume as a Function of Temperature

An example of the determination of the molal volume as a function of temperature, at 1 atmosphere, is afforded by the study of Skinner et al.,[23] of the Al_2SiO_5 polymorphs andalusite, kyanite, and sillimanite. They found, for example, that V(cm^3/mole) for andalusite is 51.525 at 0 °C and 51.624 at 100 °C, an increase of approximately 0.2 percent. At 1000 °C, V is 53.405, an increase of approximately 4 percent over the 100 °C value. The percentage changes in sillimanite and kyanite are similar and smaller than that in andalusite.

Liquids and liquid-like supercritical fluids behave somewhat like solids in that their thermal expansions and compressibilities are generally small. The pressure-temperature-volume relations of the all-important liquid, water, have been determined by Kennedy.[24]

Gases have large molal volumes, and for a reaction involving a gas, ΔV_r may be large, depending upon the particular reaction. At low pressures, the thermal expansions and the compressibilities of gases are high. With rising pressure, a gas will become more liquid-like in its properties.

Unfortunately, there exists at present no systematic up-to-date collection of molal volume data for a variety of temperatures and pressures, such as Kelley and coworkers have given in various Bureau of Mines Bulletins (*op. cit.*) for thermal data. Rather, such data as there are, are widely scattered throughout the literature.

Molal Volume and ΔF_r as a Function of Pressure

As an example of a system involving liquid water, let us consider some results obtained by Yoder and Weir[25] for the change of ΔF_r with pressure, at constant temperature, in the several reactions described in the following discussion.

Yoder and Weir were concerned with the volume changes and the free energy changes, with pressure, of reactions involving quartz, water, albite ($NaAlSi_3O_8$), analcite ($NaAlSi_2O_6 \cdot H_2O$), nepheline ($NaAlSiO_4$), and jadeite ($NaAlSi_2O_6$). For the reaction

$$2 \text{ analcite} = \text{nepheline} + \text{albite} + 2 \text{ water} \qquad (9.62)$$

$$\Delta V_r = V_{Ne} + V_{Ab} + 2V_W - 2V_{An}.$$

[23] B. J. Skinner, S. P. Clark, Jr., and D. E. Appleman, *Am. J. Sci.*, 259, 651 (1961).
[24] G. C. Kennedy, Pressure-volume-temperature relations in water at elevated temperatures and pressures: *Am. J. Sci.*, 248, 540 (1950).
[25] H. S. Yoder, Jr. and C. E. Weir, *Am. J. Sci.*, 258A, 420 (1960); idem, *Am. J. Sci.*, 249, 683 (1951).

The molal volumes, in cubic centimeters, as functions of pressure, at 25 °C, are

$$V_{Ne} = 54.120 - 1.113 \times 10^{-4}P + 0.28 \times 10^{-9}P^2$$
$$V_{Ab} = 100.399 - 2.123 \times 10^{-4}P + 2.17 \times 10^{-9}P^2$$
$$V_W = 18.016 - 4.463 \times 10^{-4}P + 14.51 \times 10^{-9}P^2$$
$$V_{An} = 97.735 - 0.843 \times 10^{-4}P - 27.04 \times 10^{-9}P^2.$$

These equations are valid for pressures of 2000 atmospheres and above. From these results, one obtains for reaction (9.62)

$$\Delta V_r = -4.919 - 10.476 \times 10^{-4}P + 85.55 \times 10^{-9}P^2 \quad cm^3/mole. \tag{9.63}$$

By virtue of equation (9.57), equation (9.63) is also equal to the rate at which the free energy of reaction changes with pressure. Recalling that 1 cc-atmosphere is equivalent to 0.02422 calorie, we multiply each numerical term in (9.63) by this conversion factor and get

$$\left(\frac{\partial \Delta F_r}{\partial P}\right)_{25 °C} = -0.119 - 0.254 \times 10^{-4}P + 2.071 \times 10^{-9}P^2 \tag{9.64}$$

where ΔF_r is in calories and P is in atmospheres.

From (9.64) we see that a large pressure change will make very little change in ΔF_r for this reaction. For example, at 3000 atmospheres

$$\left(\frac{\partial \Delta F_r}{\partial P}\right)_{25 °C} = -0.119 - 0.254 \times 3 \times 10^{-1} + 2.071 \times 9 \times 10^{-3},$$

$$\left(\frac{\partial \Delta F_r}{\partial P}\right)_{25 °C} = -0.176 \text{ cal/mole atm.}$$

Thus, every increase of 100 atmospheres (in the region of 3000 atmospheres) would result in a change in ΔF_r only of $-0.176 \times 100 = -17.6$ cal/mole.

Yoder and Weir also give similar equations for the reactions

$$analcite = jadeite + water$$
$$analcite + quartz = albite + water.$$

SUMMARY—FREE ENERGY AS A FUNCTION OF PRESSURE

The rate of change of the free energy of a reaction with pressure, at constant temperature, is given by the equation

$$\left(\frac{\partial \Delta F_r}{\partial P}\right)_T = \Delta V_r. \tag{9.57}$$

EFFECTS OF TEMPERATURE AND PRESSURE VARIATIONS

In general, the molal volume of each species entering the reaction depends upon the temperature and pressure, and therefore ΔV_r is in turn dependent upon these variables. For some reactions, ΔV_r is virtually constant, and (9.57) may be integrated directly. In other cases, ΔV_r is expressed as a polynomial series in terms of the pressure, for each temperature; (9.57) can then be integrated stepwise. Examples are given to illustrate these points. The calculation of the equilibrium pressure with varying temperature is illustrated.

FREE ENERGY AS A FUNCTION OF BOTH TEMPERATURE AND PRESSURE

Thermodynamic Relationships Concerned

For a reaction involving phases of fixed composition, when both temperature and pressure are varied, the total change in the free energy of the reaction is given by the equation

$$d\Delta F_r = \left(\frac{\partial \Delta F_r}{\partial T}\right)_P dT + \left(\frac{\partial \Delta F_r}{\partial P}\right)_T dP. \tag{9.65}$$

By virtue of equations (9.49) and (9.57), equation (9.65) becomes

$$d\Delta F_r = -\Delta S_r \, dT + \Delta V_r \, dP. \tag{9.66}$$

If the system is maintained at equilibrium, $d\Delta F_r = 0$, and equation (9.66) becomes

$$\Delta V_r \, dP - \Delta S_r \, dT = 0, \tag{9.67}$$

whence

$$\frac{dP}{dT} = \frac{\Delta S_r}{\Delta V_r}. \tag{9.68}$$

Also, by equation (9.46)

$$\Delta F_r = \Delta H_r - T\Delta S_r;$$

at equilibrium, $\Delta F_r = 0$, and

$$\Delta S_r = \frac{\Delta H_r}{T}. \tag{9.69}$$

Thus, equation (9.68) can be expressed alternatively as

$$\frac{dP}{dT} = \frac{\Delta H_r}{T\Delta V_r}. \tag{9.70}$$

Equation (9.70) is usually called the *Clapeyron equation*.

Example Involving Condensed Phases

We can illustrate the application of these equations, as well as apply some of the thermochemical considerations of earlier sections of this chapter, by considering the experimental investigation of the system hematite-goethite-water, carried out by Schmalz.[26] This investigator studied the reaction

$$\underset{\text{hematite}}{Fe_2O_3} + \underset{\text{water}}{H_2O} = \underset{\text{goethite}}{2FeOOH} \qquad (9.71)$$

at elevated temperatures and pressures. His results are shown in Figure 9.7.

Such a system provides an example of univariant equilibrium: only one degree of freedom is available to the system. At each specified temperature, for example, the equilibrium pressure is fixed. The equilibrium phase boundary found is a straight line having a positive slope, features characteristic of reactions involving condensed phases.

The slope of the phase boundary, as determined experimentally, is 22.3 bars/deg, or within the errors of the study, 22.3 atm/deg. (The conversion factor of bars to atmospheres is 1 bar = 0.98692 atm.) The volume change of reaction (9.71) was calculated for 130 °C and 1 atmosphere. It was assumed that the thermal expansion in the solids from room temperature to 130 °C would be compensatory; the volume of water was corrected to 130 °C. Thus, $\Delta V_r(130\ °C, 1\ atm) = -6.8$ cc. Then, knowing ΔV_r and dP/dT, ΔS_r can be calculated from equation (9.68), as follows

$$\frac{dP}{dT} = \frac{\Delta S'_r}{\Delta V'_r}$$

$$\Delta S'_r = -6.8 \times 22.3 = -151.64 \text{ cc-atm/deg mole}$$

$$\Delta S'_r = -151.64 \times 0.02422 = -3.7 \text{ cal/deg mole}.$$

Here, the prime refers to conditions of 130 °C and 1 atmosphere, and the value given is for 1 mole of hematite.

Schmalz has also estimated the 25 °C standard state thermodynamic properties of goethite through the following chain of reasoning. As we have previously indicated, it can safely be assumed that ΔC_P for this reaction is constant over the temperature interval from 25 °C to 130 °C. From equation (9.9) we have, for $P = 1$ and ΔC_P constant

$$\left(\frac{\partial \Delta H^°_r}{\partial T}\right)_{P=1} = \Delta C^°_P. \qquad (9.72)$$

---

[26] R. F. Schmalz, *J. Geophys. Research*, 64, 575 (1959).

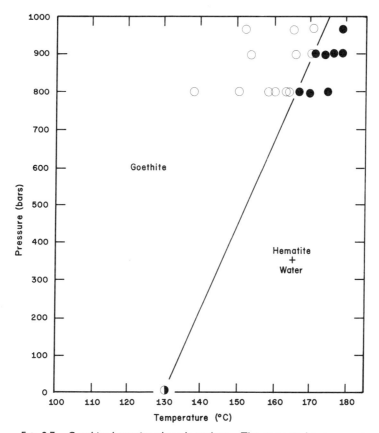

FIG. 9.7. Geothite-hematite phase boundary. The open circles represent bomb runs in which goethite was produced; filled circles, those in which hematite was produced. The half-filled circle at 130 °C represents an equilibrium run in which goethite plus hematite were found at approximately 4 atmospheres by F. D. Posnjak and H. E. Merwin, *J. Am. Chem. Soc.*, **44**, 1965 (1922). [After R. F. Schmalz, *op. cit.*]

Integration of (9.72) for ΔC_P° constant leads to

$$\Delta H_{r\,T_2}^\circ - \Delta H_{r\,T_1}^\circ = \Delta C_P^\circ (T_2 - T_1). \tag{9.72a}$$

Substituting $T_1 \Delta S_{r\,T_1}^\circ = \Delta H_{r\,T_1}^\circ$, we get

$$\Delta H_{r\,T_2}^\circ = T_1 \Delta S_{r\,T_1}^\circ + \Delta C_P^\circ (T_2 - T_1). \tag{9.73}$$

Now we set $T_2 = 298\,°\text{K}$ (25 °C) and $T_1 = 403\,°\text{K}$ (130 °C). We have just calculated ΔS_r° at T_1 and found it to be -3.7 cal/deg. The heat capacities of hematite and water are known, but that of goethite is not. However, the heat capacities of corundum, Al_2O_3, and diaspore,

AlOOH, are known. It was assumed that

$$\frac{C_P(\text{FeOOH})}{C_P(\text{Fe}_2\text{O}_3)} = \frac{C_P(\text{AlOOH})}{C_P(\text{Al}_2\text{O}_3)}.$$

This approximation yields a $C_P \cong 18.7$ cal/deg mole for goethite; finally, $\Delta C_P \cong -5.6$ cal/deg mole.

Then, putting these results into (9.73) we get

$$\Delta H^\circ_{r\,298} = (403)(-3.7) + (-5.6)(-105);$$

$$\Delta H^\circ_{r\,25\,°C} = -903 \text{ cal/mole}. \qquad (9.74)$$

Equation (9.48) states that

$$\left(\frac{\partial \Delta S_r}{\partial T}\right)_P = \frac{\Delta C_P}{T}.$$

For $P = 1$

$$\left(\frac{\partial \Delta S^\circ_r}{\partial T}\right)_{P=1} = \frac{\Delta C^\circ_P}{T}.$$

Upon integration, we get

$$\Delta S^\circ_{r\,T_2} - \Delta S^\circ_{r\,T_1} = 2.303\, \Delta C^\circ_P \log \frac{T_2}{T_1}.$$

For the temperatures of interest to us here, we have

$$\Delta S^\circ_{r\,298} = \Delta S^\circ_{r\,403} + 2.303\, \Delta C^\circ_P \log \frac{298}{403},$$

and

$$\Delta S^\circ_{r\,298} = -3.7 + (2.303)(-5.6)(-0.1311),$$

$$\Delta S^\circ_{r\,298} = -2.0 \text{ cal/deg mole}. \qquad (9.75)$$

From equation (9.1)

$$\Delta F^\circ_{r\,298} = \Delta H^\circ_{r\,298} - 298\, \Delta S^\circ_{r\,298}.$$

Thus, for this reaction

$$\Delta F^\circ_{r\,298} = -903 - (298)(-2.0) = -307 \text{ cal/mole}.$$

Since the free energies of formation of hematite and water are known, that of goethite can be calculated from the free energy of reaction, as follows

$$\Delta F^\circ_r = 2\Delta F^\circ_{f\,\text{FeOOH}} - \Delta F^\circ_{f\,\text{Fe}_2\text{O}_3} - \Delta F^\circ_{f\,\text{H}_2\text{O}_l}$$

$$-0.307 = 2\Delta F^\circ_{f\,\text{FeOOH}} - (-177.1) - (-56.69)$$

$$\Delta F^\circ_{f\,\text{FeOOH}} = -117.0 \text{ kcal/mole}.$$

Similarly,
$$\Delta H_r^\circ = 2\Delta H_{f\,\text{FeOOH}}^\circ - \Delta H_{f\,\text{Fe}_2\text{O}_3}^\circ - \Delta H_{f\,\text{H}_2\text{O}_l}^\circ$$
$$-0.903 = 2\Delta H_{f\,\text{FeOOH}}^\circ - (-196.5) - (-63.32)$$
$$\Delta H_{f\,\text{FeOOH}}^\circ = -130.4 \text{ kcal/mole}.$$

Also
$$\Delta S_r^\circ = 2S_{\text{FeOOH}}^\circ - S_{\text{Fe}_2\text{O}_3}^\circ - S_{\text{H}_2\text{O}_l}^\circ$$
$$-2.0 = 2S_{\text{FeOOH}}^\circ - 21.5 - 16.72$$
$$S_{\text{FeOOH}}^\circ = 18.1 \text{ cal/deg mole}.$$

From the approximation made in estimating C_P° for FeOOH (estimated to be ± 1 cal/mole), and the uncertainty in the slope of the experimentally determined phase boundary line, Schmalz estimated the error in $\Delta F_{f\,\text{FeOOH}}^\circ$ and $\Delta H_{f\,\text{FeOOH}}^\circ$ to be ± 1 kcal/mole, and that in S_{FeOOH}° to be ± 0.5 cal/deg mole.

Slope and Position of the P-T Curve[27]

The slope of the equilibrium P-T curve of the example just discussed is a straight line. That is

$$\frac{dP}{dT} = \frac{\Delta S_r}{\Delta V_r} = \text{constant}. \qquad (9.76)$$

The constancy of the ratio $\Delta S_r/\Delta V_r$ holds quite well at elevated temperatures and pressures for all types of reactions. At lower temperatures, for some solid-solid reactions the P-T curve may show a slight concavity toward the pressure axis. For reactions involving the production of a gas, the equilibrium curve will generally have a gentle rise of positive slope and become steeper as the gas becomes more liquid-like in its properties. Most dP/dT values are positive at higher temperatures and pressures, but there are exceptions, the ice-liquid water transition being one such exception.

The slope of the equilibrium line at 1 atmosphere can be calculated from ΔS_r° and ΔV_r°. However, this calculation gives no clue in itself about whether dP/dT will be a constant. The position of the equilibrium line on the T-axis at $P = 1$ atmosphere, can, in principle, be calculated by means of equation (9.1), i.e.,

$$0 = \Delta F_r^\circ = \Delta H_r^\circ - T\Delta S_r^\circ.$$

However, the errors in the usual experimental values of ΔH_r° and ΔS_r° are too large for T to be calculated in this way. Thompson (*op. cit.*)

[27] See J. B. Thompson, *op. cit.*

estimates that for reactions of interest here, transition temperatures calculated from equation (9.1), using ΔH_r° values determined from heat of solution measurements, will be in error by an amount of the order of $\pm 400\ ^\circ\mathrm{C}$.

The Equilibrium Constant

A fundamental relationship that we have found consistently useful in portraying the properties of equilibrium systems is that given by equation (1.11), namely,
$$\Delta F_r^\circ = -RT \ln K.$$
Because of our interest at this time in the effects of changes of temperature and pressure on the change of free energy of a reaction, ΔF_r, we wish to re-examine (1.11) and related equations.

We consider the general reaction
$$l\mathrm{A}(a_\mathrm{A}) + m\mathrm{B}(a_\mathrm{B}) = n\mathrm{C}(a_\mathrm{C}) + r\mathrm{D}(a_\mathrm{D}), \tag{9.77}$$
where the reacting species may be of any kind, and the activities a_A, a_B, a_C, and a_D may have any values whatsoever. The total pressure may have any value; similarly, the several species may be under different pressures, or in the case of gases, have varying partial pressures. It is understood that the activities of the products and reactants will be measured at the same temperature.

We shall adopt the notation $F_\mathrm{A} = \Delta F_{f\ \mathrm{A}}$, etc., and write for the change in free energy of the reaction
$$\Delta F_r = rF_\mathrm{D} + nF_\mathrm{C} - lF_\mathrm{A} - mF_\mathrm{B}, \tag{9.78}$$
where F_A is the molal free energy of a pure phase, or the partial molal free energy of a dissolved species.

Now consider the same reaction for each species taken in its standard state, i.e., at unit activity. In this case we have
$$l\mathrm{A}(a_\mathrm{A} = 1) + m\mathrm{B}(a_\mathrm{B} = 1) = n\mathrm{C}(a_\mathrm{C} = 1) + r\mathrm{D}(a_\mathrm{D} = 1). \tag{9.79}$$
The change of free energy for (9.79) is given by
$$\Delta F_r^\circ = rF_\mathrm{D}^\circ + nF_\mathrm{C}^\circ - lF_\mathrm{A}^\circ - mF_\mathrm{B}^\circ. \tag{9.80}$$

The quantities, F_A°, etc., are the free energy values of the species in their standard states. We have defined the standard state of a species to be that at 1 atmosphere (or unit fugacity in the case of a real gas), at each specified temperature. Hence, the values of F_A°, F_B°, F_C°, and F_D° depend upon the temperature, and the value of ΔF_r° is thus dependent upon the temperature. But by our definition of standard state, F_A°, etc., are independent of pressure, and consequently ΔF_r° is independent of pressure.

By virtue of equation (1.13) we may write

$$\Delta F_r = \Delta F_r^\circ + RT \ln \frac{[D]^r[C]^n}{[A]^l[B]^m},$$

where the activities may have *any values*. Since ΔF_r is a function of both temperature and pressure, and ΔF_r° a function only of temperature, then the quotient of activities will depend on both temperature and pressure. At equilibrium, $\Delta F_r = 0$, and we have equation (1.11),

$$\Delta F_r^\circ = -RT \ln \frac{[D]^r[C]^n}{[A]^l[B]^m} = -RT \ln K,$$

where the activities now have the *equilibrium values*. Since ΔF_r° is independent of pressure, then K must be independent of pressure. However, as pressure is varied, even though K does not change, the individual activities may change; it is the ratio that must remain constant.

As an illustration we shall calculate the effect of increased total pressure on the reaction

$$3\text{Fe}_c + 2\text{O}_{2\,g} = \text{Fe}_3\text{O}_{4\,c}. \tag{9.81}$$

The equilibrium constant for this reaction is given by the expression

$$K = \frac{[\text{Fe}_3\text{O}_4]}{[\text{Fe}]^3 f_{\text{O}_2}^2}. \tag{9.82}$$

Putting this in logarithmic form, we have

$$\log K = \log [\text{Fe}_3\text{O}_4] - 3 \log [\text{Fe}] - 2 \log f_{\text{O}_2}. \tag{9.83}$$

At 1 atmosphere total pressure, $[\text{Fe}_3\text{O}_4] = [\text{Fe}] = 1$, and equation (9.83) reduces to

$$\log K = -2 \log f_{\text{O}_2}. \tag{9.84}$$

We can evaluate $\log K$ by means of equation (1.11)

$$\Delta F_r^\circ = -RT \ln K,$$

by first calculating ΔF_r° from the ΔF_f°. That is

$$\Delta F_r^\circ = \Delta F_{\text{Fe}_3\text{O}_4}^\circ - 3\Delta F_{\text{Fe}}^\circ - 2\Delta F_{\text{O}_2}^\circ$$

$$\Delta F_{r\,25\,°C}^\circ = -242.4 - 3(0) - 2(0) = -242.4 \text{ kcal/mole.}$$

At 25 °C

$$\Delta F_r^\circ = -1.364 \log K$$

so that

$$\log K = 177.7. \tag{9.85}$$

From (9.84) we have

$$-2 \log f_{\text{O}_2} = 177.7$$

$$f_{\text{O}_2} = 10^{-88.9} \text{ atm.}$$

At this extremely low value of the fugacity, oxygen will behave as an ideal gas, so that we can substitute pressure for fugacity. We have found then that for reaction (9.81), at 25 °C and 1 atmosphere total pressure, the equilibrium partial pressure of oxygen is $10^{-88.9}$ atmosphere. This is, in fact, the result found in Chapter 6.

Let us now calculate the equilibrium partial pressure of oxygen for the same reaction at 25 °C and 1000 atmospheres total pressure. We start again with equation (9.83), but the activities of the solids will not be unity under these conditions. The rate of change of the activity of a pure substance with change of pressure is given by equation (2.24),[28] that is

$$\left(\frac{\partial \ln a_i}{\partial P}\right)_T = \frac{V_i}{RT}. \tag{9.86}$$

Integrating and converting to common logarithms, at constant temperature, (9.86) yields

$$\log a_{i\,P_2} - \log a_{i\,P_1} = \frac{1}{2.303RT} \int_{P_1}^{P_2} V_i \, dP. \tag{9.87}$$

Equation (9.87) can be used to evaluate the activities of Fe_3O_4 and Fe at 1000 atmospheres (25 °C) as follows: at $P = 1$ atmosphere, the activities of the solids are unity, so that for the solids involved in reaction (9.81), equation (9.87) becomes

$$\log [Fe_3O_4] - 3 \log [Fe]$$
$$= \frac{1}{2.303RT} \left\{ \int_1^{1000} V_{Fe_3O_4} \, dP - 3 \int_1^{1000} V_{Fe} \, dP \right\}. \tag{9.88}$$

Equation (9.88) may be put in the form

$$\log [Fe_3O_4] - 3 \log [Fe] = \frac{1}{2.303RT} \int_1^{1000} \Delta V_s \, dP \tag{9.89}$$

where $\Delta V_s = V_{Fe_3O_4} - 3V_{Fe}$.

It can be safely assumed that ΔV_s is constant. The compressibilities of Fe and Fe_3O_4 are not only small, but are also of the same magnitude, so that changes in volume of each of the two solids with increasing pressure is compensatory. For constant ΔV_s, (9.89) can be integrated and we get

$$\log [Fe_3O_4] - 3 \log [Fe] = \frac{999 \Delta V_s}{2.303RT}. \tag{9.90}$$

[28] Equation (2.24) is written in a general way, so that the volume is given as the partial molal volume. For a pure substance this is replaced by the molal volume.

The 25 °C molal volumes of the solids are $V_{Fe_3O_4} = 44.52$ cc, and $V_{Fe} = 7.09$ cc, whence $\Delta V_s = 23.25$ cc. Substituting numerical values into equation (9.90) we get

$$\log [Fe_3O_4] - 3 \log [Fe] = \frac{999 \times 23.25}{2.303 \times 82.06 \times 298.15}$$
$$= 0.412. \tag{9.91}$$

We now have all the data necessary to calculate f_{O_2} at 1000 atmospheres total pressure using equation (9.83), viz.,

$$\log K = \log [Fe_3O_4] - 3 \log [Fe] - 2 \log f_{O_2}.$$

Substituting numerical values from (9.85) and (9.91) in (9.83), we have

$$177.7 = 0.41 - 2 \log f_{O_2}$$
$$2 \log f_{O_2} = -177.3$$
$$f_{O_2} = 10^{-88.7} \text{ atm},$$

and consequently (since under these conditions, $\chi_{O_2} \cong 1$),

$$P_{O_2} = 10^{-88.7} \text{ atm}.$$

Thus, at 1 atmosphere total pressure and 25 °C, the equilibrium P_{O_2} is $10^{-88.9}$ atmosphere, while at 1000 atmospheres total pressure and 25 °C, P_{O_2} is $10^{-88.7}$ atmosphere; this change is extremely small.

In Figures 9.3 to 9.5 we presented several mineral stability diagrams in which the fugacity of the reacting gas, oxygen or sulfur, or the related function $\Delta F^\circ_{r,T}$, was plotted against the temperature. These diagrams hold for 1 atmosphere total pressure. By carrying out calculations of the kind just made, a set of such stability diagrams could be constructed for every pressure of interest. However, as we have just shown, the equilibrium values of the fugacities change very little with total pressure, so that higher pressure isobaric stability diagrams of this type would not differ significantly from the 1 atmosphere diagram, at least at moderately high pressures.

Ionic Equilibria

Problems involving equilibria among aqueous ions at various temperatures and pressures are treated by essentially the same methods as were used in handling the system Fe-Fe_3O_4-O_2 in the previous discussion. For example, suppose we wish to calculate the activity of hydrogen ion in pure water at 35 °C and 2000 atmospheres. For the reaction

$$H_2O_l = H^+_{aq} + OH^-_{aq},$$

the equilibrium constant $K_{35°C}$ is given by

$$K_{35°C} = \frac{[H^+][OH^-]}{[H_2O]} = 2.09 \times 10^{-14}. \tag{9.92}$$

We can calculate a_{H_2O} and a_{OH^-} at 2000 atmospheres, using equation (9.87); for dissolved species we use $\bar{V}_i$, the partial molal volume of the species in (9.87). These calculated activities can be substituted into equation (9.92) and a_{H^+} then calculated. This will be the equilibrium hydrogen ion activity at 2000 atmospheres and 35 °C.

Thus, calculations of equilibria of dissolved species do not differ from those for other types of reactions, except that a knowledge of the *partial molal volumes of ions* is required. Values of molal ionic volumes are sparse and scattered. An excellent treatment of the problem, of particular interest to geochemists, is given by Zen.[29] He considers various ways of deriving values of partial molal volumes of salts from available experimental data, and the problem of resolving these volumes into the partial molal volumes of the ions. Zen lists values for a number of ions and salts. The problem of deriving individual ionic volumes from salt volumes is analogous to the problem of deriving individual activity coefficients from mean activity coefficients—somewhere along the line an extra-thermodynamic assumption must be made.

For a salt in equilibrium with its saturated solution, we may write the equilibrium expression in the form of the ion product. For example, consider the equilibrium reaction for calcite, i.e.,

$$CaCO_{3\,c} = Ca^{++}_{aq} + CO^{--}_{3\,aq}. \tag{9.93}$$

We write

$$[Ca^{++}][CO_3^{--}] = [CaCO_3]K_{calcite} \tag{9.94}$$

and call $[Ca^{++}][CO_3^{--}]$ the *ion product*. This product is often called the *activity product* and equation (9.94) is usually written without the term $[CaCO_3]$ appearing because a pressure of 1 atmosphere is implied.

The activity of calcite can be calculated at elevated pressures, using equation (9.87), since $V_{calcite}$ is known; in addition, $K_{calcite}$ at, e.g., 25 °C, is known, so that the right-hand side of (9.94) can be evaluated for any pressure of interest. This result yields the value of the ion product for that pressure.

Owen and Brinkley[30] have calculated the ion products for a number of important equilibria at pressures to 1000 bars. They have investigated the equilibria in water, as well as in 0.725 m sodium chloride solution.

---

[29] E'an Zen, Partial molal volumes of some salts in aqueous solutions: *Geochim. et Cosmochim. Acta*, 12, 103 (1957).

[30] B. B. Owen and S. R. Brinkley, Jr., Calculation of the effect of pressure upon ionic equilibria in pure water and in salt solutions: *Chem. Revs.* 29, 461 (1941).

Their results are presented in the form of the ratio of an ion product at some pressure P atmosphere to the ion product at 1 atmosphere; some of these results are shown in Figures 9.8 and 9.9.

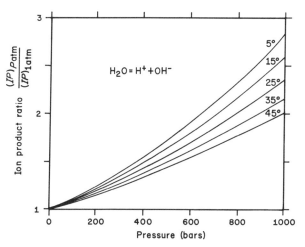

FIG. 9.8. The ion product of water, $[H^+][OH^-]$, as a function of pressure, at several temperatures. [Data from B. B. Owen and S. R. Brinkley, op. cit.]

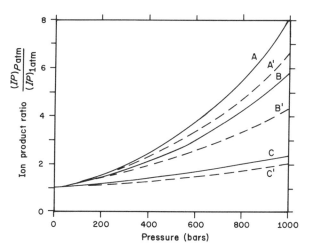

FIG. 9.9. Effect of pressure, at 25 °C, on the ion products of several substances dissolved in water (solid lines), and in 0.725m NaCl solution (dashed lines). A and A′, calcite; B and B′, $CaSO_4$; C and C′, H_2O. [Data from B. B. Owen and S. R. Brinkley, op. cit.; in Fig. 2 of Owen and Brinkley the solid and dashed lines have apparently been erroneously interchanged, judging by the numerical values these authors list for water in the absence of NaCl.]

van't Hoff Equation

The van't Hoff equation, which gives the change of the equilibrium constant with temperature, is a relationship of very general utility. We derive this equation now.

By virtue of equations (9.49) and (9.46) we have

$$\left(\frac{\partial \Delta F_r}{\partial T}\right)_P = -\Delta S_r = \frac{\Delta F_r - \Delta H_r}{T}. \qquad (9.95)$$

Keeping in mind that total pressure is constant, we rewrite (9.95) as

$$d\,\Delta F_r = \frac{(\Delta F_r - \Delta H_r)\,dT}{T}. \qquad (9.96)$$

Dividing both sides of (9.96) by T and rearranging

$$\frac{d\,\Delta F_r}{T} - \frac{\Delta F_r\,dT}{T^2} = -\frac{\Delta H_r\,dT}{T^2}. \qquad (9.97)$$

But the left-hand side of (9.97) is precisely $d(\Delta F_r/T)$. Therefore, (9.97) may be rewritten as

$$d\left(\frac{\Delta F_r}{T}\right) = -\frac{\Delta H_r\,dT}{T^2}, \qquad (9.98)$$

or alternatively as

$$d\left(\frac{\Delta F_r}{T}\right) = \Delta H_r\,d\left(\frac{1}{T}\right). \qquad (9.99)$$

Now we take the thermodynamic functions in (9.98) and (9.99) for standard conditions, i.e., for ΔF_r we substitute ΔF_r°, and for ΔH_r we substitute ΔH_r°. We can then make the further substitution, that for a system at equilibrium

$$\Delta F_r^\circ = -RT \ln K.$$

Then from (9.98) we get

$$\frac{d \ln K}{dT} = \frac{\Delta H_r^\circ}{RT^2}, \qquad (9.100)$$

and from (9.99), we get

$$-d \ln K = \frac{\Delta H_r^\circ}{R} d\left(\frac{1}{T}\right). \qquad (9.101)$$

Equation (9.100) is the *van't Hoff equation*; equation (9.101) is an alternate form. These are equations applicable to systems at equilibrium —this constraint was placed on the equations when ΔF_r° was replaced by $-RT \ln K$. The equations are extremely useful because, for many reactions, ΔH_r° is constant over an appreciable temperature range, and hence the equations can be integrated readily. For constant ΔH_r°, equation (9.101) yields (after conversion to common logarithms)

$$2.303 \log \frac{K_{T_2}}{K_{T_1}} = -\frac{\Delta H_r^\circ}{R}\left(\frac{1}{T_2} - \frac{1}{T_1}\right). \quad (9.102)$$

For constant ΔH_r°, a plot of log K vs. $1/T$ will yield a straight line of slope $-\Delta H_r^\circ/(2.303\,R)$. For example, among a number of reactions investigated at elevated temperatures and pressures by Eugster and Wones[31] in their study of the phase relations of annite, $KFe_3AlSi_3O_{10}(OH)_2$, was the reaction

$$\underset{\text{annite}}{KFe_3AlSi_3O_{10}(OH)_{2\,c}} = \underset{\text{sanidine}}{KAlSi_3O_{8\,c}} + \underset{\text{magnetite}}{Fe_3O_{4\,c}} + H_{2\,g}. \quad (9.103)$$

These investigators found that at all pressures studied for this reaction, the equilibrium constant could be represented by the equation

$$\log K = -\frac{9215}{T} + 10.99,$$

i.e., a plot of log K vs. $(1/T)$ gave a straight line of slope -9215 degrees. Thus,

$$-\frac{\Delta H_r^\circ}{2.303 R} = -9215 \text{ deg}$$

$$\Delta H_r^\circ = 2.303 \times 9215 \text{ deg} \times 1.987 \text{ cal/deg mole}$$

$$\Delta H_r^\circ = 42.2 \text{ kcal/mole}.$$

SUMMARY—FREE ENERGY AS A FUNCTION OF BOTH TEMPERATURE AND PRESSURE

When both temperature and pressure are varied, the change in free energy of reaction is given by

$$d\,\Delta F_r = \left(\frac{\partial \Delta F_r}{\partial T}\right)_P dT + \left(\frac{\partial \Delta F_r}{\partial P}\right)_T dP. \quad (9.65)$$

At equilibrium, $d\,\Delta F_r = 0$, and (9.65) yields the familiar relationships

$$\frac{dP}{dT} = \frac{\Delta S_r}{\Delta V_r} = \frac{\Delta H_r}{T \Delta V_r}. \quad (9.68;\ 9.70)$$

From an experimental P-T plot, using (9.68) and (9.70) and values of the molal volumes and heat capacities of the phases in equilibrium, it is

[31] H. P. Eugster and D. R. Wones, Stability relations of the ferruginous biotite, annite: *J. Petrology*, 3, 82 (1962).

possible to calculate the various thermodynamic functions; an example is provided. The effect of increased pressure on the activity of a condensed phase is considered; an example, involving oxygen in equilibrium with solids, is taken to show how the partial pressure of the oxygen changes with total pressure on the solids. The effect of increased pressure on ionic equilibrium is illustrated by showing how the ion product of water, and of dissolved salts, increases with total pressure. The van't Hoff equation,

$$\frac{d \ln K}{dT} = \frac{\Delta H_r^\circ}{RT^2}, \qquad (9.100)$$

is derived and an example presented to illustrate its utility.

SELECTED REFERENCES

Fyfe, W. S., F. J. Turner, and J. Verhoogen, *Metamorphic Reactions and Metamorphic Facies*. Geol. Soc. Am. Memoir 73, 1958.
  Discussion of certain aspects of the stability of mineral assemblages in metamorphic rocks, including the thermodynamic aspects.
Klotz, I., *Chemical Thermodynamics*. Englewood Cliffs, N.J., Prentice-Hall, 1950.
  Elementary presentation of basic thermodynamics.
Lewis, G. N., and Merle Randall, *Thermodynamics*, 2nd ed., revised by K. S. Pitzer and Leo Brewer. New York, McGraw-Hill, 1961.
  Generally useful for theoretical background, data, and examples of calculations.
Thompson, J. B., Jr., The thermodynamic basis for the mineral facies concept: *Am. J. Sci.*, *253*, 65 (1955).
  A comprehensive and definitive treatment of the thermodynamics of chemical equilibrium among mineral assemblages in rocks. This paper should be carefully studied and restudied by every serious student of geochemistry.

PROBLEMS

9.1. For the reaction

$$Ba_c + S_{rh} + 2O_{2\,g} = BaSO_{4\,c}$$

calculate, for 25 °C, from appropriate values in Appendix 2: (a) ΔH_r°; (b) ΔS_r°; (c) ΔF_r°, using equation (9.1). Compare result in (c) with value in Appendix 2.

9.2. Assuming that ΔC_P° for the reaction in problem 9.1 does not change appreciably with temperature, calculate ΔF_r° at 400 °K from the relation

$$\frac{\Delta(\Delta F_r^\circ)}{\Delta T} = -\Delta S_r^\circ$$

Ans. $\Delta F_r^\circ = -314.2$ kcal/mole.

EFFECTS OF TEMPERATURE AND PRESSURE VARIATIONS

9.3. Calculate ΔH_r° at 400 °C for the reaction

$$H_{2\,g} + \tfrac{1}{2}S_{2\,g} = H_2S_g.$$

[See equation (9.14a).] *Ans.* $\Delta H_r^\circ = -20.5$ kcal/mole.

9.4. From data listed in Tables 9.1 and 9.3, calculate ΔH_f° and S° for sillimanite at 1500 °K.

Ans. $\Delta H_f^\circ = -597.3$ kcal/mole; $S^\circ = 89.14$ cal/deg mole.

What additional data would be needed to calculate ΔF_f° for sillimanite at 1500 °K?

9.5. Calculate the slope and intercept of the plot of ΔF_r° vs. T for the reaction

$$3Fe_c + 2O_{2\,g} = Fe_3O_{4\,c},$$

assuming that equation (9.51) holds.

Ans. Slope = 0.0825 kcal/deg; intercept = -267.0 kcal.

9.6. The equilibrium P_{O_2} value was calculated to be $10^{-88.7}$ atmosphere at 25 °C and 1000 atmospheres total pressure, in Chapter 9, for the reaction in problem 9.5. The 25 °C, 1 bar compressibilities, $\beta_{Fe_3O_4} = 5.5 \times 10^{-7}$ bar^{-1}, and $\beta_{Fe} = 5.94 \times 10^{-7}$ bar^{-1}, were neglected in making this calculation. Show that this neglect was justified.

9.7. The density of diamond is 3.51 g/cm^3, and that of graphite is 2.55 g/cm^3. Calculate the approximate pressure at which the two forms of carbon are in equilibrium at room temperature. *Ans.* $P \simeq 22 \times 10^3$ atm.

9.8. Calculate the equilibrium constant, at 25 °C, for the reaction

$$SrCO_{3\,c} = Sr^{++}_{aq} + CO^{--}_{3\,aq}.$$

What is the value of the ion product, $[Sr^{++}][CO_3^{--}]$, at 5000 atmospheres, if $V_{SrCO_3} = 38.9$ cm^3? *Ans.* $K = 10^{-9.16}$; ion product $= 10^{-5.7}$.

9.9. For the reaction

$$H_{2\,g} + \tfrac{1}{2}S_{2\,g} = H_2S_g; \quad K = \frac{[H_2S]}{[H_2][S_2]^{1/2}}$$

the following experimental equilibrium data have been obtained

T °K	log K	T °K	log K
1023	2.025	1362	0.902
1103	1.710	1405	0.793
1218	1.305	1473	0.643
1338	0.964	1537	0.490

Construct a straight-line plot of the data by using the van't Hoff equation; determine ΔH_r°. *Ans.* $\Delta H_r^\circ \simeq -21$ kcal.

9.10. Calculate ΔF_r° at each temperature given in problem 9.9 from the corresponding value of log K. Construct a plot of ΔF_r° vs. T and determine the slope of the resulting straight line. Compare the value of this slope with ΔS_r° calculated from $\Delta F_{r\,1023}^\circ$ and ΔH_r°.

Ans. Slope $\simeq 0.0115$ kcal/deg; $-\Delta S_r^\circ = 0.0112$ kcal/deg.

Combination Diagrams

The preceding chapters, which stressed stability relations of minerals as functions of partial pressures of gases, or of Eh and pH, illustrated only two types of representations that can be used. It has been implicit in the discussion that equilibria among minerals can be described in terms of any variables that can be used to interconnect them in a balanced chemical reaction. These can include activities of dissolved molecular species or ions in aqueous solution. In the following pages, equilibria among silicates and carbonates are shown as functions of variables chosen to fit the convenience of the investigator.

DIAGRAMS INVOLVING CHIEFLY SILICATES

If the assumption is made that silicate minerals are in equilibrium with the pore waters that bathe them, the interrelations of the minerals can be shown as functions of the activities of the ions dissolved in the water. Thermodynamic data for silicates are sparse, and those that are available are uneven in their accuracy. Even so, it is possible to develop qualitative diagrams that are useful, if only to provide a graphic summary of the mineral sequences that might be expected if equilibrium were attained.

The diagrams that follow were developed in an attempt to investigate silicate mineral relations in the zone of weathering, in rocks at shallow depth permeated by ground waters, and in sediments undergoing shallow diagenesis. For these situations, diagrams constructed at 25 °C and 1 atmosphere total pressure are appropriate.

KAOLINITE-GIBBSITE RELATIONS

The essence of the problem of the origin of bauxites lies in the relation between kaolinite ($H_4Al_2Si_2O_9$) and gibbsite ($Al_2O_3 \cdot 3H_2O$). In many areas, kaolinite is clearly the mineral that precedes final dissolution of solids by the agents of weathering; elsewhere, the persistent species is gibbsite. If equilibrium between kaolinite, gibbsite, dissolved silica, and water is assumed, we can write

$$\underset{\text{kaolinite}}{H_4Al_2Si_2O_{9\,c}} + \underset{\text{water}}{5H_2O_l} = \underset{\text{gibbsite}}{Al_2O_3 \cdot 3H_2O_c} + \underset{\text{dissolved silica}}{2H_4SiO_{4\,aq}}. \quad (10.1)$$

Under the severe leaching conditions generally accepted as required for the formation of gibbsite, the pore water is reasonably pure, so its activity is fixed at unity. Then the equilibrium constant for equation (10.1) is

$$[H_4SiO_4]^2 = K. \quad (10.2)$$

Applying free-energy values to equation (10.1)

$$\Delta F_r^\circ = \Delta F_{f\,\text{gibbsite}}^\circ + 2\Delta F_{f\,H_4SiO_4}^\circ - \Delta F_{f\,\text{kaolinite}}^\circ - 5\Delta F_{f\,H_2O}^\circ$$

$$\Delta F_r^\circ = -554.6 + 2(-300.3) - (-884.5) - 5(-56.69)$$

$$\Delta F_r^\circ = 12.7 \text{ kcal} = -1.364 \log [H_4SiO_4]^2 \quad (10.3)$$

$$K^{1/2} = [H_4SiO_4] = 10^{-4.7}. \quad (10.4)$$

Consequently, equilibrium is attained between gibbsite and kaolinite at a fixed value of the activity (= concentration) of dissolved silica of about 2 ppm (as H_4SiO_4; 1 ppm as SiO_2). At any value of dissolved silica below this value, kaolinite should tend to dissolve incongruently to leave a residuum of gibbsite, with silica disappearing in solution. But what controls the silica content of the soil solution?

Additional information can be obtained by recognizing that if gibbsite is in equilibrium with pore water, it is in equilibrium with its various dissociation products, i.e.

$$Al_2O_3 \cdot 3H_2O_c + 6H_{aq}^+ = 2Al_{aq}^{3+} + 6H_2O_l \quad (10.5)$$

$$Al_2O_3 \cdot 3H_2O_c = 2H_{aq}^+ + 2AlO_{2\,aq}^- + 2H_2O_l. \quad (10.6)$$

Equilibrium constants for these reactions can be obtained in the usual way from ΔF_f° values. For reaction (10.5)

$$K_1 = \frac{[Al^{3+}]}{[H^+]^3} \quad (10.7a)[1]$$

$$\log K_1 = 5.7 = \log [Al^{3+}] + 3\,pH. \quad (10.7b)$$

For equation (10.6)

$$K = [H^+][AlO_2^-]$$

$$\log K_2 = -14.6 = \log [AlO_2^-] - pH. \quad (10.8)$$

[1] K_1 as given for equation (10.7a) is actually the square root of the equilibrium constant for the reaction as written. In this and similar cases, such a simplification will be made.

Figure 10.1 is a plot in terms of equations (10.7b) and (10.8), and shows how a field of stability of gibbsite can be delineated in terms of $[Al^{3+}]$ and $[AlO_2^-]$. The relations in Figure 10.1 illustrate the general behavior to be expected from gibbsite; it should dissolve to yield aluminum ion under acid conditions, and aluminate ion under alkaline conditions. The ion activities are equal at pH 5.1. This should not be looked upon as an accurate *solubility* diagram, because aluminum forms a variety of ions in aqueous solution, especially on the acid side, where the small Al^{3+} tends to form complexes with many anions. However, the implication is clear that in dilute solutions, aluminum content must be very low to prevent precipitation of aluminum under most pH conditions.

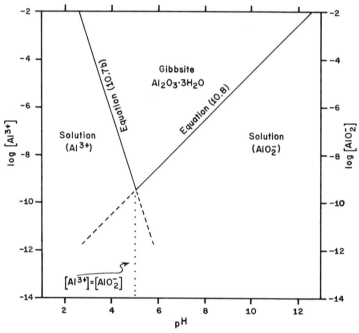

FIG. 10.1. Stability of gibbsite expressed in terms of pH, and activities of aluminum ion and aluminate ion, at 25 °C and 1 atmosphere. Equation numbers refer to the text.

Gibbsite is almost invariably accompanied by one or both of the dimorphs boehmite and diaspore $(Al_2O_3 \cdot H_2O)$. Their relations to gibbsite and to each other are pertinent to the weathering problem, even though they are metastable in pure water. For the reaction at standard conditions

$$Al_2O_3 \cdot H_2O = Al_2O_3 \cdot H_2O \qquad \Delta F_r^\circ \cong -0.1 \text{ kcal.}$$
$$\underset{\text{boehmite}}{} \quad \underset{\text{diaspore}}{}$$

The free-energy change is very small, but is in favor of diaspore by about 0.1 kilocalorie. This is in accord with the formation and persistence of both phases during weathering.

Gibbsite and diaspore are related by the reaction

$$Al_2O_3 \cdot 3H_2O_c = Al_2O_3 \cdot H_2O_c + 2H_2O_1.$$

In the presence of pure water, this reaction runs to the left, with a driving force of several kilocalories, indicating that diaspore should hydrate to gibbsite. The reaction is exceedingly slow at room temperature; we can guess that the existence of gibbsite and diaspore may be controlled by two factors. One may be the metastable formation and persistence of diaspore; the other, the formation of diaspore during periods of drying, in which the activity of water in the environment is lowered far below that of pure water, followed by metastable persistence of diaspore during wetter conditions.

Kaolinite stability can be described similarly to that of gibbsite. The dissociation reactions are

$$H_4Al_2Si_2O_9 + 6H^+_{aq} = 2Al^{3+}_{aq} + 2H_4SiO_{4\ aq} + H_2O_1 \qquad (10.9)$$

$$H_4Al_2Si_2O_9 + 3H_2O_1 = 2H^+_{aq} + 2AlO_2^- + 2H_4SiO_{4\ aq}. \qquad (10.10)$$

The equilibrium constant for reaction 10.9 is

$$K_1 = \frac{[Al^{3+}][H_4SiO_4]}{[H^+]^3}$$

$$\log K_1 = 1.0 = \log [Al^{3+}] + \log [H_4SiO_4] + 3\,\mathrm{pH}. \qquad (10.11)$$

The equilibrium constant for reaction (10.10) is

$$K_2 = [H^+][AlO_2^-][H_4SiO_4]$$

$$\log K_2 = -19.3 = \log [AlO_2^-] + \log [H_4SiO_4] - \mathrm{pH}. \qquad (10.12)$$

Equations (10.11) and (10.12), describing the dissociation of kaolinite, are analogous to equations (10.7b) and (10.8), describing the dissociation of gibbsite, except that each contains an additional term for dissolved silica. Because of this added term, stability relations for kaolinite must be depicted as a volume, rather than as an area. Figure 10.2 shows this volume, which is included between the intersecting planes defined by equations (10.11) and (10.12). These planes are labeled with the equation number that pertains to them. The method of construction was to fix the silica activity at 10^{-8}, which permitted drawing of the relations for the front face of the block; and then to fix it at $10^{-2.0}$, which fixed the geometry of the back face. The rest of the construction follows directly.

The front face of Figure 10.2, at constant $[H_4SiO_4]$, has the same geometry as Figure 10.1, which represents the stability of gibbsite. With increasing $[H_4SiO_4]$, kaolinite stability increases, reaching a maximum in solutions saturated with amorphous silica. The saturation value obtained from the observed solubility of H_4SiO_4 is about $10^{-2.6}$, which can be equated to activity in dilute solutions.

So far, gibbsite and kaolinite have been considered separately. But if it is recognized that the stability of gibbsite can be shown as a volume on a diagram using pH, Al^{3+}, AlO_2^-, and H_4SiO_4 as variables, even though its stability alone is independent of $[H_4SiO_4]$, it becomes possible to combine Figures 10.1 and 10.2 into Figure 10.3. The boundary

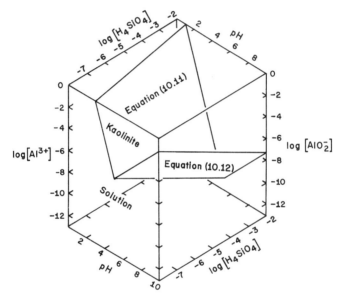

FIG. 10.2. Stability field of kaolinite as defined by its dissociation into Al^{3+}, AlO_2^-, H^+, and H_4SiO_4, at 25 °C.

between gibbsite and kaolinite is a plane representing a fixed activity of H_4SiO_4, in accord with equations (10.1) and (10.2). At H_4SiO_4 activities greater than this value, kaolinite can be expected to dissolve congruently, and thus to leave no gibbsite residue. The prevalence of kaolinite in the A horizons of most soils formed under strong leaching conditions, leads to the prediction that most soil solution compositions are toward the rear of Figure 10.3.

Figure 10.4 is an enlarged diagram showing the gibbsite-kaolinite relations as functions of $[Al^{3+}]$ rather than as a function of both Al^{3+} and AlO_2^-, with the pH axis of Figure 10.3 reversed to clarify solution-solid relations in acid solutions. The dashed arrows show the change in solution composition resulting if kaolinite dissolves at fixed pH. Under such conditions the ratio of H_4SiO_4 to Al^{3+} in solution is maintained at unity [see equation (10.9)]. If a kaolinite soil is subjected to leaching by rain water, with pH fixed at 4, perhaps by living plants or decomposing organic materials which release CO_2 to the soil water, the solution composition will change along the line indicated by arrow no. 1, and the solid field intersected is that of kaolinite. In other words, kaolinite dissolves congruently to leave a kaolinite residue. On the other hand, if

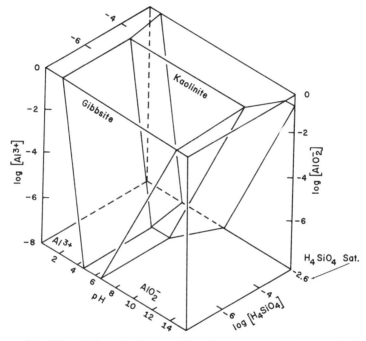

FIG. 10.3. Gibbsite-kaolinite stability fields expressed in terms of pH and activities of the dissociation products Al^{3+}, AlO_2^-, and H_4SiO_4, at 25 °C and 1 atmosphere.

there is no pH buffering, H^+ will be used up when kaolinite dissolves, and the solution composition will change along a path similar to that shown by arrow no. 2, indicating incongruent solution of kaolinite to produce a gibbsite residue.

These diagrams thus lead to a hypothesis concerning some major controls of the formation of bauxites, and serve to show the utility of such methods of representation as stimuli to the collection of data about relations that might not otherwise be deduced. Most soils in areas of high rainfall and concomitant high leaching rates are kaolinitic; they

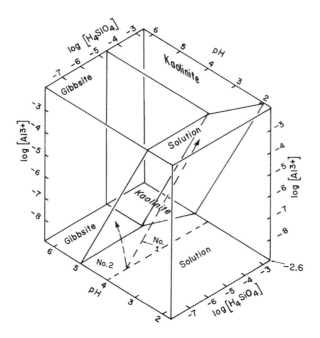

FIG. 10.4. Gibbsite-kaolinite-solution relations at 25 °C and 1 atmosphere, for [H+] values on the acid side. The direction of the pH scale is reversed from that of Figure 10.3. Arrow 1 shows that a solution maintained at pH 4, and increasing in Al^{3+} and $H_4SiO_{4\ aq}$ by dissolving kaolinite will intersect the kaolinite stability field; arrow 2 that lack of pH buffering results in a gibbsite residue.

also tend to have low pH values. The low pH apparently results in part from a high CO_2 content in the soil atmosphere, an effect induced by decomposition of organic materials in the soil, and in part from pH lowering at the surfaces of roots. Perhaps under exceptional conditions of high rainfall and high temperature, bacterial decomposition is so rapid that CO_2 resulting from organic decomposition passes directly into the atmosphere, and the soils are so heavily leached that plant ground cover is relatively sparse. If these conditions occur, rainwater descending through the soil will be relatively unbuffered, and a gibbsite residue could result. Whatever the details of the explanation, it does appear that the requirement of abundant unbuffered soil water for bauxite formation is in harmony with the occurrence of low silica-high alumina soil residues. This conclusion is bolstered by analyses of ground waters in Jamaica that drain a terrain composed of a mixture of kaolinite and bauxite.[2] The silica content of these waters is 3 to 6 ppm, near the value of 1 ppm (as SiO_2) predicted for equilibrium between gibbsite and kaolinite.

[2] V. G. Hill and A. C. Ellington, *Econ. Geol.*, **56**, 533 (1961).

GIBBSITE-KAOLINITE-MICA-FELDSPAR RELATIONS

Figures 10.2 and 10.3 are good examples to show how increase in the number of phases of a system of fixed components restricts the field of stable occurrence of a given phase, and hence increases its usefulness as an indicator of environment. Figure 10.2 shows the result of considering kaolinite alone; Figure 10.3 shows how addition of gibbsite to the diagram restricts the occurrence of kaolinite. Figure 10.3 shows that kaolinite can occur stably only between solution compositions with sufficient H_4SiO_4 to silicate gibbsite and those saturated with amorphous silica. The geochemist's ideal mineral is one that has a small field of stable existence, but which has that field in the range of conditions commonly encountered. By adding one more component, namely, K_2O, to the Al_2O_3-SiO_2-H_2O system, two more phases of interest in low-temperature systems can be investigated. These phases are K-feldspar and K-mica.

Before choosing the dissolved species that are most useful in delineating relations in a given system, it is usually instructive and helpful to make a

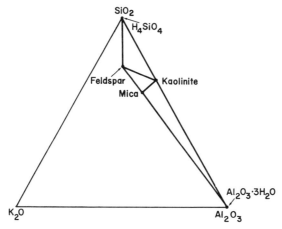

FIG. 10.5. Assumed associations in the system involving feldspar, mica, kaolinite, gibbsite, and amorphous silica at 25 °C and 1 atmosphere. The association feldspar-kaolinite presumably would be replaced by mica-quartz in the stable system with quartz rather than H_4SiO_4.

composition diagram and to show compatibilities as a guide to writing reactions. Figure 10.5 is a triangular diagram showing the system K_2O-Al_2O_3-SiO_2-H_2O. Because nearly pure water is assumed as an associate of all minerals in this system, the K_2O-Al_2O_3-SiO_2 triangle is all that is required, with the composition of the hydrated phases projected from their positions on the tetrahedron required for complete representation. The compatibilities, as estimated from the higher temperature relations are K-feldspar–K-mica; K-feldspar–kaolinite; K-mica–kaolinite; K-mica–gibbsite; kaolinite-gibbsite; K-feldspar–H_4SiO_4; and kaolinite-H_4SiO_4. In high-temperature work the last two compatibilities are with quartz, rather than with H_4SiO_4 (amorphous silica). However, as shown by Siever,[3] the maximum silica solubility at low temperature is controlled by amorphous silica rather than by quartz. The rate of crystallization of quartz is so slow that amorphous silica, a metastable phase, should be looked upon as the upper limit of dissolved silica content of natural waters for most low-temperature processes.

When reactions are written for the mineral compatibilities just given, it is discovered that only three variables need be considered: [K^+], [H^+], and [H_4SiO_4]. Furthermore, the ratio of [K^+] to [H^+] in the various equilibrium constants is always unity. Therefore, all the mineral relations can be described in two-dimensional representations involving the ratio of [K^+] to [H^+] as one axis, and the activity of H_4SiO_4 as the other. The reactions and their equilibrium constants are as follows

$$3\underset{\text{K-feldspar}}{KAlSi_3O_{8\,c}} + 2H^+_{aq} + 12H_2O_l$$

$$= \underset{\text{K-mica}}{KAl_3Si_3O_{10}(OH)_{2\,c}} + 6\underset{\substack{\text{dissolved}\\\text{silica}}}{H_4SiO_{4\,aq}} + 2K^+_{aq} \quad (10.13)$$

$$K_1 = \frac{[K^+][H_4SiO_4]^3}{[H^+]}$$

$$\log K_1 = -4.9 = \log \frac{[K^+]}{[H^+]} + 3 \log [H_4SiO_4] \quad (10.14)$$

$$2\underset{\text{K-feldspar}}{KAlSi_3O_{8\,c}} + 2H^+_{aq} + 9H_2O_l = \underset{\text{kaolinite}}{H_4Al_2Si_2O_{9\,c}} + 2K^+_{aq} + 4H_4SiO_{4\,aq}$$
$$(10.15)$$

$$K_2 = \frac{[K^+][H_4SiO_4]^2}{[H^+]}$$

$$\log K_2 = -1.0 = \log \frac{[K^+]}{[H^+]} + 2 \log [H_4SiO_4] \quad (10.16)$$

$$2\underset{\text{K-mica}}{KAl_3Si_3O_{10}(OH)_{2\,c}} + 2H^+_{aq} + 3H_2O_l = 3\underset{\text{kaolinite}}{H_4Al_2Si_2O_{9\,c}} + 2K^+_{aq}$$
$$(10.17)$$

$$K_3 = \frac{[K^+]}{[H^+]}$$

$$\log K_3 = +6.5 = \log \frac{[K^+]}{[H^+]} \quad (10.18)$$

[3] R. Siever, *Am. Mineral*, 42, 826 (1957).

$$2\text{KAl}_3\text{Si}_3\text{O}_{10}(\text{OH})_{2\,\text{c}} + 2\text{H}^+_{\text{aq}} + 18\text{H}_2\text{O}_1$$
$$\underset{\text{K-mica}}{}$$

$$= 3\text{Al}_2\text{O}_3\cdot 3\text{H}_2\text{O}_{\text{c}} + 2\text{K}^+_{\text{aq}} + 6\text{H}_4\text{SiO}_{4\,\text{aq}} \quad (10.19)$$
$$\underset{\text{gibbsite}}{} \qquad \underset{\text{dissolved silica}}{}$$

$$K_4 = \frac{[\text{K}^+][\text{H}_4\text{SiO}_4]^3}{[\text{H}^+]}$$

$$\log K_4 = -7.6 = \log\frac{[\text{K}^+]}{[\text{H}^+]} + 3\log[\text{H}_4\text{SiO}_4] \quad (10.20)$$

$$\text{H}_4\text{Al}_2\text{Si}_2\text{O}_{9\,\text{c}} + 5\text{H}_2\text{O}_1 = \text{Al}_2\text{O}_3\cdot 3\text{H}_2\text{O}_{\text{c}} + 2\text{H}_4\text{SiO}_{4\,\text{aq}} \quad (10.21)$$
$$\underset{\text{kaolinite}}{} \qquad \underset{\text{gibbsite}}{} \qquad \underset{\text{dissolved silica}}{}$$

$$K_5 = [\text{H}_4\text{SiO}_4]$$
$$\log K_5 = -4.7 = \log[\text{H}_4\text{SiO}_4]. \quad (10.22)$$

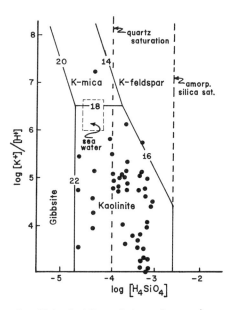

FIG. 10.6. Stability relations of some phases in the system K_2O-Al_2O_3-SiO_2-H_2O at 25 °C and 1 atmosphere, as functions of $[K^+]/[H^+]$ and $[H_4SiO_4]$. Numbers in boundary lines refer to equations in text. Solid circles represent analyses of waters in arkosic sediments.

Figure 10.6 shows the preceding relations as a plot of $\log[K^+]/[H^+]$ vs. $\log[H_4SiO_4]$. Among the aspects of interest is that most ground and stream waters fall into the kaolinite field of stability. Also, they have dissolved silica contents ranging between the solubility of quartz and that of amorphous silica (solid circles). Mica and kaolinite are predicted as intermediate phases in the bauxitization of K-feldspar. The association kaolinite and K-feldspar, observed so commonly in examination of weathering products, is indicated here as a metastable association permitted only by the reluctance of quartz to crystallize at low temperatures. The position of the bulk of the analyses is in harmony with the "aggressiveness" of most stream and ground waters—they tend to kaolinize rock minerals. If "pure" rain water (H_2O in equilibrium with $10^{-3.5}$ atmosphere CO_2 pressure) were permitted to move very slowly through a pure potassium feldspar arkose, the water composition might move from nearly pure water to the gibbsite-kaolinite boundary, then up that boundary to the gibbsite-kaolinite-mica intersection, and then across to the kaolinite-mica-feldspar intersection. It would be of interest to make a series of analyses of ground waters in a micaeous arkose as a function of distance from the area of catchment. Such analyses should give valuable information on the relative rates of equilibration of feldspar and mica. The presence of silica in most waters in excess of that predicted for reactions with K-mica and K-feldspar can be attributed to the presence of other minerals that equilibrate more rapidly than K-feldspar.

ALBITE-ANALCITE-MONTMORILLONITE-KAOLINITE-GIBBSITE RELATIONS

The relations deduced for the system K_2O-Al_2O_3-SiO_2-H_2O raise immediately in the investigator's mind questions of parallel relations for phases in the system Na_2O-Al_2O_3-SiO_2-H_2O. Figure 10.7 has been constructed by writing reactions similar to those for phases in the system containing K_2O. However, free-energy data are not available for albite or analcite, so that although the general geometry is reliable, dictated by the equilibrium constants, the actual numbers on the axis are estimated. The diagram is adapted from an original by Kurt Linn,[4] who based his estimate of room temperature compatibilities on an extrapolation of the experimental relations at higher temperatures and pressures. It was necessary to use a three-dimensional diagram because the presence of Na-montmorillonite yields equilibrium constants for which the ratio $[Na^+]/[H^+]$ is not unity. The position of the analcite-albite boundary is placed beyond the value for saturation with H_4SiO_4 because of the common occurrence of analcite and chalcedony in sedimentary rocks, which implies an initial association of analcite and amorphous silica. The presence of authigenic albite, however, indicates that the reaction,

[4] Kurt Linn, personal communication, 1959.

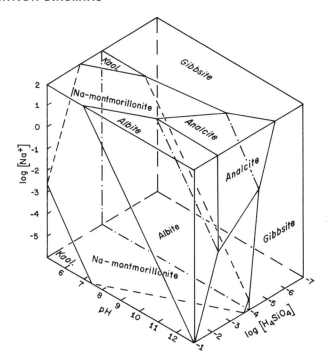

FIG. 10.7. Stability relations of some phases in the system $Na_2O-Al_2O_3-SiO_2-H_2O$ at 25 °C and 1 atmosphere, as functions of $[Na^+]$, pH, and $[H_4SiO_4]$. The diagram shown is strictly an estimated one based on extrapolation of experimental relations at higher temperatures and pressures, and the numerical values are approximate ones only. [After K. Linn, personal communication, 1959.]

analcite + amorphous silica = albite + water, which is in the go-no-go category, probably goes to the left at room temperature, and to the right at a slightly higher temperature. At any rate, analcite is seen as a mineral expected at high $[Na^+]/[H^+]$ ratio at room temperature, a mineral to be found in alkaline sodic waters.

$K_2O-MgO-Al_2O_3-SiO_2-H_2O$ RELATIONS

As in the case of systems containing Na_2O, free-energy data are not available for systems containing MgO. As before, it is possible to express mineral stabilities qualitatively as functions of the activities of ions in the aqueous environment. Linn[5] has made an estimate of the stability relations at room temperature in the system involving pure magnesium chlorite, phlogopite, kaolinite, K-mica, and K-feldspar.

[5] Kurt Linn, personal communication, 1959.

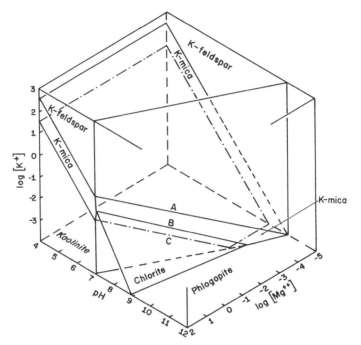

FIG. 10.8. Preliminary diagram showing suggested stability relations in part of the system K_2O-MgO-FeO-Al_2O_3-SiO_2-H_2O at 25 °C and 1 atmosphere. Quartz, "FeO," and H_2O are assumed to be present in excess. The lines marked with letters are lines of three-phase equilibria, as follows: (A) K-feldspar–phlogopite–K-mica, (B) K-mica–phlogopite–chlorite, (C) kaolinite–K-mica–chlorite. [After K. Linn, personal communication, 1959.]

The results are shown in Figure 10.8. Although the numbers on the axes are not trustworthy, the general configuration is comparable to that determined experimentally at high temperature.[6] Perhaps the most interesting aspect of the diagram is illustrated by the stability relations shown clearly on the front face, where chlorite, Mg-mica, kaolinite, K-feldspar, and K-mica all are stable within a relatively small range of $[K^+]$ and pH at a given value of $[Mg^{++}]$. Perhaps this corresponds to the relations in seawater, in which detrital minerals seem to persist for long periods of time with little change.

In a sense, seawater can be considered to be a solution that tries to "please" all the minerals dumped into it. Although this is patently impossible, the composition of seawater may be so slightly removed from that required for stability of a great many phases that there is little chemical potential available for equilibration; hence rates of change of unstable species may well be vanishingly low.

[6] J. J. Hemley, personal communication, 1960.

Figure 10.8 also gives some insight into the problem of hydrothermal alteration of felsic rocks adjacent to sulfide ore deposits. The general sequence of phases outward from the vein (sericite, kaolinite, chlorite, feldspar, and biotite) can reasonably result from a continuous decrease in the ratio of [K$^+$] to [H$^+$], accompanied by an increase in [Mg^{++}].

SILICATE DIAGRAMS AT ELEVATED TEMPERATURES

Diagrams illustrating relations among silicates can be constructed for many temperatures at which equilibrium constants can be obtained. For example, Hemley et al.[7] studied the relations among phases in the systems Na_2O-Al_2O_3-SiO_2-H_2O and K_2O-Al_2O_3-SiO_2-H_2O as a function of temperature and the ratio of aqueous NaCl to HCl or KCl to HCl, keeping the activity of silica constant at a given temperature by having quartz present in all experiments. Figure 10.9 is taken from their work. For a reaction such as a change from albite to paragonite

$$3NaAlSi_3O_{8\ c} + 2HCl_{aq}$$
albite

$$= NaAl_3Si_3O_{10}(OH)_{2\ c}$$
paragonite

$$+ 2NaCl_{aq} + 6SiO_{2\ c,\ qtz}.$$

(10.23)

The equilibrium constant is

$$K = \frac{[NaCl]}{[HCl]}.$$

Hemley et al. (op. cit.) actually measured the ratio m_{NaCl}/m_{HCl}, but it is probable that this ratio is close to the ratio of [NaCl]/[HCl] or [Na$^+$]/[H$^+$] (see Chapter 2). At any rate, Figure 10.9 shows that equilibrium

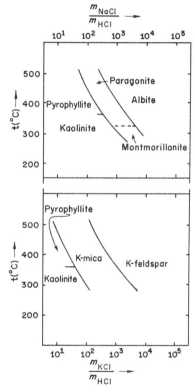

FIG. 10.9. Some stability relations in the systems Na_2O-Al_2O_3-SiO_2-H_2O and K_2O-Al_2O_3-SiO_2-H_2O as functions of temperature and the ratios m_{NaCl}/m_{HCl} and m_{KCl}/m_{HCl}. Total pressure is 15,000 psi and quartz is present in excess. [After Hemley et al., op. cit.]

---

[7] J. J. Hemley, Charles Meyer, and D. H. Richter, U.S. Geol. Sur. Prof. Paper 424-D, 338 (1961).

relations among silicates can be expressed in terms of temperature and properties of an aqueous solution coexisting with the minerals.

By combining the work of Orville[8] on the alkali ion exchange between vapor and feldspar phases and that of Hemley et al., it is possible to construct isothermal diagrams for some important silicate phases at high temperatures, and to express the relations among the silicates in terms of solution composition.

Figure 10.10 is a diagram for 300 °C and a total pressure of 15,000 psi. It describes the stability relations of the feldspars, micas, and of kaolinite in terms of m_{NaCl}, m_{KCl}, and m_{HCl}. The data used to construct the

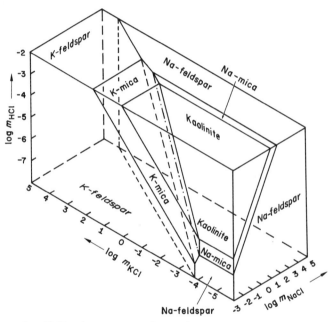

FIG. 10.10. Approximate phase relations in the system Na_2O-K_2O-Al_2O_3-SiO_2-H_2O at 300 °C and 15,000 psi; quartz is assumed to be present in excess. [Diagram constructed from data taken from Hemley et al., op. cit., and from Orville, op. cit., by O. R. Ekstrand and R. M. Garrels, personal communication. Previously published in Harold C. Helgeson, *Complexing and Hydrothermal Ore Deposition*, New York, Pergamon, 1964.]

diagram, and the application of graphical methods to check the data, warrant detailed discussion.

Let us consider details of the construction of the top surface of the block shown in Figure 10.10. This surface is at a fixed concentration of HCl. The experiments of Hemley et al. provide us with the relations of

[8] P. M. Orville, *Am. J. Sci.*, *261*, 201 (1963).

K-feldspar, K-mica, and kaolinite, as well as Na-feldspar, Na-mica, and kaolinite, as functions of the ratio of KCl or NaCl to HCl at a given temperature. Therefore, if a value of HCl is stipulated, we can obtain values of the concentration of NaCl or KCl at which various pairs of these minerals are in equilibrium. Orville (*op. cit.*) determined the ratio of KCl to NaCl for equilibrium between Na-feldspar and K-feldspar as a function of temperature, so from his data it is possible to obtain the equilibrium ratio at a given temperature, such as the one of interest (300 °C). Then we can write the equations relating the solid phases, $NaCl_{aq}$ and KCl_{aq}, and obtain equilibrium boundaries between the phases. If the two sets of data are consistent, the lines should all join to form a complete equilibrium diagram for 300 °C, with the assumption that quartz is ubiquitous and that HCl concentration is constant at $10^{-2} m$. (In addition, it is assumed that equilibrium is attained at 300 °C between Na-feldspar and Na-mica, rather than between Na-feldspar and montmorillonite. The metastable extension of the Na-feldspar–Na-mica curve in Figure 10.9 must lie very close to the boundary given for Na-feldspar–Na-montmorillonite.)

Considering the reactions individually at 300 °C, and reading the values of the equilibrium constants from Figure 10.9

$$3\underset{\text{Na-feldspar}}{NaAlSi_3O_8}{}_c + 2HCl_{aq}$$
$$= \underset{\text{Na-mica}}{NaAl_3Si_3O_{10}(OH)_2}{}_c + 2NaCl_{aq} + 6\underset{\text{quartz}}{SiO_2}{}_c$$

$$K_1 = \frac{[NaCl]}{[HCl]} \simeq \frac{m_{NaCl}}{m_{HCl}} \simeq \frac{[Na^+]}{[H^+]} \simeq 10^{3.8}.$$

If $m_{HCl} = 10^{-2}$, then m_{NaCl} is $10^{1.8}$ \hfill (10.24)

$$2\underset{\text{Na-mica}}{NaAl_3Si_3O_{10}(OH)_2}{}_c + 2HCl_{aq} + 3H_2O = 3\underset{\text{kaolinite}}{H_4Al_2Si_2O_9}{}_c + 2NaCl_{aq}$$

$$K_2 = \frac{[NaCl]}{[HCl]} \simeq \frac{m_{NaCl}}{m_{HCl}} \simeq \frac{[Na^+]}{[H^+]} \simeq 10^{3.1}.$$

If $m_{HCl} = 10^{-2}$, then m_{NaCl} is $10^{1.1}$. \hfill (10.25)

$$3\underset{\text{K-feldspar}}{KAlSi_3O_8}{}_c + 2HCl_{aq} = \underset{\text{K-mica}}{KAl_3Si_3O_{10}(OH)_2}{}_c + 2KCl_{aq} + 6\underset{\text{quartz}}{SiO_2}{}_c$$

$$K_3 = \frac{[KCl]}{[HCl]} \simeq \frac{m_{KCl}}{m_{HCl}} \simeq \frac{[K^+]}{[H^+]} \simeq 10^{3.6}.$$

If $m_{HCl} = 10^{-2}$, then m_{KCl} is $10^{1.6}$. \hfill (10.26)

$$3H_2O + 2\underset{\text{K-mica}}{KAl_3Si_3O_{10}(OH)_2}{}_c + 2HCl_{aq} = 3\underset{\text{kaolinite}}{H_4Al_2Si_2O_9}{}_c + 2KCl_{aq}$$

$$K_4 = \frac{[KCl]}{[HCl]} \simeq \frac{m_{KCl}}{m_{HCl}} \simeq \frac{[K^+]}{[H^+]} \simeq 10^{2.0}.$$

If $m_{HCl} = 10^{-2}$, then m_{KCl} is 10^0. \hfill (10.27)

These values of m_{NaCl} and m_{KCl} are shown on Figure 10.11, and are numbered as in the text. From Orville's work, we obtain, for 300 °C

$$\underset{\text{K-feldspar}}{KAlSi_3O_{8\ c}} + NaCl_{aq} = \underset{\text{Na-feldspar}}{NaAlSi_3O_{8\ c}} + KCl_{aq}$$

$$K = \frac{[KCl]}{[NaCl]} \cong \frac{m_{KCl}}{m_{NaCl}} \cong \frac{[K^+]}{[Na^+]} \cong 10^{-0.8}. \tag{10.28}$$

Therefore the boundary between Na-feldspar and K-feldspar must be fixed on Figure 10.11 at this ratio and the line must have a 1:1 slope. A line showing this ratio and slope is labeled on the diagram with the equation number.

Also, we can determine the slope of the boundary between Na-mica and K-mica from the reaction

$$\underset{\text{Na-mica}}{NaAl_3Si_3O_{10}(OH)_{2\ c}} + KCl_{aq} = \underset{\text{K-mica}}{KAl_3Si_3O_{10}(OH)_{2\ c}} + NaCl_{aq}$$

$$K = \frac{[NaCl]}{[KCl]} \cong \frac{m_{NaCl}}{m_{KCl}} \cong \frac{[Na^+]}{[K^+]}. \tag{10.29}$$

The equilibrium boundary between these phases must be a line on Figure 10.11 with unit slope, but its position can be determined only by relations to other equilibria.

Finally, the slope of the boundary between K-mica and Na-feldspar can be obtained

$$\underset{\text{K-mica}}{KAl_3Si_3O_{10}(OH)_{2\ c}} + \underset{\text{quartz}}{6SiO_{2\ c}} + 3NaCl_{aq} = \underset{\text{Na-feldspar}}{3NaAlSi_3O_8}$$
$$+ 2HCl_{aq} + KCl_{aq}$$

$$K = \frac{[KCl][HCl]^2}{[NaCl]^3} \cong \frac{m_{KCl}m_{HCl}^2}{m_{NaCl}^3} \cong \frac{[K^+][H^+]^2}{[Na^+]^3}. \tag{10.30}$$

Figure 10.12 shows the completed diagram, with metastable extensions of lines of Figure 10.11 removed. A small inconsistency in the data is shown by the difference in slope of line (10.30) from the slope predicted by the equation (dashed line). However, this difference is no more than would be expected from the accuracy of reading values for equilibrium constants from Figure 10.9.

The preceding discussion shows how the experimental data were used to construct the top surface of Figure 10.10. Any other section, representing a different m_{HCl}, can be constructed similarly, and the composite of such sections gives the block diagram.

The success of this amalgamation of data from Hemley *et al.* and from Orville has some far-reaching implications. First, it shows that use of molalities of the solution constituents gives geometric relations predicted on the basis of activities. In other words, the molality ratios of HCl,

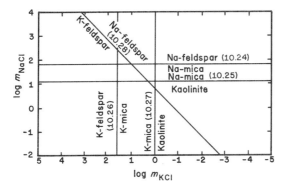

FIG. 10.11. Plot of experimental values of m_{KCl} and m_{NaCl} for equilibrium between various solid phases (and solution) at 300 °C and 15,000 psi. Quartz is present in excess and $m_{HCl} = 10^{-2}$. The numbers in parentheses are equation numbers. [Data from Hemley et al., op. cit., and Orville, op. cit.]

NaCl, and KCl must be nearly the same as their activity ratios; activity coefficients of these salts must be similar under the experimental conditions. Second, and perhaps even more important, the data from Hemley et al. were from experiments in which the solids contained only Na or K, whereas those of Orville involved coexisting Na and K feldspars, each of which contained a few mole percent of the other component. All of the equilibrium constants used in the preceding discussion were derived on the assumption that the activity of the solid phases was unity—i.e., each was in its pure state. Corrections for the deviations from unit

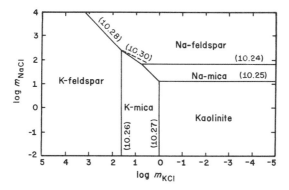

FIG. 10.12. Phase relations in a portion of the system Na_2O-K_2O-Al_2O_3-SiO_2-H_2O at 300 °C and 15,000 psi. Quartz is present in excess and $m_{HCl} = 10^{-2}$. The numbers in parentheses are equation numbers. The dashed line (10.30) indicates the calculated slope of the boundary between K-mica and Na-feldspar.

activity of Orville's feldspars should be negligible unless their behavior as solid solutions deviated widely from ideality. Clearly, they do not. This conclusion can be corroborated by a study of Orville's extensive data on solution and solid compositions. The same considerations apply to the micas; in the presence of K and Na, the actual phases formed are not the pure end members studied by Hemley *et al.*

It is of interest to note that the standard free-energy changes for all the reactions between the phases shown on Figure 10.12 can be calculated to within a few tenths of a kilocalorie. It should not be long before enough standard free energies of formation of compounds are known at elevated temperatures to permit us to make stability diagrams showing the relations of solids to their aqueous environment with the same facility that we now can make them at room temperature. Once ΔF_f° values are obtained for a few key compounds, the rest can be obtained from reactions like those just described.

DIAGRAMS INVOLVING CARBONATES AND SULFATES

Only a few diagrams showing stability relations among carbonates and sulfates will be presented here. The reader is referred to Schmitt's compilation[9] for more comprehensive coverage. The ones given here have been chosen to illustrate various combinations of variables that can be used as the basis for graphic description of phase relations.

SOME RELATIONS IN THE SYSTEM Na_2O-CO_2-H_2O

None of the systems considered so far has been one in which the activity of water deviated markedly from unity. But in the system Na_2O-CO_2-H_2O, which is a first approximation to the composition of some saline lakes, the concentration of dissolved material is so high that water activity is a critical factor in determining the stable solids. The activity of water is taken here as the ratio of P_{H_2O} of the system to the vapor pressure of pure liquid water under the same P, T conditions. This definition covers aqueous solutions and also those environments in which a liquid phase is not present. Water activity, defined this way, and relative humidity are synonymous (when relative humidity is considered as ratio rather than percentage).

Some of the important minerals in carbonate-rich lakes are nahcolite, $NaHCO_3$; trona, $NaHCO_3 \cdot Na_2CO_3 \cdot 2H_2O$; natron, $Na_2CO_3 \cdot 10H_2O$; and thermonatrite, $Na_2CO_3 \cdot H_2O$.

[9] Harrison H. Schmitt (ed.), *Equilibrium Diagrams for Minerals*. Cambridge, Mass., The Geological Club of Harvard, 1962.

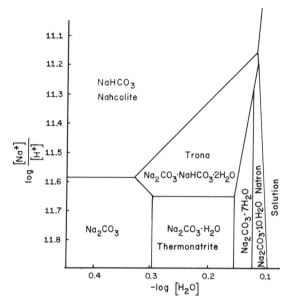

FIG. 10.13. Relations among some phases in the system Na_2O-CO_2-H_2O at 25 °C and 1 atmosphere, as functions of $[H_2O]$ and $[Na^+]/[H^+]$. System is in equilibrium with solution. [Adapted from figure by A. Truesdell, 1959, in *Equilibrium Diagrams for Minerals*, Harrison H. Schmitt (ed.). Cambridge, Mass., The Geological Club of Harvard, 1962, p. 4; by permission.]

Figure 10.13 shows relations among these species at 25 °C as a function of $[H^+]/[Na^+]$ vs. $[H_2O]$. For the progressive hydration of Na_2CO_3, shown across the bottom of the diagram, the activity of water is the only variable in the equilibrium constant. For example

$$Na_2CO_3 \cdot H_2O_c + 6H_2O_l = Na_2CO_3 \cdot 7H_2O_c$$

$$K_1 = \frac{1}{[H_2O]^6}.$$

Therefore, the boundaries between these hydrates are vertical lines, independent of the ordinate.

On the other hand, for the reaction from thermonatrite to trona

$$2Na_2CO_3 \cdot H_2O_c + H^+_{aq} = Na_3H(CO_3)_2 \cdot 2H_2O_c + Na^+_{aq}$$

$$K_2 = \frac{[Na^+]}{[H^+]}.$$

The boundary between these phases is a horizontal line independent of $[H_2O]$.

For the boundary between trona and nahcolite

$$Na_3H(CO_3)_2 \cdot 2H_2O_c + H^+_{aq} = 2NaHCO_{3\,c} + Na^+_{aq} + 2H_2O_{aq}$$

$$K_3 = \frac{[Na^+][H_2O]^2}{[H^+]}$$

and the boundary is a sloping line if log $[Na^+]/[H^+]$ is plotted vs. log $[H_2O]$.

It could have been predicted from the phase rule that equilibrium between these phases needs only two descriptive variables. If three solid phases are present in a three-component system at fixed temperature

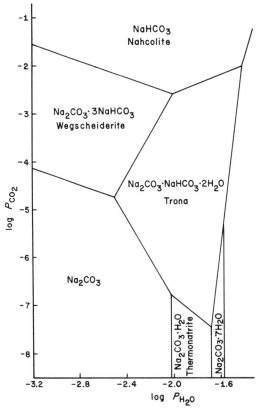

FIG. 10.14. A portion of the phase diagram for the system Na_2O-CO_2-H_2O at 25 °C and 1 atmosphere total pressure, as functions of P_{H_2O} and P_{CO_2}. The partial pressures are given in atmospheres. In constructing Figures 10.13 and 10.14, some of the ΔF°_f values used were slightly different. [Adapted from a figure by Malcolm Ross, 1961; personal communication, 1963.]

and pressure, an invariant point must result, such as the point at which Na_2CO_3, thermonatrite, and trona coexist. If two phases are present, the boundary is a line; if only one phase exists, it occupies an area on a two-variable plot.

A plot such as Figure 10.13 is useful to the investigator who wants to know the effect of adding acid, perhaps from volcanic emanations, to a soda lake; or the effect of increasing NaCl content. It is also extremely convenient for the field geochemist who can measure directly $[Na^+]$ and $[H^+]$ with glass electrodes, and can determine the vapor pressure of the lake water. He can tell immediately whether the solids in a given lake are in equilibrium with the supernatant solution.

In Figure 10.14, a different pair of descriptive variables—vapor pressure of water and pressure of CO_2—have been used to describe a portion of the same phase diagram as in Figure 10.13. The new diagram also includes the phase wegscheiderite, which persists metastably from higher temperatures to 25 °C.

SOME RELATIONS AMONG COPPER CARBONATES AND SULFATES

Another example of a choice of descriptive variables to fit the investigator's intent is the work of Silman[10] on stability relations among the copper carbonates and sulfates. Silman was interested in the amount of copper that could be carried in solution in natural waters, in connection with geochemical prospecting for copper by water analysis. From a water analysis one can easily obtain total dissolved CO_2 ($CO_3^{--} + HCO_3^- + H_2CO_3$) and SO_4^{--}, as well as pH. Silman reasoned that the maximum amount of dissolved copper in a stream or lake probably was that in equilibrium with various basic sulfates and carbonates, and that he might be able to establish certain conditions of water composition that would have equilibrium copper contents too small to be useful in prospecting.

Figure 10.15 shows the stability fields of the oxidized copper minerals as a function of SO_4^{--}, pH, and total dissolved carbonate. This initial diagram serves nicely to define the copper minerals that should form in a given stream or lake. Because Silman was interested in *solubility* of *copper*, rather than *activity* of *cupric ion*, he measured actual concentration of dissolved copper in equilibrium with the various mineral species, and then correlated total dissolved copper with the various dissolved species into which it was partitioned.

Figure 10.16 is a diagram constructed by Schmitt[11] from Silman's

[10] J. R. Silman, Ph.D. thesis, Department of Geological Sciences, Harvard University, 1958.
[11] Harrison H. Schmitt, in Harrison H. Schmitt (ed.), *op. cit.*, p. 117.

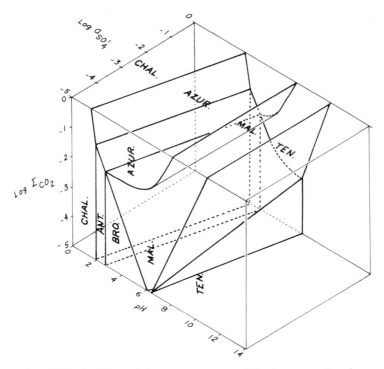

FIG. 10.15. Stability relations among some oxidized copper minerals at 25 °C as a function of the properties of stream and lake waters. [From J. R. Silman, thesis, 1958, in Schmitt (ed.), *op. cit.*, p. 119; by permission.]

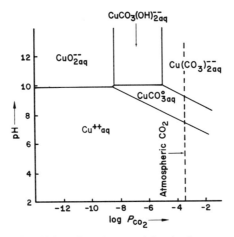

FIG. 10.16. Distribution of dissolved copper species at 25 °C and 1 atmosphere total pressure, as a function of pH and P_{CO_2} (in atmospheres). [Adapted from figure by H. H. Schmitt, in Schmitt (ed.), *op. cit.*, p. 116; by permission.]

COMBINATION DIAGRAMS

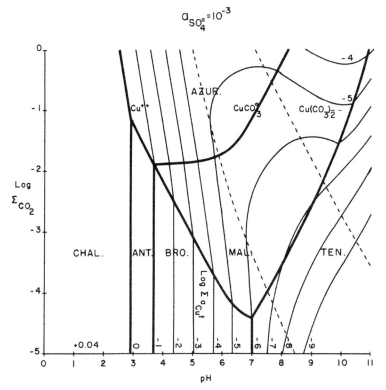

FIG. 10.17. Section at $a_{SO_4^{--}} = 10^{-3}$ of Figure 10.15. Shown are: (a) the stability fields of some oxidized copper minerals, (b) fields of dominance of some dissolved copper species, and (c) contours of total dissolved copper activity from 10^0 to 10^{-9}. $\Sigma CO_2 = [H_2CO_3] + [HCO_3^-] + [CO_3^{--}]$; $\Sigma a_{Cu^t} = [Cu^{++}] + [CuCO_{3\,aq}] + [Cu(CO_3)_2^{--}]$. [From J. R. Silman, thesis, 1958, in Schmitt (ed.) op. cit., p. 123; by permission.]

data, that shows clearly the various copper species considered by Silman, and their environments of occurrence; Figure 10.17 is a diagram by Silman that is a vertical section of Figure 10.15 at a fixed SO_4^{--} activity. Contours of the various dissolved copper species are superimposed upon the mineral stability fields.

From Silman's diagrams it is possible to determine the copper mineral in equilibrium with a water from a standard water analysis, the maximum dissolved copper that can exist in equilibrium with the mineral, and the kinds and properties of dissolved species that make up the total copper solubility.

SOME RELATIONS AMONG CALCIUM AND MAGNESIUM CARBONATES

The reason for including a diagram concerning calcium and magnesium carbonates is not to show new kinds of variables that are useful as descriptive coordinates, but to illustrate the use of equilibrium diagrams involving metastable phases. Even though a phase is metastable with respect to another phase, or phases, it may well be possible to establish equilibrium between such a phase and its environment. It is possible, for example, to measure the solubility of aragonite, approaching the saturation value from both under- and oversaturation. A diagram showing aragonite-solution relations can be an equilibrium diagram insofar as solid and solution are concerned, but a nonequilibrium diagram in terms of solid-solid relationships.

Carpenter[12] has constructed an interesting diagram that shows metastable and stable equilibria among the calcium and magnesium carbonates (Figure 10.18). The metastable phases involved are aragonite and

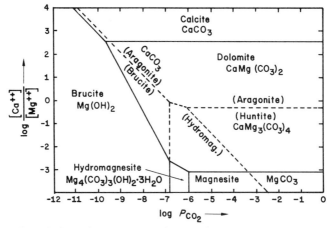

FIG. 10.18. Relations among calcium and magnesium carbonates at 25 °C and 1 atmosphere total pressure, as a function of P_{CO_2} (in atmospheres) and $[Ca^{++}]/[Mg^{++}]$. Solid lines show equilibrium relations between stable phases; dashed lines show equilibrium relations between metastable phases, and between metastable and stable phases. [From a figure by A. B. Carpenter, in Schmitt (ed.), *op. cit.*, p. 26; by permission.]

huntite; the boundaries between these phases and between these and stable phases are calculated from ΔF_f° values in exactly the same way as for stable phases alone. The diagram as presented has been useful to Carpenter in interpreting paragenetic relations at Crestmore, California, where supergene veinlets containing aragonite, huntite, brucite, dolomite, magnesite, hydromagnesite, and calcite are found. The early mineral

[12] A. B. Carpenter, in Harrison H. Schmitt (ed.), *op. cit.*, p. 26.

compatibilities are those that would be deduced from the metastable (dashed lines) diagram; these early minerals apparently eventually equilibrate to give the phase relations demonstrated by the solid lines. The diagram has the added virtue of illustrating a basic relation in geology—"the older the rock, the less initial environmental information available from it." Note the relative restriction of ordinate and abscissa values for the fields of the minerals in the diagram involving the metastable phases, as opposed to the greater environmental range of the stable solids. Time and the attendant water percolation through sediments tend to diminish the number of mineral phases initially present, and to leave only minerals with broad ranges of stability.

SUMMARY

In this chapter, emphasis is on the use of variables chosen for the purposes of the investigator in depicting stability relations among minerals. Using silicates, carbonates, and sulfates as examples, we show that in addition to Eh, pH, and partial pressure, various combinations of these and other parameters can be used to suit the needs of the investigator. Diagrams are reproduced that use individual ion activities, molalities of salts, activity of water, ratios of ions, and temperature as descriptive parameters. Also, emphasis is placed on the utility of diagrams involving metastable phases.

SELECTED REFERENCES

Schmitt, Harrison H., (ed.), *Equilibrium Diagrams for Minerals.* Cambridge, Mass., The Geological Club of Harvard, 1962.

Sillén, Lars Gunnar, Graphic presentation of equilibrium data, chap. 8, in *Treatise on Analytical Chemistry*, Part I, Vol. 1, I. M. Kolthoff and P. J. Elving (eds.). New York, Interscience, 1959.

PROBLEMS

10.1. Write the reaction for huntite's changing to dolomite in such a way that the equilibrium constant is expressed in terms of
    a. Mg_{aq}^{++} and $CO_{3\,aq}^{--}$
    b. Mg_{aq}^{++}, H_{aq}^{+}, and $HCO_{3\,aq}^{-}$
    c. Mg_{aq}^{++}, H_{aq}^{+}, H_2O_l, and $CO_{2\,g}$
    d. Ca_{aq}^{++} and $CO_{3\,aq}^{--}$

Answers
a. $CaMg_3(CO_3)_{4\,c} = CaMg(CO_3)_{2\,c} + 2Mg^{++}_{aq} + 2CO^{--}_{3\,aq}$
b. $CaMg_3(CO_3)_{4\,c} + 2H^+_{aq} = CaMg(CO_3)_{2\,c} + 2Mg^{++}_{aq} + 2HCO^-_{3\,aq}$
c. $CaMg_3(CO_3)_{4\,c} + 4H^+_{aq} = CaMg(CO_3)_{2\,c} + 2Mg^{++}_{aq} + 2H_2O_l + 2CO_{2\,g}$
d. $CaMg_3(CO_3)_{4\,c} + 2Ca^{++}_{aq} + 2CO^{--}_{3\,aq} = 3CaMg(CO_3)_{2\,c}$

10.2. The vapor pressure of pure water at 100 °C is 760.0 millimeters; that of a 6 molar solution is 583.5 millimeters. What is the activity of water in the NaCl solution, at that temperature? *Ans. $a = 0.768$.*

10.3. The equilibrium vapor pressure of pure water at 25 °C is 23.8 millimeters. What is the approximate vapor pressure of water in a solution in equilibrium with $Na_2CO_3 \cdot 10H_2O_c$ and $Na_2CO_3 \cdot 7H_2O_c$? (See Figure 10.13.) *Ans. $P_{H_2O} \cong 18$ mm.*

10.4. What is the standard free energy of the reaction $3H_2O_l + Na_2CO_3 \cdot 7H_2O_c = Na_2CO_3 \cdot 10H_2O_c$ at 25 °C? (Use only results of question 10.3.) *Ans. $\Delta F°_r \cong 0.5$ kcal (per mole of 10-hydrate formed).*

10.5. Can gibbsite, kaolinite, and quartz be brought into equilibrium at 25 °C by controlling the activity of water in the system?
Ans. No. For equilibrium in the reaction
$$H_4Al_2Si_2O_{9\,c} + H_2O_l = Al_2O_3 \cdot 3H_2O_c + 2SiO_{2\,c,\,quartz}$$
the activity of water is given by
$$K = \frac{1}{[H_2O]}.$$

Show that $[H_2O]$ is a large number, essentially unattainable, regardless of pressure.

CHAPTER 11

Some Geological Applications of Mineral Stability Diagrams

This chapter is a review of some of the ways that mineral stability diagrams involving solids and solutions have been applied to geological problems. It should serve as a partial guide to the literature, and should also suggest ways in which the preceding diagrams may be used. The examples are selected because they illustrate the use of diagrams showing stability relations of solids. There is a large additional literature, especially in Russian, that reports measurements of solution properties and discusses the relations of these properties to the solids present. However, the criterion used for inclusion here is presentation on a graphical basis, simply as a convenient method of preventing this book from becoming unwieldy, and as a practice in harmony with the preoccupation with diagrammatic relations in the preceding chapters. Also, emphasis here is on Eh-pH diagrams because they have been the most widely used. In a few years it is likely that the variety of descriptive variables used will increase markedly.

NATURAL LIMITS OF Eh AND pH

The range of Eh and pH that must be considered in constructing diagrams is determined by the limits of values observed in nature. Baas Becking, Kaplan, and Moore,[1] in an important paper on this subject, have accumulated essentially all the measurements available in the literature. Figure 11.1, from their work, shows these measurements, and serves admirably to illustrate both range and frequency of values. These investigators discuss in detail the controls of Eh and pH, with emphasis on biologic aspects, and present a comprehensive bibliography. Figure 11.2 is an attempt to show the position of the greatest frequency of

[1] L. G. M. Baas Becking, I. R. Kaplan, and D. Moore, *Jour. Geol.*, 68, 243 (1960).

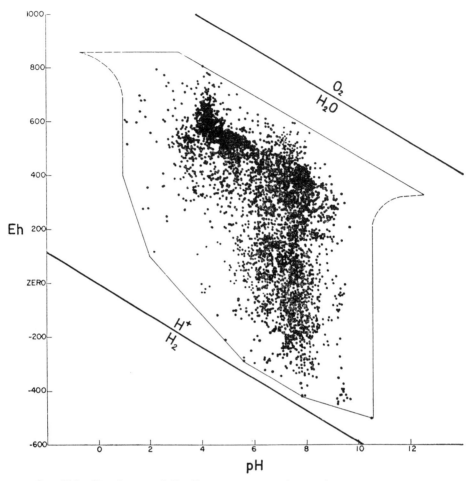

Fig. 11.1. Distribution of Eh-pH measurements of natural aqueous environments. [After Baas Becking et al., op. cit., p. 276.]

readings for several different types of natural waters. Note that measurements on most aerated surface waters (mine waters, rain, streams, normal ocean water on Figure 11.2) fall along a line well below, but roughly parallel to, the upper boundary of water stability. This line is the "irreversible oxygen potential," and has been explained by Sato[2] as being the H_2O_2-O_2 stability boundary on the basis that oxidation by oxygen has to go through a rate-controlling hydrogen peroxide step. If the measured potentials are a true guide to the effect of the oxygen of the atmosphere, then we see that natural waters behave as if they contained a very small amount of dissolved oxygen. On the other hand, for example,

[2] Motoaki Sato, Econ. Geol., 55, 928 (1960).

GEOLOGICAL APPLICATIONS OF MINERAL STABILITY DIAGRAMS 381

oxidized vanadium ores contain the mineral navajoite, which is stable only at potentials near 1.0 volt and pH values of the order of 1.5 to 2.0. Figure 11.3 shows that values from mines fall over a large range of

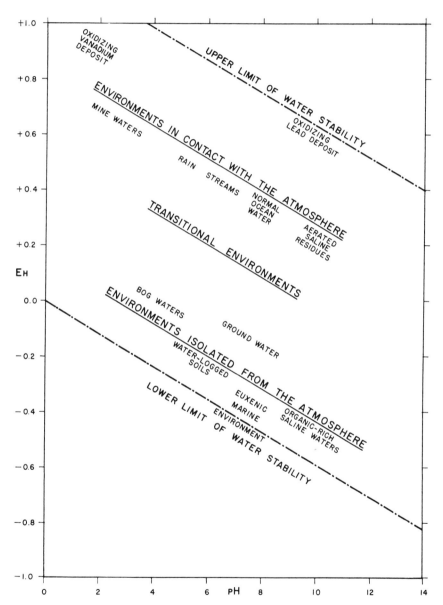

FIG. 11.2. Approximate position of some natural environments as characterized by Eh and pH.

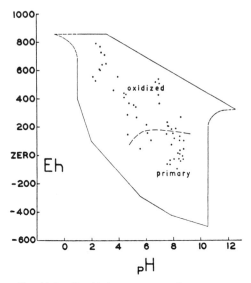

FIG. 11.3. Eh-pH characteristics of mine waters. An attempt is made to distinguish between the water coming from the primary ore and that draining from the oxidized ore. [After Baas Becking et al., op. cit., p. 255.]

conditions, with those samples protected from the atmosphere (primary zone) falling into the transition zone of Figure 11.2. Inasmuch as many rocks—sedimentary, igneous, and metamorphic—contain both magnetite and hematite, it may be that the potentials of most underground waters approach the magnetite-hematite boundary (Figure 7.3).

The general picture that emerges is that only surface waters with good circulation are oxidizing, whereas confined waters rapidly lose their oxygen content, whether confinement results from fixation in rock or soil pores, or by prevention of overturn of open waters. Organic-free waters lose their oxidizing character by reaction with silicates containing ferrous iron, such as biotite, chlorite, amphiboles, pyroxenes, or by contact with sulfides or ferrous iron-containing carbonates. There are many other possible inorganic reducing agents, but the iron compounds are probably the most important quantitatively. The pH tends to rise because of hydrolysis of the silicates, so that the environment becomes alkaline as well as reducing. In environments containing organic matter, biochemical reactions quickly remove oxygen, commonly with marked increase in CO_2, and with production of hydrogen sulfide. The influence of bacteria is paramount,[3] and deoxygenation tends to be accompanied

[3] The reader is referred to Baas Becking, Kaplan, and Moore, op. cit., for detailed authoritative discussion of the role of microorganisms in controlling environment.

by pH lowering as CO_2 and H_2S are generated. Some bacteria even release hydrogen, and potentials may sink close to the water stability limit. Under inorganic conditions many ions, notably sulfate, may persist metastably during long periods of time, but the presence of organic material and the accompanying bacteria almost guarantees approach to equilibrium, as denoted by the measured potentials. Below the water table the environment can be assumed to be alkaline and reducing, except in local instances of high rates of water flow. The best evidence is the persistence of sulfides in ores for millions of years, without destruction of the finest details of surface characteristics.

SEDIMENTARY IRON ORES

The classic paper on sedimentary iron ores, which ties the environmental characteristics to the observed facies, is the study by James.[4] His abstract (p. 236) summarizes the relations as succinctly as possible:

The sedimentary iron formations of Precambrian age in the Lake Superior region can be divided on the basis of the dominant original iron mineral in four principal facies: sulfide, carbonate, oxide, and silicate. As chemical sediments, the rocks reflect certain aspects of the depositional environments. The major control, at least for the sulfide, carbonate, and oxide types, probably was the oxidation potential. The evidence indicates that deposition took place in restricted basins, which were separated from the open sea by thresholds that inhibited free circulation and permitted development of abnormalities in oxidation potential and water composition.

... The *sulfide facies* is represented by black slates in which pyrite makes up as much as 40 percent of the rock. The free carbon content of these rocks typically ranges from 5 to 15 percent, indicating that ultra-stagnant conditions prevailed during deposition. Locally, the pyrite rocks contain layers of iron-rich carbonate. The *carbonate facies* consists, in its purer form, of interbedded iron-rich carbonate and chert. It is a product of an environment in which oxygen concentration was sufficiently high to destroy most of the organic material but not high enough to permit formation of ferric compounds. The *oxide facies* is found as two principal types, one characterized by magnetite and the other by hematite. Both minerals appear to be of primary origin. The magnetite-banded rock is one of the dominant lithologies in the region; it consists typically of magnetite interlayered with chert, carbonate, or iron silicate, or combinations of the three. Its mineralogy and association suggest origin under weakly oxidizing to moderately reducing conditions, but the mode of precipitation of magnetite is not clearly understood. The hematite-banded rocks consist of finely crystalline hematite interlayered with chert or jasper. Oolitic structure is common. This facies doubtless accumulated in a

[4] H. L. James, *Econ. Geol.*, 49, 235 (1954).

strongly oxidizing probably near-shore environment similar to that in which younger hematitic ironstones such as the Clinton were deposited. The *silicate facies* contain one or more of the hydrous ferrous silicates (greenalite, minnesotaite, stilpnomelane, chlorite) as a major constituent.... The most common association of the silicate rocks is with either carbonate or magnetite-bearing rocks, which suggests that the optimum conditions for deposition ranged from slightly oxidizing to slightly reducing....

The relation of James' conclusion to equilibrium stability relations is clear from Figure 7.23. With slight undersaturation with amorphous silica, magnetite can replace ferrous silicates; with diminution of $\Sigma\, CO_2$, magnetite or ferrous silicate replaces siderite; with increase in sulfide sulfur, pyrite takes over; with oxygenation, hematite holds sway. Furthermore, all these facies lie within a narrow range of pH and Eh values. Castano and Garrels[5] and Huber and Garrels[6] demonstrated experimentally that the iron minerals precipitate in the laboratory according to the metastable relations shown in Figure 7.4, and that the response of freshly precipitated iron "hydroxides" and carbonates to a given Eh-pH environment is rapid. White[7] has done an excellent job of deciphering the relation of the iron mineral facies to the environment of deposition, and correlates facies with aeration and distance from shore. Carroll[8] suggested that in many deposits transportation of iron took place as a shell of iron oxide at the surface of clay minerals, and that removal and concentration of the iron might take place after initial sedimentation if the environment became one that permitted reduction and dissolution of ferric oxides. She performed experiments in which it was demonstrated that bacterial activity could result in a change of Eh and pH conditions to those appropriate to the solution of iron oxide, and that solution actually was accomplished.

Our understanding of the chemistry of iron in natural waters has been advanced greatly by the recent studies of Hem[9] and his coworkers and

---

[5] J. R. Castano and R. M. Garrels, *Econ. Geol.*, **45**, 755 (1950).

[6] N. K. Huber and R. M. Garrels, *Econ. Geol.*, **48**, 337 (1953).

[7] D. A. White, The stratigraphy and structure of the Mesabi Range, Univ. of Minnesota, *Minnesota Geol. Survey, Bull.* 38, 1954.

[8] Dorothy Carroll, *Geochim. et Cosmochim. Acta*, **14**, 1 (1958).

[9] J. D. Hem and W. H. Cropper, A survey of ferrous-ferric chemical equilibria and redox potentials: *U.S. Geol. Survey Water-Supply Paper* 1459-A, 1–31 (1959).

J. D. Hem, Restraints on dissolved ferrous iron imposed by bicarbonate, redox potential, and pH: *U.S. Geol. Survey Water-Supply Paper* 1459-B, 33–55 (1960).

J. D. Hem, Some chemical relationships among sulfur species and dissolved ferrous iron: *U.S. Geol. Survey Water-Supply Paper* 1459-C, 57–73 (1960).

J. D. Hem, Complexes of ferrous iron with tannic acid: *U.S. Geol. Survey Water-Supply Paper* 1459-D, 75–94 (1960).

J. D. Hem and M. W. Skougstad, Coprecipitation effects in solutions containing ferrous, ferric, and cupric ions: *U.S. Geol. Survey Water-Supply Paper* 1459-E, 95–110 (1960).

by Stumm and Lee.[10] By their detailed studies of natural waters and their laboratory analogs, they have shown, in essence, that Eh-pH measurements, when combined with adequate thermodynamic data, can actually be used to determine the concentrations and species of dissolved iron. We need many more studies like these on other elements. Both Hem and his coworkers and Stumm and Lee present numerous Eh-pH diagrams involving solids and solutions.

In summary, the study of natural occurrences of iron minerals shows that there is sufficient approach to equilibrium with the environment of deposition to permit use of relations based on thermochemical considerations. The diagrams tell nothing about the rate at which equilibrium might have been attained, nor the path by which it was approached, but if we know what should happen, we are in a position to use the stable associations as a guide to interpreting processes as indicated by replacement sequences among the minerals. We can thus put reasonable limits on the amount of dissolved CO_2 required to have a siderite facies, and can demonstrate that oxygen must be to all intents and purposes absent during formation of siderite, pyrite, magnetite, or silicate facies of iron formation. It is also important to know that any sulfide sulfur in an iron-rich environment will be reflected in pyrite, and that the presence of both pyrite and pyrrhotite puts rather definite limits on the total dissolved sulfur that was in equilibrium with these compounds. Our major gap in knowledge is in thermal data pertaining to the primary iron silicates in sedimentary iron ores—it was necessary to use $FeSiO_3$ in the calculations, and no mineral of this composition is known to exist. There is a nasty problem in the coexistence of chert and magnetite as primary minerals on the one hand, and iron silicates on the other. The fact that the real silicates are aluminous must enter into the actual equations.

SEDIMENTARY MANGANESE DEPOSITS

Eh-pH relations have been applied to sedimentary deposits of manganese by Krauskopf[11] and by Marchandise.[12] Figure 11.4, taken from Krauskopf's paper, shows his plot of the stability boundaries among anhydrous manganese compounds. Note that Krauskopf prefers to plot equations relating mineral pairs, rather than delimiting and labeling

[10] W. Stumm and G. F. Lee, *Schweizerische Zeits. für Hydrologie*, 22, 295 (1960).
[11] K. B. Krauskopf, *Geochim. et Cosmochim. Acta*, 12, 61 (1957).
[12] H. Marchandise, Symposium Sobre Yacimientos de Manganese, *XX Intern. Geol. Congr. Mexico*, 1, 107 (1956).

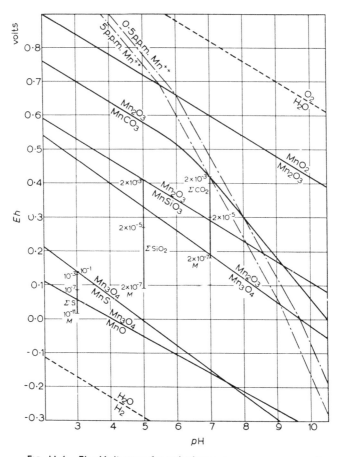

FIG. 11.4. Eh-pH diagram for anhydrous manganese compounds. Solid lines are boundaries of stability fields; each line separates the field of an oxidized form (above) from that of a reduced form (below). Crossbars on vertical lines show positions of field boundaries at lower concentrations of carbonate, sulfide, and silica. Dashed lines are limits of possible redox potentials in water solution. Dashed-dot lines are "isoconcentration" lines, drawn through points where the concentration of Mn^{++} in equilibrium with the oxides is 5 ppm and 0.5 ppm, respectively. [K. B. Krauskopf, *op. cit.*, p. 63.]

mineral fields, as generally used in this text. Figure 11.5 shows his diagram for the manganese hydroxides. Again the best summary of the relation of the diagrams to the observed associations comes from the author's abstract (Krauskopf, *op. cit.*, p. 61):

Thermochemical data on compounds of manganese and iron are in general agreement with reported mineral associations, provided that the more complex mineral compounds, for which data do not exist, be assumed to have somewhat

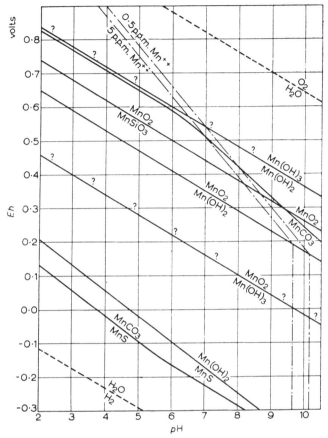

FIG. 11.5. Eh-pH diagram for manganese hydroxides. Symbols same as in Figure 11.4. Question marks along stability boundaries for $Mn(OH)_3$ indicate that under these conditions $Mn(OH)_3$ is metastable. [K. B. Krauskopf, op. cit., p. 66.]

larger stability fields than their nearest simple chemical equivalents. The data show that the iron compounds to be expected in nature are uniformly less soluble than the corresponding manganese compounds, and that ferrous ion is more easily oxidized than manganous ion under any naturally occurring pH-Eh conditions. Thus, inorganic processes should always lead to precipitation of iron before manganese from a solution containing both metals, unless the Mn/Fe ratio is very high.

The oxidation of manganous and ferrous ions by atmospheric oxygen takes place by slow reactions which can be utilized as an energy source by bacteria. Selective oxidation and precipitation by different species of bacteria can lead to partial separation of the metals, but this is probably not a major factor in the formation of large, nearly pure deposits of manganese or iron compounds.

Selective dissolution of the metals from igneous rocks, a mechanism of separation often postulated in the literature, was tested by treating basaltic andesite with a number of solvents at temperatures ranging from 25° to 300 °C. The ratios of Mn to Fe in the resulting solutions were all approximately the same as in the original rock, showing that this assumed process of separation is ineffective.

Isolation of manganese in solution can be accomplished by precipitating the iron first. This is most effectively done by adding alkali gradually to a solution containing both metals, keeping the solution in contact with atmospheric oxygen. The reaction can be demonstrated in the laboratory under conditions similar to those in nature by letting dilute acid percolate through crushed lava and then through limestone: iron dissolved from the lava is precipitated in the limestone, and the solution is left with a high Mn/Fe ratio. This suggests a possible explanation for the origin of many manganese deposits, especially those associated with lavas and tuffs, but it requires that iron oxide in amounts many times that of the manganese be deposited in the rocks through which the solutions have passed.

OXIDATION AND SECONDARY ENRICHMENT OF ORES

Eh-pH and related diagrams are especially useful in studies of oxidation and secondary enrichment of ore deposits. The temperatures and pressure of the processes are close to 25 °C and 1 atmosphere total pressure, and the range of values of Eh and pH encountered is extreme. Also, the concentration of readily oxidizable or reducible material is high compared to most other near-surface environments, and the variety of minerals is great.

Weathering of Sandstone Type Uranium Deposits

Uranium deposits of the so-called sandstone type, which occur in continental sediments as pods, lenses, and blankets, provide the student of mineral associations with an endless variety of relations for investigation. Prior to about 1950, when the best-known deposits in the United States were in southwestern Colorado, the chief ore mineral was carnotite, a potassium uranyl vanadate $[K_2(UO_2)_2V_2O_8 \cdot 0\text{-}3H_2O]$, and it was believed by many to be a primary sedimentary mineral, precipitated during or shortly after the deposition of the enclosing sediments.[13] The mineral is certainly fully oxidized, inasmuch as the uranium is in the sexivalent state, and the vanadium quinquevalent. But discovery of increasing amounts of uraninite in deeper deposits began to raise a valence problem, since at least part of the uranium of uraninite is quadrivalent. The possibility began to develop that uraninite is an early uranium

[13] For example, see R. P. Fischer, Vanadium deposits of Colorado and Utah: *U.S. Geol. Survey Bull.*, 936-P, 363 (1942).

mineral and that carnotite is an oxidation product. This suggestion was further strengthened by the identification of the mineral montroseite [VO(OH)], which contains trivalent vanadium. To complicate matters still further, other vanadium minerals containing both V^{3+} and V^{4+}, V^{4+} alone, and both V^{4+} and V^{5+} were identified.

Today it is generally concluded that carnotite is a weathering product. This view is a result of a great deal of work by a great many people—here we shall consider only those aspects relating occurrence of minerals and theoretical stability relations. The specific problem that had to be attacked can be stated somewhat as follows: if an unoxidized uranium ore consists of uraninite, montroseite, and pyrite, with some sphalerite and galena, will oxidation yield the complex mineral assemblages observed? What oxidation products can coexist stably? Which minerals will oxidize first under near-equilibrium conditions? The use of an Eh-pH framework for representation of stability relations is indicated, so that the various elements can be compared by superposition of individual diagrams.

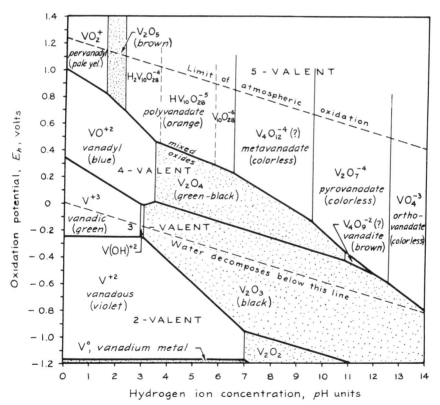

FIG. 11.6a. Stability of some vanadium compounds and ions in water at 25 °C and 1 atmosphere total pressure. [From H. T. Evans, Jr., and R. M. Garrels, op. cit.]

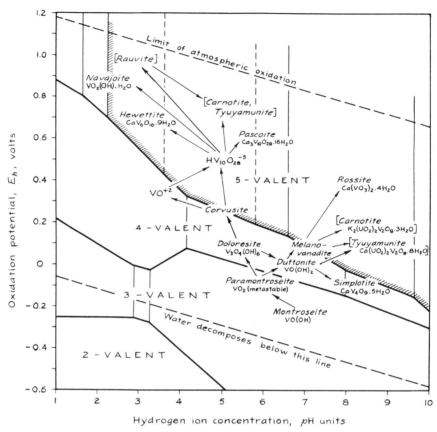

FIG. 11.6b. Eh-pH diagram showing the oxidation relations of some naturally occurring vanadium compounds. [From H. T. Evans, Jr., and R. M. Garrels, *op. cit.*]

The first diagrams prepared were rather crude preliminary ones for vanadium and uranium.[14] They did, however, suffice to show that uraninite and montroseite can coexist under strongly reducing conditions, but that montroseite oxidizes at a lower potential than uraninite in water, so that uraninite can coexist with vanadium oxides containing trivalent or quadrivalent vanadium. These crude diagrams were superseded by a pair of diagrams developed chiefly by Evans[15] which compared theoretical relations among a large variety of vanadium-oxygen compounds and ions with the observed sequences of vanadium minerals in Colorado Plateau deposits. The new diagrams are shown in Figures 11.6a,b. Not only can the minerals be fitted into the theoretical framework, but also the calculated diagram has been a useful predictive device

[14] R. M. Garrels, *Am. Mineral.*, 38, 1251 (1953); *ibid.*, 40, 1004 (1955).
[15] H. T. Evans, Jr. and R. M. Garrels, *Geochim. et Cosmochim. Acta*, 15, 131 (1958).

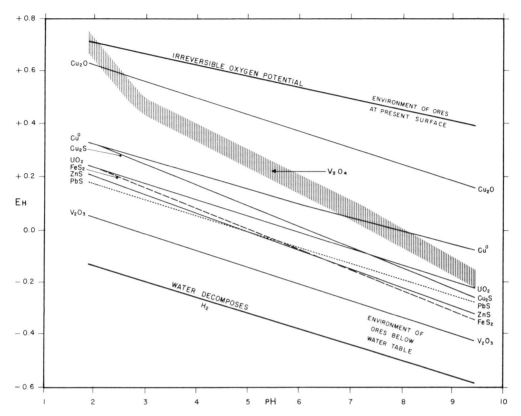

FIG. 11.7a. Stability limits of some of the minerals of the Colorado Plateau uranium ores. [R. M. Garrels, Am. Inst. Chem. Engrs., Nuclear Eng. and Sci. Congr., Preprint 250 (1955). Reproduced with permission of the American Geological Institute.]

in searching for new minerals. A composite diagram showing oxidation boundaries for various compounds has also been prepared,[16] and the interesting relation has been found that the observed sequence of minerals during oxidation of a primary uraninite, montroseite, metal sulfide ore corresponds closely to the sequence predicted from the assumption of progressive oxidation of such an ore with near-equilibrium being maintained (Figures 11.7a,b,c). Hostetler and Garrels[17] used Eh-pH diagrams to illustrate fields of "solubility" and the solids in the complex system U-V-K-O_2-CO_2-H_2O in attempting to illustrate the conditions under which uranium might be transported in underground waters in the presence of vanadium. Figures 11.8a,b,c show some of their deductions concerning the various fields of stability of solids and the

[16] R. M. Garrels, Am. Inst. Chem. Engrs., Nuclear Eng. & Sci. Congr., Preprint 250 (1955).
[17] P. B. Hostetler and R. M. Garrels, Econ. Geol., 57, 137 (1962).

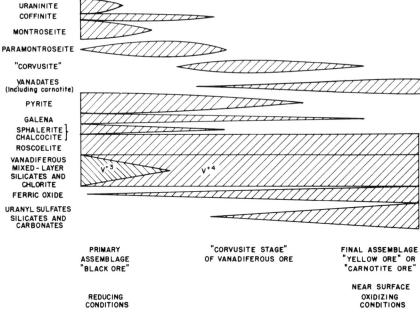

FIG. 11.7b. Approximate relations of some ore minerals in the oxidation sequence of the uranium ores of the Colorado Plateau. [Same reference as in Figure 11.7a.]

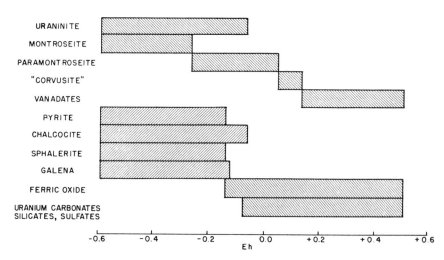

FIG. 11.7c. Predicted oxidation sequence of Colorado Plateau uranium ores if equilibrium were maintained during continuous rise of oxidation potential at pH 7. [Same reference as in Figure 11.7a.]

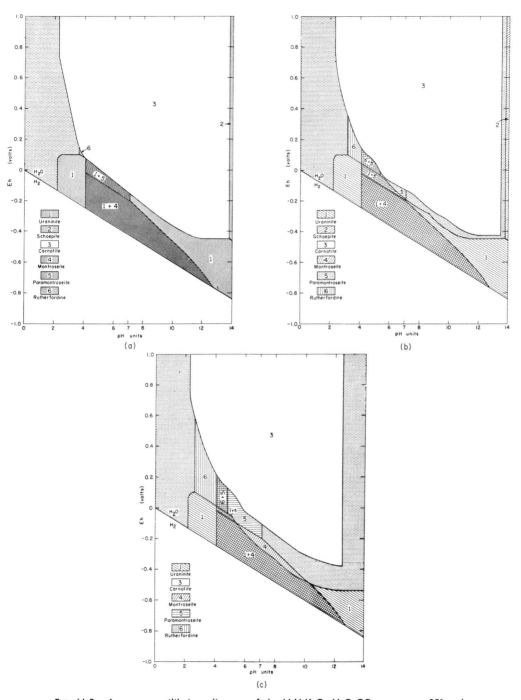

FIG. 11.8. Aqueous equilibrium diagram of the U-V-K-O_2-H_2O-CO_2 system at 25° and 1 atmosphere. Stability fields of solid phases are numbered and areas indicating soluble conditions are stippled. The boundary between solid phases and solution is fixed for $\Sigma U = 10^{-4}, \Sigma V = 10^{-3}, \Sigma K = 10^{-3}$. In (a) $\Sigma CO_2 = 10^{-3}$, in (b) $\Sigma CO_2 = 10^{-2}$, and in (c) $\Sigma CO_2 = 10^{-1}$. [From P. B. Hostetler and R. M. Garrels, op. cit., by permission.]

environments in which transportation of significant concentrations of vanadium and uranium could take place.

Coleman and Deleveaux,[18] used an Eh-pH diagram in explaining the behavior of selenium during weathering.

Thus again the relations calculated from thermochemical data have been useful in studying the behavior of natural deposits.

Oxidation of Sulfide Ores

A composite diagram showing the Eh-pH conditions for the oxidation of the common metal sulfides has been prepared.[19] This is shown in Figure 11.9. The nature of the oxidation products and their general behavior in the zone of weathering, as calculated, correspond fairly well with observed behavior, with the reservation that thermochemical data are not available for many phases known to exist. The diagram is useful in giving a quantitative method of describing the boundary between oxidation and stability of sulfides, and in classifying some individual relationships such as the stable coexistence of oxidized zinc minerals and copper sulfides such as chalcocite. The diagram shown is rather crude; those developed by Natarajan and Garrels (Figures 7.24 and 7.25) are much more nearly complete and are capable of even greater refinement as better data become available. An interesting comparison of real and theoretical relations has been made by Frondel and Ito,[20] who showed that the germaniferous sulfide ores of the Tsumeb mine oxidize essentially as would be predicted.

Kelly and Cloke[21] studied the controls of the solubility of gold in the zone of oxidation of ores, and showed that gold is appreciably soluble at low pH, high oxidation potential, and high chloride content of ground water. They described the behavior of gold diagrammatically, using Eh, pH, and activity of chloride ion as variables.

THE ENVIRONMENT OF MARINE CHEMICAL SEDIMENTS

There is, of course, a voluminous literature on the environment of marine sediments, but so far little use has been made of Eh-pH diagrams in interpreting the interrelations of the chemical precipitates. On the other hand, there is considerable qualitative use of Eh and pH in

---

[18] R. G. Coleman and Maryse Deleveaux, *Econ. Geol.*, 52, 499 (1957).
[19] R. M. Garrels, *Geochim. et Cosmochim. Acta*, 5, 153 (1953).
[20] C. Frondel and Jun Ito, *Am. Mineral.*, 42, 743 (1957).
[21] W. C. Kelly and P. L. Cloke, *Mich. Acad. Sci., Arts, and Letters*, 46, 19 (1961).

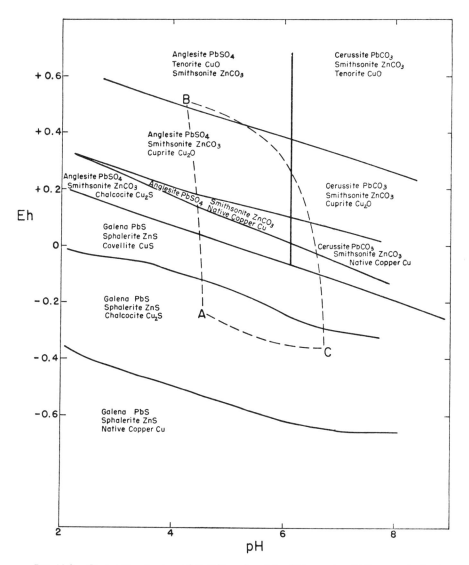

Fig. 11.9. Composite diagram of stability of metal sulfides and oxidation products at 25 °C and 1 atmosphere total pressure in the presence of total dissolved carbonate = $10^{-1.5}$, total dissolved sulfur = 10^{-1}. [From R. M. Garrels, Geochim. et Cosmochim. Acta, 5, 165, (1953).]

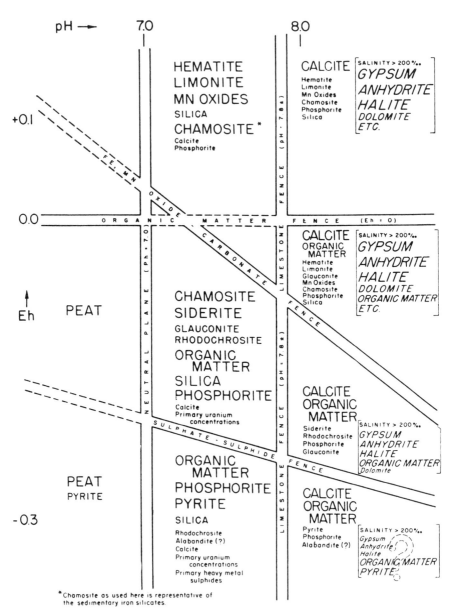

FIG. 11.10. Sedimentary chemical end-member associations in their relations to environmental limitations imposed by selected Eh and pH values. Associations in brackets refer to hypersaline solutions. [From W. C. Krumbein and R. M. Garrels, op. cit., p. 26.]

describing the environments.[22] Krauskopf[23] used stability diagrams to illustrate the activities of ions in equilibrium with various possible stable solids in his study of rare elements in the sea, and Krumbein and Garrels[24] qualitatively classified a variety of chemical sediments within an Eh-pH framework (Figure 11.10). Unfortunately, so many chemical compounds are so nearly in equilibrium with sea water, metastability is so common, and thermal data so crude, that it will be some time before it is possible to illustrate what should exist at equilibrium on the sea floor. On the other hand, if adequate diagrams showing stability relations could be constructed, we would have a powerful tool in our attempts to decipher the importance of reaction rates, adsorption phenomena as controls of ionic content of ocean waters as opposed to equilibrium with solid phases, the role of various types of soluble complexes in influencing solubility, and the critical questions concerning the ability of organisms to overcome a given set of external environmental conditions as they precipitate mineral substances.

THE ENVIRONMENT OF ORE DEPOSITION

The study of the relations between minerals and solutions occurring with them is of great importance in the study of ore deposition, but progress has been slow, largely because of the difficulties of experimentation at high temperatures and pressures. It has required ingenuity, skill, and extreme care to obtain accurate analyses of the aqueous solutions coexisting with various solids. A purely theoretical approach has been hampered by the small amount of thermochemical data available. However, there is much current interest in the properties of aqueous solutions at high temperatures and pressures, not only on the part of geochemists, who have done much of the pioneering work, but by physical chemists as well. Rapid progress can be expected.

Some of the possibilities that await have been suggested by Barton[25] who, using the mineral associations in veins, has done a remarkable job of corralling the nature of the ore-forming environment. Figure 11.11 is taken from his paper, and shows his deductions concerning the Eh and pH of formation of some ores. He makes clear the limitations because of

---

[22] For example, see K. O. Emery and S. C. Rittenberg, Early diagenesis of California basin sediments: *Bull. Am. Assoc. Petrol. Geol.*, 36, 735 (1952); G. I. Theodorovich, Siderite geochemical facies of seas and saline waters in general as oil producing: *Doklady Akad. Nauk S.S.S.R.*, 69, 227 (1949).
[23] K. B. Krauskopf, *Econ. Geol., 50th Anniv. Volume*, 411 (1955).
[24] W. C. Krumbein and R. M. Garrels, *Jour. Geol.*, 60, 1 (1952).
[25] Paul B. Barton, Jr., *Econ. Geol.*, 52, 333 (1957).

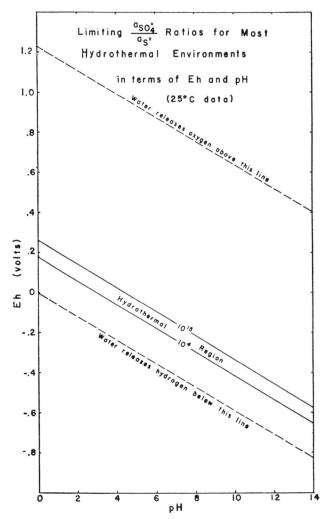

FIG. 11.11. Environment of hydrothermal ore deposition. [From P. B. Barton, Jr., *op. cit.*, p. 345.]

lack of free energy data at the temperatures of actual ore deposition. Barnes and Kullerud,[26] with additional data, constructed a number of pH-P_{O_2} diagrams representing temperatures up to 250 °C. Figure 11.12 is typical of their work.

Hemley and his coworkers[27] opened a new era with their systematic

[26] H. L. Barnes and G. Kullerud, *Econ. Geol.*, **56**, 648 (1961).
[27] J. J. Hemley, Some mineralogical equilibria in the system K_2O-Al_2O_3-SiO_2-H_2O: *Am. J. Sci.*, **257**, 241 (1959).
J. J. Hemley, C. Meyer, and D. H. Richter, Some alteration reactions in the system Na_2O-Al_2O_3-SiO_2-H_2O: *U.S. Geol. Survey Prof. Paper* 424-D, 338 (1961).

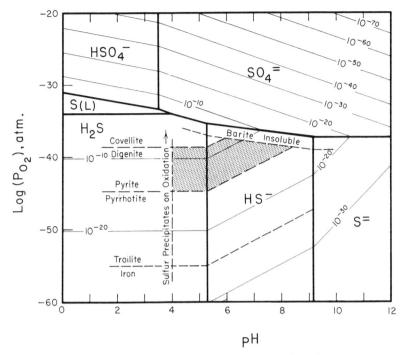

FIG. 11.12. The common range of acidity and oxidation state of sphalerite-depositing ore solutions at $\Sigma S = 0.1m$ and 250 °C. [From H. L. Barnes and G. Kullerud, *op. cit.*, by permission.]

studies of solution-solid relations among the feldspars, micas, and aluminum silicates. Their work has been used by Helgeson,[28] in conjunction with his studies of lead sulfide solubility, to estimate compositional changes in vein solutions with changing temperature (Figure 11.13).

Helgeson (*op. cit.*), and Barnes and Helgeson[29] have summarized the data available for ionization constants in aqueous solutions at high temperatures.

The literature utilizing partial pressures of gases as descriptive variables for equilibrium relations of ore minerals and their gangues is now rather extensive. Some key publications are the study of Krauskopf[30] of gas-solid equilibria at 600 °C; the investigation by Eugster and Wones[31] on the stability relations of annite, in which they controlled the composition of iron-bearing phases by "buffering" the system with mineral pairs of

[28] H. C. Helgeson, *Complexing and Hydrothermal Ore Deposition*, New York, Pergamon, 1964.
[29] H. L. Barnes and H. C. Helgeson, Ionization constants in aqueous solution, *Handbook of Physical Constants*, S. P. Clark, Jr. (ed.), *Geol. Soc. Amer. Spec. Paper 36* (in press).
[30] K. B. Krauskopf, *Econ. Geol.*, 52, 786 (1957).
[31] H. P. Eugster and D. R. Wones, *J. Petrology*, 3, 82 (1962).

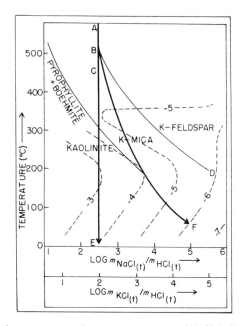

FIG. 11.13. Use of temperature and concentration ratios of NaCl, HCl, and KCl to describe the environment of ore deposition. Dashed lines are calculated isosolubility curves for PbS expressed as log $m_{Pb(t)}$ (note minus sign before the numbers); light solid lines are boundaries of mineral stability fields. Heavy line ABCE shows the effect of cooling a hydrothermal solution of initial composition A, without reaction with wall rock. Heavy line ABF shows the path of solution composition if reaction takes place with felsic wall rocks. In this instance the solubility of PbS decreases steadily. [From H. C. Helgeson, *Complexing and Hydrothermal Ore Deposition*, New York, Pergamon, 1964, p. 103.]

known equilibrium oxygen pressure at a given temperature; and the work of Holland[32] on the delineation of the gas pressures of ore deposition by consideration of the controls exerted by typical mineral associations.

The age-old question concerning ore deposition, of course, is the *composition* of the ore-forming fluids. Judging from the large quantity of data we need to predict compositions of waters from mineral assemblages at low temperatures, it may take many experiments to provide us with data of comparable usefulness at high temperatures. There is some suggestion, however, that aqueous chemistry may be simpler at 400 °C than it is at 25 °C.

[32] H. D. Holland, *Econ. Geol.*, 54, 184 (1959).

APPENDIX I

Symbols and Physical Constants

Symbols

A, B	Constants in Debye-Hückel equation	a,b,c	Constants in heat-capacity equations
B	Constant in regular solution equation	a	Activity
C	Heat capacity	c_P	Gram-atomic heat capacity
E	Potential of galvanic cell; EMF	e	Electron
		f	Fugacity; formality
Eh	Oxidation potential	k	Henry's law constant
F	Free energy (Gibbs)	m	Molality
$\mathscr{F}$	Faraday constant	n	Number of moles
H	Heat content (enthalpy)	q	Quantity of heat absorbed
I	Ionic strength	t	Temperature, °C
K	Equilibrium constant	z_+, z_-	Charge on a positive, negative ion
M	Molarity		
N	Mole fraction	α	Coefficient of thermal expansion
P	Pressure	β	Coefficient of compressibility
Q	Reaction quotient	γ	Activity coefficient on molal basis
R	Gas constant	δ	Increment
S	Entropy	λ	Activity coefficient on mole fraction basis
T	Absolute temperature; °K		
V	Volume	ν	Stoichiometric number
$\bar{V}$	Partial molal volume	χ	Gas activity coefficient

In addition, a Greek letter is sometimes used to denote the polymorphic form of a crystalline substance, as, e.g., α-SiO_2.

Superscripts:
  °   Standard state   *   Pure substance

Subscripts:

1	Solvent	P	Constant pressure
2	Solute	r	Reaction
f	Reaction of formation from the elements	T	Constant temperature
		V	Constant volume
j, i	Constituent of the j'th, i'th kind		

Values of Some Physical Constants and Numerical Factors

Temperature of ice point, 0 °C ....273.150 °K
Thermochemical calorie (defined)..........4.1840 joules
Atmosphere (760 mm Hg, defined) ..1,013,250 dynes/cm^2
Bar (10^6 dynes/cm^2, defined)...............0.98692 atm
R = 1.98726 cal/deg mole
   = 8.31470 joules/deg mole
   = 82.0597 cc-atm/deg mole
$\mathscr{F}$ = 96,493.5 coulombs/equiv (or joules/volt equiv)
   = 23,062.3 cal/volt equiv
$\ln x$ = 2.302585 $\log x$

1 cal/mole = 4.1840 joules/mole = 41.292 cc-atm/mole
1 cc-atm/mole = 0.024218 cal/mole = 0.10133 joule/mole
1 joule/mole = 0.23901 cal/mole = 9.8692 cc-atm/mole

APPENDIX 2

Tables of $\triangle H_f^°$, $S°$, and $\triangle F_f^°$ Values at 25 °C

In the following tables are listed values for the standard free energies of formation, $\Delta F_f^°$, the standard heats of formation, $\Delta H_f^°$, and the standard entropies, $S°$, for a number of substances, including gases and aqueous ions, at 25 °C. The values for $\Delta F_f^°$ and $\Delta H_f^°$ are given in kilocalories per mole of the species; values of $S°$ are given in calories per degree per mole. The standard states on which these values are based are the usual ones, and these are discussed in Chapter 9 and in Reference 1 of Appendix 2. No attempt has been made to assess the probable errors of the values, although it is known that the precision of these values varies from excellent to poor. All data are subject to constant revision.

Aluminum

Formula	Description	State	$\Delta H_f^°$ (kcal)	$S°$ (cal/deg)	$\Delta F_f^°$ (kcal)	Source
Al	Metal	c	0	6.769	0	1
Al^{3+}		aq	−125.4	−74.9	−115.0	1
AlO_2^-		aq			−200.7	3
$H_2AlO_3^-$		aq			−257.4	3
$Al(OH)_3$		amorph.	−304.9	17.	−271.9	2
Al_2O_3	Corundum	c	−399.09	12.186	−376.77	1
$Al_2O_3 \cdot H_2O$	Boehmite	c	−471.	23.15	−435.0	1
$Al_2O_3 \cdot H_2O$	Diaspore	c			−435.1	15
$Al_2O_3 \cdot 3H_2O$	Gibbsite	c		33.51	−554.6	3
$Al_2O_3 \cdot 3H_2O$	Bayerite	c			−552.5	3
$Al_2Si_2O_5(OH)_4$	Kaolinite	c			−884.5	16
Al_2S_3		c	−121.6	23.	−117.7	1
$Al_2(SO_4)_3$		c	−820.98	57.2	−738.99	1

Antimony

Formula	Description	State	ΔH_f° (kcal)	S° (cal/deg)	ΔF_f° (kcal)	Source
Sb		g	60.8	43.06	51.1	1
Sb	Metal	c	0	10.5	0	1
Sb_2		g	52.	60.9	40.	1
SbO^+		aq			−42.0	1
SbO_2^+		aq			−65.5	3
SbO_2^-		aq			−82.5	2
SbO_3^-		aq			−122.9	3
Sb_2O_4		c	−193.3		−165.9	2
Sb_4O_6	Cubic senarmontite	c, II	−336.8	58.8	−298.0	1
Sb_4O_6	Orth. valentinite	c			−294.0	3
Sb_2O_5		c	−234.4	29.9	−200.5	1
SbH_3		g	34.	53.	35.3	2
SbO_2^-		aq			−82.5	2
$HSbO_2$		aq			−97.5	2
SbF_3		c	−217.2	25.2	−199.8	2
$SbCl_3$		g	−75.2	80.8	−72.3	1
$SbCl_3$		c	−91.34	44.5	−77.62	1
SbS_3^-		aq			−32.0	2
Sb_2S_3	amorph.		−36.0	30.3	−32.0	2
SbS_2^-		aq			−13.	2

Arsenic

Formula	Description	State	ΔH_f° (kcal)	S° (cal/deg)	ΔF_f° (kcal)	Source
As		g	60.64	41.62	50.74	1
As	α, gray metal	c	0	8.4	0	1
As	β	amorph.	1.0			2
As	γ, yellow	c	3.53			1
As_2		g	29.6	57.3	17.5	1
As_4		g	35.7	69.	25.2	1
AsO		g	4.79			1
AsO^+		aq			−39.1	1
AsO_2^-		aq			−83.7	1
AsO_4^{3-}		aq	−208.	−34.6	−152.	1
As_2O_5		c	−218.6	25.2	−184.6	1
$As_2O_5 \cdot 4H_2O$		c	−500.3	62.6	−411.1	2
As_4O_6	Octahedral	c, II	−313.94	51.2	−275.36	1
AsH_3		g	41.0	52.	42.0	2
$HAsO_4^{--}$		aq	−214.8	0.9	−169.	1
$H_2AsO_3^-$		aq	−170.3		−140.4	2
$H_2AsO_4^-$		aq	−216.2	28.	−178.9	1
H_3AsO_3		aq	−177.3	47.0	−152.9	1
H_3AsO_4		aq	−214.8	49.3	−183.8	1
AsF_3		g	−218.3	69.08	−214.7	1
$AsCl_3$		g	−71.5	78.2	−68.5	1
As_2S_2		g	−4.22			1
As_2S_2		c	−31.9	32.9	−32.15	2
As_2S_3		c	−35.0	26.8	−32.46	2

Barium

Formula	Description	State	ΔH_f° (kcal)	S° (cal/deg)	ΔF_f° (kcal)	Source
Ba	Metal	c	0	16.	0	1
Ba^{++}		aq	−128.67	3.	−134.0	1
BaO		c	−133.4	16.8	−126.3	1
BaO$_2$		c	−150.5	15.7	−135.8	2
BaO$_2 \cdot$H$_2$O		c	−223.5	25.1	−195.0	2
Ba(OH)$_2$		c	−226.2	22.7	−204.7	2
Ba(OH)$_2 \cdot$8H$_2$O		c	−799.5		−666.8	2
BaF$_2$		c	−286.9	23.1	−272.5	2
BaS		c	−106.0	18.7	−104.5	2
BaSO$_4$		c	−350.2	31.6	−323.4	1
BaSeO$_4$		c	−280.0	36.1	−253.8	2
Ba$_3$(PO$_4$)$_2$		c	−998.0	85.1	−944.4	2
BaCO$_3$	Witherite	c, II	−291.3	26.8	−272.2	1
BaSiO$_3$		c	−359.5	24.2	−338.7	2
Ba$_2$SiO$_3$		c	−496.8	46.4	−470.6	2
BaMnO$_4$		c	−282.	36.8	−257.	2
BaMoO$_4$		c	−373.8	38.7	−349.3	2
BaWO$_4$		c	−407.7	41.	−373.6	2
BaCl$_2$		c	−205.56	30.	−193.8	2
Ba(NO$_3$)$_2$		c	−237.06	51.1	−190.0	2

Beryllium

Formula	Description	State	ΔH_f° (kcal)	S° (cal/deg)	ΔF_f° (kcal)	Source
Be	Metal	c	0	2.28	0	1
Be^{++}		aq	−93.0	−55.	−85.2	2
BeO	Hexagonal	c	−146.0	3.37	−139.0	1
BeO$_2^{--}$		aq	−187.8	−27.	−155.3	2
Be$_2$O$_3^-$		aq			−298.	2
Be$_2$O^{++}		aq			−218.	2
Be(OH)$_2$	α	c	−216.8	13.3	−196.2	2
Be(OH)$_2$	β	c	−216.1	13.3	−195.5	2
BeO·Be(OH)$_2$	Pptd.	c	−366.2	16.7	−338.	2
BeS		c	−55.9	9.3	−55.9	2
BeCl$_2$		c	−122.3	20.5	−111.8	2
BeSO$_4$		c	−286.0	21.5	−260.2	2
BeH		g	78.1	40.84	71.3	1

Bismuth

Formula	Description	State	ΔH_f° (kcal)	S° (cal/deg)	ΔF_f° (kcal)	Source
Bi		g	49.7	44.67	40.4	1
Bi	Metal	c	0	13.6	0	1
Bi_2		g	59.4	65.4	48.0	1
BiO		c	−49.85	17.	−43.5	2
Bi^{3+}		aq			14.83	3
BiO^+		aq			−34.54	1
$BiOH^{++}$		aq			−39.13	3
Bi_2O_3		c	−137.9	36.2	−118.7	1
Bi_2O_4		c			−109.0	2
Bi_2O_5		c			−91.57	3
Bi_4O_7		c			−232.75	3
BiOOH		g			−88.4	2
$Bi(OH)_3$		amorph.	−169.6	24.6	−137.	2
BiCl		g	10.7	58.9	5.2	1
$BiCl_3$		g	−64.7	85.3	−62.2	1
$BiCl_3$		c	−90.61	45.3	−76.23	1
BiOCl		c	−87.3	20.6	−77.0	1
BiBr		g	12.7	61.6	3.8	1
BiI		g	16.	63.4	11.	1
$BiCl_4^-$		aq			−114.2	2
Bi_2S_3		c	−43.8	35.3	−39.4	1
BiH_3		g			55.34	3

Boron

Formula	Description	State	ΔH_f° (kcal)	S° (cal/deg)	ΔF_f° (kcal)	Source
B	Crystalline	c	0	1.56	0	1
BO_2^-		aq	−183.5(?)	20.	−169.6(?)	2
B_2O_3		c	−302.0	12.91	−283.0	1
B_2O_3		glass	−297.6	18.8	−280.4	1
$B_{10}H_{14}$		c			65.	3
$B_4O_7^{--}$		aq			−616.(?)	2
$H_2BO_3^-$		aq	−251.8	7.3	−217.63	1
HBO_3^{--}		aq			−200.29	3
BO_3^{3-}		aq			−181.48	3
H_3BO_3		c	−260.2	21.41	−230.2	1
H_3BO_3		aq			−230.16	3
BF_3		g	−265.4	60.70	−261.3	1
BF_4^-		aq	−365.	40.	−343.	1
B_2S_3		c	−57.0	13.7	−53.3	2
B_2H_6		g	7.5	55.66	19.8	1
B_5H_9		g	15.0	65.88	39.6	1
$B_{10}H_{14}$		g			71.	3
BH		g			112.6	3
BO		g			−19.5	3

Bromine

Formula	Description	State	ΔH_f° (kcal)	S° (cal/deg)	ΔF_f° (kcal)	Source
Br_2		l	0	36.4	0	1
Br^-		aq	−28.90	19.29	−24.574	1
Br_2		g	7.34	58.639	0.751	1
Br_2		aq	−1.1		0.977	2
Br_3^-		aq	−32.0	40.(?)	−25.27	2
HBrO		aq			−19.9	2
BrO^-		aq			−8.0	2
HBr		g	−8.66	47.437	−12.72	1
$HBrO_7^-$		aq			−628.27	3
BrO_3^-		aq			2.30	3

Cadmium

Formula	Description	State	ΔH_f° (kcal)	S° (cal/deg)	ΔF_f° (kcal)	Source
Cd	Metal, α	c	0	12.3	0	1
Cd	Metal, γ	c			0.140	1
Cd^{++}		aq	−17.30	−14.6	−18.58	1
CdO	Cubic	c	−60.86	13.1	−53.79	1
$Cd(OH)_2$	"Active"	c	−133.26	22.8	−112.46	1
$Cd(OH)_2$	"Inactive"	c			−113.13	3
$HCdO_2^-$		aq			−86.5	3
CdF_2		c	−164.9	27.	−154.8	1
$CdCl_2$		c	−93.00	28.3	−81.88	1
$CdCl^+$		aq		5.6	−51.8	2
$CdCl_2^\circ$	Un-ionized	aq		17.	−84.3	2
$CdBr_2$		c	−75.15	31.9	−70.14	1
$CdCl_3^-$		aq		50.7	−115.9	2
CdS		c	−34.5	17.	−33.6	1
$CdSO_4$		c	−221.36	32.8	−195.99	1
$CdSO_4 \cdot H_2O$		c	−294.37	41.1	−254.84	1
$CdSO_4 \cdot \tfrac{8}{3}H_2O$		c	−411.82	57.9	−349.63	1
CdTe		c	−24.30	22.6	−23.82	1
$CdCO_3$		c	−178.7	25.2	−160.2	1
$Cd(NH_3)_4^{++}$		aq			−53.73	2
$Cd(CN)_4^{--}$		aq			111.	2
CdH		g	62.54	50.76	55.73	1

Calcium

Formula	Description	State	ΔH_f° (kcal)	S° (cal/deg)	ΔF_f° (kcal)	Source
Ca	Metal	c, II	0	9.95	0	1
Ca^{++}		aq	−129.77	−13.2	−132.18	1
$CaOH^+$		aq			−171.55	7
CaO	Cubic	c	−151.9	9.5	−144.4	1
CaO_2		c	−157.5	10.3	143.	2
$Ca(OH)_2$	Rhombic	c	−235.80	18.2	−214.22	1
CaF_2		c	−290.3	16.46	−277.7	1
$CaCl_2$		c	−190.0	27.2	−179.3	1
$CaBr_2$		c	−161.3	31.	−156.8	1
CaI_2		c	−127.8	34.	−126.6	1
CaS		c	−115.3	13.5	−114.1	1
$CaCO_3$	Calcite	c	−288.45	22.2	−269.78	1
$CaCO_3$	Aragonite	c	−288.49	21.2	−269.53	1
$CaSiO_3$	Pseudo-wollastonite	c	−377.4	20.9	−357.4	1
$CaSiO_3$	Wollastonite	c	−378.6	19.6	−358.2	1
Ca_2SiO_4	β	c	−538.0	37.6	−512.7	2
Ca_2SiO_4	γ	c	−539.0	37.6	−513.7	2
$CaSO_4$	Anhydrite	c	−342.42	25.5	−315.56	1
$CaSO_4$	Soluble, α	c	−340.27	25.9	−313.52	1
$CaSO_4$	Soluble, β	c	−339.21	25.9	−312.46	1
$CaSO_4 \cdot \frac{1}{2}H_2O$	α	c	−376.47	31.2	−343.02	1
$CaSO_4 \cdot \frac{1}{2}H_2O$	β	c	−375.97	32.1	−342.78	1
$CaSO_4 \cdot 2H_2O$		c	−483.06	46.36	−429.19	1
$Ca_3(PO_4)_2$	α	c	−986.2	57.6	−929.7	1
$Ca_3(PO_4)_2$	β	c	−988.9	56.4	−932.0	1
$CaHPO_4$		c	−435.2	21.	−401.5	1
$CaHPO_4 \cdot 2H_2O$		c	−576.0	40.	−514.6	1
$Ca(H_2PO_4)_2$	Precipitated	c	−744.4	45.3	−672.	2
$CaWO_4$		c	−392.5	36.1	−368.7	2
$CaSO_4^\circ$	Un-ionized	aq			−312.67	7
$CaCO_3^\circ$	Un-ionized	aq			−262.76	7
$CaHCO_3^+$		aq			−273.67	7
$CaMg_3(CO_3)_4$	Huntite (natural)	c			−1007.7	14
$CaBa(CO_3)_2$	Alstonite (natural)	c			−543.0	14
$CaBa(CO_3)_2$	Barytocalcite (natural)	c			−542.9	14
$CaMn(CO_3)_2$	Kutnahorite (natural)	c			−466.2	14
$CaMg(CO_3)_2$	Dolomite (natural)	c			−520.5	14

Carbon

Formula	Description	State	ΔH_f° (kcal)	S° (cal/deg)	ΔF_f° (kcal)	Source
C	Diamond	c	0.4532	0.5829	0.6850	1
C	Graphite	c	0	1.3609	0	1
CO		g	−26.4157	47.301	−32.8079	1
CO_2		g	−94.0518	51.061	−94.2598	1
CO_2		aq	−98.69	29.0	−92.31	1
CH_4		g	−17.889	44.50	−12.140	1
C_2H_2		g	54.194	47.997	50.000	1
H_2CO_3		aq	−167.0	45.7	−149.00	1
HCO_3^-		aq	−165.18	22.7	−140.31	1
CO_3^{--}		aq	−161.63	−12.7	−126.22	1
COS		g	−32.80	55.34	−40.45	1
CS_2		g	27.55	56.84	15.55	1
CF_4		g	−162.5	62.7	−151.8	1
$H_2C_2O_4$		aq	−195.57		−166.8	2
$HC_2O_4^-$		aq	−195.7		−165.12	2
$C_2O_4^{--}$		aq	−195.7	10.6	−159.4	2
CH_3OH		aq	−58.77	31.6	−41.88	1
HCHO		aq			−31.0	2
HCO_2H		aq	−98.0	39.1	−85.1	1
HCO_2^-		aq	−98.0	21.9	−80.0	1

Cerium

Formula	Description	State	ΔH_f° (kcal)	S° (cal/deg)	ΔF_f° (kcal)	Source
Ce	Metal	c, III	0	13.8	0	1
Ce^{3+}		aq	−173.8	−44.	−170.5	1
CeO_2		c	−233.	15.8	−219.	2
$Ce(OH)^{3+}$		aq			−187.7	1
$Ce(OH)_2^{++}$		aq			−244.0	1
$Ce(OH)_3$		c			−311.63	2
CeS_2		c	−153.9	18.8	−151.5	2
Ce_2S_3		c	−298.7	31.5	−293.1	2

Cesium

Formula	Description	State	ΔH_f° (kcal)	S° (cal/deg)	ΔF_f° (kcal)	Source
Cs	Metal	c	0	19.8	0	1
Cs^+		aq	−59.2	31.8	−67.41	1
CsOH		c, II	−97.2	18.6	−84.9	2
Cs_2O		c	−75.9	29.6	−65.6	2
Cs_2O_2		c	−96.2	28.2	−78.2	2
Cs_2O_3		c	−111.2	28.7	−86.1	2
Cs_2O_4		c	−124.2	31.2	−92.5	2

Chlorine

Formula	Description	State	ΔH_f° (kcal)	S° (cal/deg)	ΔF_f° (kcal)	Source
Cl^-		aq	−40.023	13.17	−31.350	1
Cl_2		g	0	53.286	0	1
Cl_2		aq			1.65	2
HCl		g	−22.063	44.617	−22.769	1
HCl		aq	−40.023	13.16	−31.350	1
HClO		aq	−27.83	31.	−19.110	2
ClO^-		aq		10.0	− 8.9	2
$HClO_2$		aq	−13.68	42.	0.07	2
ClO_2^-		aq	−17.18	24.1	2.74	2
$HClO_3$		aq	−23.50	39.0	− 0.62	2
ClO_3^-		aq	−23.50	39.0	− 0.62	1
$HClO_4$		aq	−31.41	43.2	− 2.47	2
ClO_4^-		aq	−31.41	43.5	− 2.57	1

Chromium

Formula	Description	State	ΔH_f° (kcal)	S° (cal/deg)	ΔF_f° (kcal)	Source
Cr	Metal	c	0	5.68	0	1
Cr^{++}		aq	− 33.2(?)		− 42.1	2
Cr^{3+}	$[Cr(6H_2O)]^{3+}$	aq	− 61.2(?)	−73.5	− 51.5	2
Cr_2O_3		c	−269.7	19.4	−250.2	1
$Cr_2O_7^-$		aq	−364.0	51.1	−315.4	2
$Cr(OH)^{++}$	$[Cr(5H_2O)(OH)]^{++}$	aq	−113.5	−16.4	−103.0	2
H_2CrO_4		aq			−185.92	3
$HCrO_4^-$		aq	−220.2	16.5	−184.9	2
CrO_4^{--}		aq	−213.75	9.2	−176.1	2
CrO_2^-		aq			−128.0	3
$Cr(OH)_2$		c			−140.5	2
$Cr(OH)_3$		c	−247.1	19.2	−215.3	2
$Cr(OH)_3$	Hydrous, probably $[Cr(5H_2O)(OH)](OH)_2$	c	−236.6	19.6	−205.5	2
$Cr(OH)_4$		c			−242.4	3
$Cr(OH)_2^+$		aq			−151.2	3
CrO_3^{3-}		aq			−144.2	3

Cobalt

Formula	Description	State	ΔH_f° (kcal)	S° (cal/deg)	ΔF_f° (kcal)	Source
Co	Metal	c, III	0	6.8	0	1
Co^{++}		aq	− 14.2	−27.	− 12.8	2
Co^{3+}		aq			28.9	2
CoO		c	− 55.2	10.5	− 49.0	2
CoO_2		c			− 51.84	3
Co_3O_4		c			−167.8	3

Cobalt (continued)

Formula	Description	State	ΔH_f° (kcal)	S° (cal/deg)	ΔF_f° (kcal)	Source
Co(OH)$_2$		c	−129.3	19.6	−109.0	2
Co(OH)$_3$		c	−174.6	20.	−142.6	2
CoS	α, Precipitated	c	−19.3	16.1	−19.8	2
Co$_2$S$_3$		c	−47.(?)			2
CoSO$_4$		c	−205.5	27.1	−180.1	2
CoCO$_3$		c			−155.57	2
Co(NO$_3$)$_2$		c			−55.1	3
HCoO$_2$		aq			−82.97	3

Copper

Formula	Description	State	ΔH_f° (kcal)	S° (cal/deg)	ΔF_f° (kcal)	Source
Cu	Metal	c	0	7.96	0	1
Cu$^+$		aq	12.4	−6.3	12.0	1
Cu^{++}		aq	15.39	−23.6	15.53	1
CuO		c	−37.1	10.4	−30.4	1
HCuO$_2^-$		aq			−61.42	2
CuO$_2^{--}$		aq			−43.3	1
Cu$_2$O		c	−39.84	24.1	−34.98	1
Cu(OH)$_2$		c	−106.1	19.	−85.3	2
CuS		c	−11.6	15.9	−11.7	1
Cu$_2$S		c, II	−19.0	28.9	−20.6	1
Cu$_2$SO$_4$		c	−179.2	43.6	−156.	2
CuSO$_4$		c	−184.00	27.1	−158.2	1
CuSO$_4 \cdot$H$_2$O		c	−259.00	35.8	−219.2	1
CuSO$_4 \cdot$3H$_2$O		c	−402.27	53.8	−334.6	1
CuSO$_4 \cdot$5H$_2$O	Chalcanthite	c, II	−544.45	73.0	−449.3	1
CuSe		c	−6.6	22.2	−7.9	2
Cu$_2$(OH)$_2$CO$_3$	Malachite	c			−216.44	9
Cu$_3$(OH)$_2$(CO$_3$)$_2$	Azurite	c			−345.8	9
Cu$_4$(OH)$_6$SO$_4$	Brochantite	c			−434.62	9
Cu$_4$(OH)$_6$SO$_4 \cdot$1.3H$_2$O	Langite	c			−505.5	9
Cu$_3$(OH)$_4$SO$_4$	Antlerite	c			−345.5	9
CuCO$_3^\circ$	Un-ionized	aq			−119.9	9
Cu(CO$_3$)$_2^{--}$		aq			−250.5	9
CuCl		c	−32.2	21.9	−28.4	1
CuCO$_3$		c	−142.2	21.	−123.8	1
CuCl$_2$		c	−52.3	26.8	−42.	2

Fluorine

Formula	Description	State	ΔH_f° (kcal)	S° (cal/deg)	ΔF_f° (kcal)	Source
F		g	18.3	37.917	14.2	1
F_2		g	0	48.6	0	1
F_2O		g	5.5	58.95	9.7	2
F^-		aq	-78.66	-2.3	-66.08	1
HF		g	-64.2	41.47	-64.7	1
HF		aq	-78.66	26.	-70.41	2
HF_2^-		aq	-153.6	0.5	-137.5	2

Germanium

Formula	Description	State	ΔH_f° (kcal)	S° (cal/deg)	ΔF_f° (kcal)	Source
Ge		c	0	10.14	0	1
GeO	Hydrated, a	c			-69.9	3
GeO	Hydrated, b	c			-62.7	3
GeO_2		c			-136.1	3
GeO_2	Precipitated	c			$-132.$	3
Ge^{++}		aq			0	3
H_2GeO_3		aq			186.8	3
$HGeO_3^-$		aq			-175.2	3
GeO_3^{--}		aq			-157.9	3
$HGeO_2^-$		aq			-92.1	3

Gold

Formula	Description	State	ΔH_f° (kcal)	S° (cal/deg)	ΔF_f° (kcal)	Source
Au	Metal	c	0	11.4	0	1
Au^+		aq			39.0	2
Au^{3+}		aq			103.6	2
AuO_2		c			48.0	3
Au_2O_3		c	19.3	30.	39.0	1
H_3AuO_3		aq			-61.8	1
$H_2AuO_3^-$		aq			-45.8	1
$HAuO_3^{--}$		aq			-27.6	1
AuO_3^{3-}		aq			-5.8	1
$Au(OH)_3$		c	-100.0	29.	-69.3	1

Hydrogen

Formula	Description	State	ΔH_f° (kcal)	S° (cal/deg)	ΔF_f° (kcal)	Source
H^+		aq	0	0	0	2
H_2		g	0	31.211	0	1

Iodine

Formula	Description	State	ΔH_f° (kcal)	S° (cal/deg)	ΔF_f° (kcal)	Source
I^-		aq	13.37	26.14	-12.35	1
I_3^-		aq	-12.4	41.5	-12.31	1
I_5^-		aq			-6.9	3
I_2		g	14.876	62.280	4.63	1
I_2		c	0	27.9	0	1
I_2		aq	5.0		3.926	2
IO_3^-		aq	-55.0	27.7	-32.4	1
H_2IO^+		aq			-25.4	13
HIO		aq	$-38.(?)$		-23.5	2
IO^-		aq	$-34.(?)$		-8.5	2
HIO_4		aq			-15.02	3
IO_4^-		aq			-12.7	3
$H_4IO_6^-$		aq			-123.88	3
HIO_5^{--}		aq			-58.11	3
IO_5^{3-}		aq			-43.11	3
ICl		c	-8.03	24.5	-3.24	2
ICl_3		c	-21.1	41.1	-5.36	1
HIO_3		aq			-33.34	3
ICl		aq			-4.0	2

Iron

Formula	Description	State	ΔH_f° (kcal)	S° (cal/deg)	ΔF_f° (kcal)	Source
Fe		c	0	6.49	0	1
Fe^{++}		aq	-21.0	-27.1	-20.30	1
Fe^{3+}		aq	-11.4	-70.1	-2.52	1
$Fe_{0.95}O$	"FeO" Wüstite	c	-63.7	12.9	-58.4	1
Fe_2O_3	Hematite	c	-196.5	21.5	-177.1	1
Fe_3O_4	Magnetite	c	-267.0	35.0	-242.4	1
$Fe(OH)^{++}$		aq	-67.4	-23.2	-55.91	1
$Fe(OH)_2$		c	-135.8	19.	-115.57	1
$Fe(OH)_2^+$		aq			-106.2	1
$Fe(OH)_3$		c	-197.0	23.	-166.0	2
$FeCl^{++}$		aq	-42.9	$-22.$	-35.9	1
FeO_2H^-		aq			-90.6	3
FeS	α	c	-22.72	16.1	-23.32	1
FeS_2	Pyrite	c			-36.00	10
$FePO_4$		c	-299.6	22.4	$-272.$	2
$FeCO_3$	Siderite	c	-178.70	22.2	-161.06	1
FeSe	Precipitated	c	-16.5		-13.9	1, 2
$FeSiO_3$		c	$-276.$	20.9	$-257.$	2
Fe_2SiO_4		c	-343.7	35.4	-319.8	1
$FeMoO_4$		c	-257.5	33.4	-234.8	2
$FeWO_4$		c	-274.1	35.4	-250.4	2
$FeSO_4$		c	-220.5	27.6	-198.3	2
$FeCl_3$		c	-96.8	31.1	-80.4	2
FeO_4^{--}		aq			$-111.$	3

Lanthanum

Formula	Description	State	ΔH_f° (kcal)	S° (cal/deg)	ΔF_f° (kcal)	Source
La	Metal	c, III	0	13.7	0	1
La^{3+}		aq	−176.2	−39.	−174.5	2
La_2O_3		c	−458.	29.1	−426.9	2
$La(OH)_3$		c	−345.0	25.	−313.2	2
LaS_2		c	−156.7	18.8	−154.7	2
La_2S_3		c	−306.8	31.5	−301.2	2

Lead

Formula	Description	State	ΔH_f° (kcal)	S° (cal/deg)	ΔF_f° (kcal)	Source
Pb	Metal	c	0	15.51	0	1
Pb^{++}		aq	0.39	5.1	−5.81	1
Pb^{4+}		aq			72.3	2
PbO	Red	c, II	−52.40	16.2	−45.25	1
PbO	Yellow	c, I	−52.07	16.6	−45.05	1
$HPbO_2^-$		aq			−81.0	2
PbO_3^{--}		aq			−66.34	3
PbO_4^{4-}		aq			−67.42	3
$Pb(OH)_2$		c	−123.0	21.	−100.6	1
PbO_2		c	−66.12	18.3	−52.34	1
Pb_3O_4		c	−175.6	50.5	−147.6	1
PbF_2		c	−158.5	29.	−148.1	1
$PbCl_2$		c	−85.85	32.6	−75.04	1
PbS		c	−22.54	21.8	−22.15	1
PbS_2O_3		c	−150.1	35.4	−134.0	2
$PbSO_4$		c, II	−219.50	35.2	−193.89	1
$PbSO_4 \cdot PbO$		c	−282.5	48.7	−258.9	2
PbSe		c	−18.0	26.9	−15.4	2
$PbSeO_4$		c	−148.	37.	−122.	2
PbTe		c	−17.5	27.6	−18.1	2
$Pb_3(PO_4)_2$		c	−620.3	84.45	−581.4	1
$PbHPO_3$		c	−234.5	31.9	−208.3	2
$PbCO_3$		c	−167.3	31.3	−149.7	1
$PbO \cdot PbCO_3$		c	−220.0	48.5	−195.6	1
$2PbO \cdot PbCO_3$		c	−273.	65.	−242.	1
$Pb_3(OH)_2(CO_3)_2$		c			−406.0	8
$PbCrO_4$		c	−225.2	36.5	−203.6	2
$PbMoO_4$		c	−265.8	38.5	−231.7	2
$PbSiO_3$		c	−258.8	27.	−239.0	1
Pb_2SiO_4		c	−312.7	43.	−285.7	1
Pb_2O_3		c			−98.42	3
$PbBr_2$		c	−66.21	38.6	−62.24	1
PbI_2		c	−41.85	42.3	−41.53	1
PbH_2		g			69.5	3

Lithium

Formula	Description	State	ΔH_f° (kcal)	S° (cal/deg)	ΔF_f° (kcal)	Source
Li	Metal	c	0	6.70	0	1
Li$^+$		aq	-66.554	3.4	-70.22	1
LiH		c	-21.61	5.9	-16.72	1
LiOH		c	-116.45	12.	-106.1	1
Li$_2$O		c	-142.4	9.06	-133.9	2
Li$_2$O$_2$		c	-151.7	8.	$-135.$	2
LiCl		c	-97.70	13.2	-91.7	2
Li$_2$SO$_4$		c, II	-342.83	27.	-316.6	2
LiNO$_3$		c	-115.279	25.2	-93.1	2
Li$_2$CO$_3$		c	-290.54	21.60	-270.66	1
LiOH		aq	-121.511	0.9	-107.82	1
LiCl		aq	-106.577	16.6	-101.57	1
Li$_2$SO$_4$		aq	-350.01	10.9	-317.78	1
LiNO$_3$		aq	-115.926	38.4	-96.63	1
Li$_2$CO$_3$		aq	-294.74	-5.9	-266.66	1
LiH		g	30.7	40.77	25.2	1

Manganese

Formula	Description	State	ΔH_f° (kcal)	S° (cal/deg)	ΔF_f° (kcal)	Source
Mn	α	c, IV	0	7.59	0	1
Mn	γ	c, II	0.37	7.72	0.33	1
Mn^{++}		aq	-53.3	$-20.$	-54.4	2
Mn^{3+}		aq	$-27.$		-19.6	2
MnO		c	-92.0	14.4	-86.8	1
HMnO$_2^-$		aq			-120.9	2
MnO$_2$	Pyrolusite	c	-124.2	12.7	-111.1	2
MnO$_4^-$		aq	129.7	45.4	-107.4	2
MnO$_4^{--}$		aq			-120.4	2
Mn$_2$O$_3$		c	-232.1	22.1	-212.3	2
Mn$_3$O$_4$	Hausmannite	c			-306.2	13
Mn(OH)$_2$	Precipitated (identical to pyrochroite)	c			-147.34	13
Mn(OH)$_3$		c	$-212.$	23.8	$-181.$	2
MnS	Green	c, I	-48.8	18.7	-49.9	1
MnS	Precipitated	c			-53.3	2
Mn$_3$(PO$_4$)$_2$	Precipitated	c	$-771.$	71.6	$-683.$	2
MnCO$_3$		c	-213.9	20.5	-195.4	1
MnCO$_3$	Precipitated	c	$-212.$	23.8	-194.3	2
MnSiO$_3$		c	-302.5	21.3	-283.3	1
MnOOH	"γ Mn$_2$O$_3$"	c			-132.2	13
MnOOH	γ, Manganite	c			-133.3	13
MnO$_2$	δ, Birnessite, MnO$_{1.70}$ to MnO$_2$ (contains hydroxyl)	c			-108.3	13
MnO$_2$	γ, Nsutite, MnO$_{1.75}$ to MnO$_2$ (contains hydroxyl)				-109.1	13
MnCO$_3$		aq	-213.9	-32.7	-179.6	1
MnSO$_4$		c	-254.24	26.8	-228.48	1
MnCO$_3$	Rhodochrosite (natural)	c			-195.7	14

Magnesium

Formula	Description	State	ΔH_f° (kcal)	S° (cal/deg)	ΔF_f° (kcal)	Source
Mg	Metal	c	0	7.77	0	1
Mg^{++}		aq	−110.41	−28.2	−108.99	1
MgO		c	−143.84	6.4	−136.13	1
MgO	Finely divided	c	−142.95	6.66	−135.31	1
$Mg(OH)_2$		c	−221.00	15.09	−199.27	1
MgS		c	−83.	12.6	−83.6	2
$Mg_3(PO_4)_2$		c	−961.5	56.8	−904.	2
$Mg_3(AsO_4)_2$		c	−731.3	53.8	−679.3	2
$MgCO_3$		c	−266.	15.7	−246.	1
$MgNH_4PO_4$		c			−390.	2
$MgCl_2$		c	−153.40	21.4	−141.57	1
MgOHCl		c	−191.3	19.8	−175.0	1
$MgBr_2$		c	−123.7	29.4	−119.3	2
$MgSO_4$		c	−305.5	21.9	−280.5	1
$MgSO_4^\circ$	Un-ionized	aq			−289.55	7
$MgCO_3^\circ$	Un-ionized	aq			−239.85	7
$MgHCO_3^+$		aq			−250.88	7
$MgOH^+$		aq			−150.10	7
MgH		g	41.	47.61	34.	1
$Mg_4(CO_3)_3(OH)_2 \cdot 3H_2O$	Hydromagnesite (natural)	c			−1108.3	14

Mercury

Formula	Description	State	ΔH_f° (kcal)	S° (cal/deg)	ΔF_f° (kcal)	Source
Hg	Metal	l	0	18.5	0	1
Hg^{++}		aq	41.59	−5.4	39.38	2
Hg_2^{++}		aq			36.35	2
HgO	Red	c, II	−21.68	17.2	−13.990	1
HgO	Yellow	c, I	−21.56	17.5	−13.959	1
$Hg(OH)_2$		aq			−65.70	2
$HHgO_2^-$		aq			−45.42	1
HgCl		g	19.	62.2	14.	1
$HgCl_2$		c	−55.0	34.5	−44.4	2
Hg_2Cl_2		c	−63.32	46.8	−50.350	1
$HgCl_4^{--}$		aq			−107.7	2
$HgBr_4^{--}$		aq	−99.9	84.(?)	−88.0	2
Hg_2Br_2		c	−49.42	50.9	−42.714	1
$HgBr_2$		c	−40.5	37.2	−35.22	1
HgI		g	33.	67.1	23.	1
Hg_2I_2	Yellow	c	−28.91	57.2	−26.60	1
HgS	Red, cinnabar	c, II	−13.90	18.6	−11.67	1
HgS	Black, metacinnabar	c, I	−12.90	19.9	−11.05	1
HgS_2^{--}		aq			11.6	2
$HgSO_4$		c	−168.3	32.6	−141.0	2
Hg_2SO_4		c	−177.34	47.98	−149.12	1
Hg_2CO_3		c			−105.8	2
Hg_2CrO_4		c			−155.75	2
HgH		g	58.06	52.42	52.60	1
Hg		g	14.54	41.80	7.59	1

Molybdenum

Formula	Description	State	ΔH_f° (kcal)	S° (cal/deg)	ΔF_f° (kcal)	Source
Mo	Metal	c	0	6.83	0	1
Mo^{3+}		aq			−13.8	3
$HMoO_4^-$		aq			−213.6	3
MoO_2		c			−120.0	3
MoO_3		c	−180.33	18.68	−161.95	1
MoO_4		aq	−173.5	40.0	−154.	2
MoO_4^-		aq			−205.42	3
H_2MoO_4	Probably complex	aq			−227.	2
MoS_2		c	−55.5	15.1	−53.8	1
MoS_3		c	−61.2	18.	−57.6	2
$MoO_3 \cdot H_2O$		c			−283.69	3

Nickel

Formula	Description	State	ΔH_f° (kcal)	S° (cal/deg)	ΔF_f° (kcal)	Source
Ni	Metal	c, II	0	7.20	0	1
Ni^{++}		aq	−15.3(?)		−11.53	2
$HNiO_2^-$		aq			−83.46	3
NiO_2		c			−47.5	2
$NiO_2 \cdot 2H_2O$		c			−164.8	3
$Ni_3O_4 \cdot 2H_2O$		c			−283.53	3
$Ni_2O_3 \cdot H_2O$		c			−169.96	3
$Ni(OH)_2$		c	−128.6	19.	−108.3	1
$Ni(OH)_3$		c	−162.1	19.5	−129.5	2
NiS	α	c			−17.7	2
NiS	γ	c			−27.3	2
NiO		c			−51.3	3
$NiSO_4$		c	−213.0	18.6	−184.9	1
$NiSO_4 \cdot 6H_2O$	Green	c, I	−644.98			1
$NiSO_4 \cdot 6H_2O$	Blue	c, II	−642.5	73.1	−531.0	1
$NiCO_3$		c	−158.7	21.9	−147.0	2

Niobium

Formula	Description	State	ΔH_f° (kcal)	S° (cal/deg)	ΔF_f° (kcal)	Source
Nb	Metal	c	0	8.3	0	1
NbO		c			−90.5	3
NbO_2		c			−176.	3
Nb_2O_4		c	−387.8	29.2	−362.4	2
Nb_2O_5		c			−422.0	3
Nb^{3+}		aq			−76.	2

Nitrogen

Formula	Description	State	ΔH_f° (kcal)	S° (cal/deg)	ΔF_f° (kcal)	Source
N_2		g	0	45.767	0	1
NO		g	21.600	50.339	20.719	1
NO_2		g	8.091	57.47	12.390	1
NO_2^-		aq	−25.4	29.9	− 8.25	2
NO_3^-		aq	−49.372	35.0	−26.43	2
$N_2O_2^{--}$		aq	− 2.59	6.6	33.0	1
N_2O_4		g	2.309	72.73	23.491	1
NH_3		g	−11.04	46.01	− 3.976	1
NH_3		aq	−19.32	26.3	− 6.37	1
NH_4^+		aq	−31.74	26.97	−19.00	1
HNO_3		l	−41.404	37.19	−19.100	1
HNO_3		aq	−49.372	35.0	−26.41	1
NH_4OH		aq	−87.64	43.0	−63.05	2
$NH_2(OH)_2^+$		aq			−13.54	3
NH_2OH		aq	−21.7	40.	− 5.60	2
N_2		aq			2.994	3
N_2O		g	19.49	52.58	24.76	1
N_2O_4		g	2.309	72.73	23.491	1
NOCl		g	12.57	63.0	15.86	1
NOBr		g	19.56	65.16	19.70	1

Oxygen

Formula	Description	State	ΔH_f° (kcal)	S° (cal/deg)	ΔF_f° (kcal)	Source
O_2		g	0	49.003	0	1
OH^-		aq	−54.957	− 2.519	−37.595	1
H_2O		g	−57.7979	45.106	−54.6357	1
H_2O		l	−68.3174	16.716	−56.690	1
H_2O_2		aq	−45.68		−31.470	2
O_2^-		aq			13.0	2
HO_2^-		aq			−15.610	2

Palladium

Formula	Description	State	ΔH_f° (kcal)	S° (cal/deg)	ΔF_f° (kcal)	Source
Pd	Metal	c	0	8.9	0	1
Pd^{++}		aq			45.5	2
PdO		c	− 20.4	13.2	− 14.4	2
$Pd(OH)_2$		c	− 92.1	21.7	− 72.	2
$Pd(OH)_4$		c	−169.4(?)	24.7	−126.2(?)	2
Pd_2H		c			− 1.097	3
PdO_3		c			24.1	3

Phosphorus

Formula	Description	State	ΔH_f° (kcal)	S° (cal/deg)	ΔF_f° (kcal)	Source
P	White	c, III	0	10.6	0	1
P	Red	c, II	− 4.4	7.0	− 3.3	2
P	Black	c, I	− 10.3			1
PO_4^{3-}		aq	− 306.9	− 52.	− 245.1	1
$H_2PO_2^-$		aq			− 122.4	2
HPO_3^{--}		aq	− 233.8		− 194.0	2
HPO_4^{--}		aq	− 310.4	− 8.6	− 261.5	1
$H_2PO_3^-$		aq		19.	− 202.35	2
$H_2PO_4^-$		aq	− 311.3	21.3	− 271.3	1
H_3PO_4		aq	− 308.2	42.1	− 274.2	2
H_3PO_2		aq	− 145.6	38.	− 125.1	2
$H_2PO_2^-$		aq			− 122.4	2
H_3PO_3		aq	− 232.2	40.	− 204.8	2
$H_2PO_3^-$		aq		19.	− 202.35	2
HPO_3^{--}		aq	− 233.8		− 194.0	2
$H_4P_2O_6$		aq			− 392.	2
$H_3P_2O_6^-$		aq			− 389.0	3
$H_2P_2O_6^{--}$		aq			− 385.2	3
$HP_2O_6^{3-}$		aq			− 375.3	3
$P_2O_6^{4-}$		aq			− 361.7	3
HPO_3		aq	− 234.8(?)	36.	− 215.8	2
PH_3		g	2.21	50.2	4.36	1
P_2		g	33.82	52.13	24.60	1
P_4		g	13.12	66.90	5.82	1
PCl_3		g	− 73.22	74.49	− 68.42	1
PCl_5		g	− 95.35	84.3	− 77.59	1

Platinum

Formula	Description	State	ΔH_f° (kcal)	S° (cal/deg)	ΔF_f° (kcal)	Source
Pt	Metal	c	0	10.0	0	1
Pt^{++}		aq			54.8	2
$Pt(OH)_2$		c	− 87.2	26.5	− 68.2	1
PtS		c	− 20.8(?)	20.2	− 21.6(?)	2
PtS_2		c	− 27.8	17.8	− 25.6	2

Potassium

Formula	Description	State	ΔH_f° (kcal)	S° (cal/deg)	ΔF_f° (kcal)	Source
K	Metal	c	0	15.2	0	1
K$^+$		aq	− 60.04	25.5	− 67.466	1
KCl		c	− 104.175	19.76	− 97.592	1
KCl		aq	− 100.06	37.7	− 98.816	1
K$_2$O		c	− 86.4	20.8	− 46.2	2
KOH		c	− 101.78	14.2	− 89.5	2
K$_2$S		c	− 100.	26.6	− 96.6	2
K$_2$CO$_3$		c	− 273.93	33.6	− 255.5	2
KAlSi$_3$O$_8$	Feldspar	c			− 856.0	8
KAl$_3$Si$_3$O$_{10}$(OH)$_2$	Mica	c			− 1300.	8
KBr		c	− 93.73	23.05	− 90.63	1
KI		c	− 78.31	24.94	− 77.03	1
KNO$_3$		c	− 117.76	31.77	− 93.96	1
KOH		aq	− 115.00	22.0	− 105.061	1
KSO$_4^-$		aq			− 246.11	7

Rubidium

Formula	Description	State	ΔH_f° (kcal)	S° (cal/deg)	ΔF_f° (kcal)	Source
Rb	Metal	c, I	0	16.6	0	1
Rb$^+$		aq	− 58.9	29.7	− 67.45	1
Rb$_2$S		c	− 83.2	32.	− 80.6	2
Rb$_2$CO$_3$		c	− 269.6	23.3	− 249.3	2
RbH		c			− 7.3	3
RbOH		c, II	− 98.9	16.9	− 87.1	2
Rb$_2$O		c	− 78.9	26.2	− 69.5	2
Rb$_2$O$_2$		c	− 101.7	24.8	− 83.6	2
Rb$_2$O$_3$		c	− 116.7	25.3	− 92.4	2
Rb$_2$O$_4$		c	− 126.2	27.8	− 94.6	2
RbOH		aq	− 113.9	27.2	− 105.05	1

Scandium

Formula	Description	State	ΔH_f° (kcal)	S° (cal/deg)	ΔF_f° (kcal)	Source
Sc	Metal	c	0	8.	0	2
Sc^{3+}		aq	− 148.8	− 56.	− 143.7	2
ScOH^{++}		aq			− 193.7	3
Sc(OH)$_3$		c			− 293.5	2

Selenium

Formula	Description	State	ΔH_f° (kcal)	S° (cal/deg)	ΔF_f° (kcal)	Source
Se		g	48.37	42.21	38.77	1
Se	I, gray hexagonal	c, I	0	10.0	0	1
Se^{--}		aq	31.6	20.0	37.2	1
Se_2		g	33.14	60.22	21.15	1
SeO_2		c	−55.00	13.6	−41.5	2
SeO_3^{--}		aq	−122.39	3.9	−89.33	1
SeO_4^{--}		aq	−145.3	5.7	−105.42	1
HSe^-		aq	24.6	42.3	23.57	1
H_2Se		g	20.5	52.9	17.0	1
H_2Se		aq	18.1	39.9	18.4	1
$HSeO_3^-$		aq	−123.5	30.4	−98.3	1
$HSeO_4^-$		aq	−143.1	22.0	−108.2	1
H_2SeO_3		aq	−122.39	45.7	−101.8	1
H_2SeO_4		aq	−145.3	5.7	−105.42	1
SeF_6		g	−246.	75.10	−222.	1

Silicon[a]

Formula	Description	State	ΔH_f° (kcal)	S° (cal/deg)	ΔF_f° (kcal)	Source
Si	Metal	c	0	4.47	0	1
Si		g	88.04	40.120	77.41	1
SiO_2	Quartz, II	c	−205.4	10.00	−192.4	1
SiO_2	Cristobalite, II	c	−205.0	10.19	−192.1	1
SiO_2	Tridymite, IV	c	−204.8	10.36	−191.9	1
SiO_2	Vitreous	glass	−202.5	11.2	−190.9	1
SiH_4		g	−14.8	48.7	−9.4	1
SiF_6^{--}		aq	−558.5	−12.(?)	−511.	2
H_4SiO_4		aq			−300.3	5
$H_3SiO_4^-$		aq			−286.8	6
SiF_4		g	−370.	68.0	−360.	1
$SiCl_4$		g	−145.7	79.2	−136.2	1

[a] The values for silicon species are in the process of drastic changes, but the values given here are internally consistent, and are consistent with the values for silicates given in the other tables of Appendix 2.

Silver

Formula	Description	State	ΔH_f° (kcal)	S° (cal/deg)	ΔF_f° (kcal)	Source
Ag	Metal	c	0	10.206	0	1
Ag^+		aq	25.31	17.67	18.430	1
Ag^{++}		aq			64.1	2
AgO^+		aq			53.9	2
AgO^-		aq			− 5.49	2
Ag_2O		c	− 7.306	29.09	− 2.586	1
AgO		c	− 6.0		2.6	2
Ag_2O_3		c			20.8	2
AgCl		c	− 30.362	22.97	− 26.224	1
AgBr		c, II	− 23.78	25.60	− 22.930	1
AgI		c, II	− 14.91	27.3	− 15.85	1
Ag_2S	Rhombic, α	c, II	− 7.60	34.8	− 9.62	1
Ag_2S	β	c, I	− 7.01	35.9	− 9.36	1
Ag_2SO_4		c, II	− 170.50	47.8	− 147.17	1
Ag_2SeO_4		c	− 94.7	43.3	− 68.5	1
Ag_2CO_3		c	− 120.97	40.0	− 104.48	1
$AgMoO_4$		c			− 196.4	2
Ag_2WO_4		c			− 206.0	2
Ag_2CrO_4		c	− 176.2	51.8	− 154.7	2
AgOH		c			− 21.98	3
$AgNO_3$		c, II	− 29.43	33.68	− 7.69	1
$Ag(S_2O_3)_2^{3-}$		aq	− 285.5		− 247.6	2
$Ag(SO_3)_2^{3-}$		aq			− 225.4	2
$Ag(NH_3)_2^+$		aq	− 26.724	57.8	− 4.16	2
$Ag(CN)_2^-$		aq	64.5	49.0	72.05	1

Sodium

Formula	Description	State	ΔH_f° (kcal)	S° (cal/deg)	ΔF_f° (kcal)	Source
Na	Metal	c	0	12.2	0	1
Na^+		aq	− 57.279	14.4	− 62.589	1
NaCl		aq	− 97.302	27.6	− 93.939	1
NaCl		c	− 98.232	17.30	− 91.785	1
Na_2S		c	− 89.2	23.2	− 86.6	2
Na_2CO_3		c	− 270.3	32.5	− 250.4	1
Na_2CO_3		aq			− 251.4	2
$NaHCO_3$		c	− 226.5	24.4	− 203.6	1
$NaHCO_3$		aq	− 222.5	37.1	− 202.89	2
$NaHCO_3^\circ$	Un-ionized	aq			− 202.56	7
Na_2SiO_3		c	− 363.	27.2	− 341.	1
$NaOH \cdot H_2O$		c	− 175.17	20.2	− 149.00	1
NaOH		c, II	− 101.99	12.5	− 90.1	2
$NaOH^\circ$	Un-ionized	aq			− 99.23	7
Na_2O		c	− 99.4	17.4	− 90.0	1
$NaCO_3^-$		aq			− 190.54	7
$NaSO_4^-$		aq			− 240.91	7
NaF		c	− 136.0	14.0	− 129.3	1
NaBr		c	− 86.030	20.5	− 83.1	2
NaI		c	− 68.84	22.1	− 56.7	2

Sodium (continued)

Formula	Description	State	ΔH_f° (kcal)	S° (cal/deg)	ΔF_f° (kcal)	Source
Na_2SO_4		c, II	-330.90	35.73	-302.78	1
$NaNO_3$		c, II	-111.54	27.8	-87.45	1
$Na_2CO_3 \cdot 10H_2O$		c			-819.546	17
$Na_2CO_3 \cdot 7H_2O$		c			-649.12	17
$Na_2CO_3 \cdot H_2O$		c			-307.493	17
$NaHCO_3 \cdot Na_3CO_3 \cdot 2H_2O$	Trona	c			-570.40	18

Strontium

Formula	Description	State	ΔH_f° (kcal)	S° (cal/deg)	ΔF_f° (kcal)	Source
Sr	Metal	c	0	13.0	0	1
Sr^{++}		aq	-130.38	-9.4	-133.2	1
SrO		c	-141.1	13.0	-133.8	1
SrO_2		c	-153.6	13.	139.	2
$Sr(OH)_2$		c	-229.3	21.	-207.8	2
SrF_2		c	-290.3	21.4	-277.8	2
SrS		c	-108.1	17.	-97.4	2
$SrSO_4$		c	-345.3	29.1	-318.9	1
$Sr_3(PO_4)_2$		c	-987.3	70.	-932.1	2
$SrHPO_4$		c	-431.3	31.2	-399.7	2
$SrCO_3$	Strontianite	c, II	-291.2	23.2	-271.9	1
$SrSiO_3$		c	-371.2	22.5	-350.8	2
Sr_2SiO_4		c	-520.6	43.	-495.7	2
$Sr(WO_4)$		c	-398.3	37.8	-366.5	2
$SrCl_2$		c	-198.0	28.	-186.7	1

Sulfur

Formula	Description	State	ΔH_f° (kcal)	S° (cal/deg)	ΔF_f° (kcal)	Source
S	Rhombic	c, II	0	7.62	0	1
S	Monoclinic	c, I	0.071	7.78	0.023	1
S		g	53.25	40.085	43.57	1
S^{--}		aq			21.96	3
S$_2$		g	29.86		19.13	4
S$_2^-$		aq			19.75	3
S$_3^-$		aq			17.97	3
S$_4^{--}$		aq			16.61	3
S$_5^{--}$		aq			15.69	3
SO$_2$		g	$-$ 70.96	59.40	$-$ 71.79	1
SO$_3$		g	94.45	61.24	$-$ 88.52	1
SO$_3^{--}$		aq	$-$151.9	$-$ 7.	$-$116.1	2
SO$_4^{--}$		aq	$-$216.90	4.1	$-$177.34	1
S$_2$O$_3^{--}$		aq	$-$154.	29.	$-$127.2	1
S$_2$O$_4^{--}$		aq	$-$164.	57.	$-$138.	1
S$_2$O$_5^{--}$		aq	$-$232.	25.	$-$189.	2
S$_2$O$_6^{--}$		aq	$-$280.4	30.	$-$231.	2
S$_2$O$_8^{--}$		aq	$-$324.3	35.	$-$262.	2
S$_3$O$_6^{--}$		aq	$-$279.	33.	$-$229.	2
S$_4$O$_6^{--}$		aq	$-$290.	62.	$-$246.3	1
S$_5$O$_6^{--}$		aq	$-$281.	40.	$-$228.5	2
HS$^-$		aq	$-$ 4.22	14.6	3.01	1
H$_2$S		g	$-$ 4.815	49.15	$-$ 7.892	1
H$_2$S		aq	$-$ 9.4	29.2	$-$ 6.54	1
HSO$_3^-$		aq	$-$150.09	31.64	$-$126.03	1
HSO$_4^-$		aq	$-$211.70	30.32	$-$179.94	1
H$_2$SO$_3$		aq	$-$145.5	56.	$-$128.59	2
H$_2$SO$_4$		aq	$-$216.90	4.1	$-$177.34	1
HS$_2$O$_4^-$		aq			$-$141.4	3
H$_2$S$_2$O$_4$		aq	$-$164.		$-$140.0	2
H$_2$S$_2$O$_8$		aq	$-$324.3	35.	$-$262.	2
SF$_6$		g	$-$262.	69.5	$-$237.	1
S$_2$Cl$_2$		l	$-$ 14.4	40.	$-$ 5.9	2
SO$_2$Cl$_2$		g			$-$ 73.6	2
H$_2$S$_2$O$_3$		aq			$-$129.9	3
HS$_2$O$_3^-$		aq			$-$129.5	3
SO		g	19.02	53.04	12.78	1

Tantalum

Formula	Description	State	ΔH_f° (kcal)	S° (cal/deg)	ΔF_f° (kcal)	Source
Ta	Metal	c	0	9.9	0	1
Ta$_2$O$_5$		c	$-$499.9	34.2	$-$470.6	1

Tellurium

Formula	Description	State	ΔH_f° (kcal)	S° (cal/deg)	ΔF_f° (kcal)	Source
Te		c, II	0	11.88	0	1
Te^{--}		aq			52.7	2
Te$_2^{--}$		aq			38.75	2
Te$_2$		g	41.0	64.07	29.0	1
HTe$^-$		aq			37.7	2
H$_2$Te		g	36.9	56.	33.1	1
H$_2$Te		aq			34.1	2
TeO$_2$		c			-65.32	3
TeO$_3^-$		aq			-93.79	3
H$_6$TeO$_6$		c			-245.04	3
H$_2$TeO$_3$		c			-114.36	3
TeOOH$^+$		aq			-61.78	2
TeF$_6$		g	-315.0	80.67	$-292.$	1
TeCl$_4$		c	-77.2	50.	-56.7	2
TeCl$_6^{--}$		aq			-137.4	2
Te^{4+}		aq			52.38	3
HTeO$_2^+$		aq			-62.51	3
HTeO$_3^-$		aq			-104.34	3
HTeO$_4^-$		aq			-123.27	3
TeO$_4^{--}$		aq			-109.1	3
H$_2$TeO$_4$		aq			-131.66	3

Thorium

Formula	Description	State	ΔH_f° (kcal)	S° (cal/deg)	ΔF_f° (kcal)	Source
Th	Metal	c	0	13.6	0	1
Th^{4+}		aq	-183.0	$-75.$	-175.2	2
ThO$_2$		c	$-292.$	16.9	-278.4	2
Th(OH)$_4$	"Soluble"	c	-421.5	32.	$-379.$	2
Th$_2$S$_3$		c	-262.0	35.7	-257.7	2

Tin

Formula	Description	State	ΔH_f° (kcal)	S° (cal/deg)	ΔF_f° (kcal)	Source
Sn	Gray	c, III	0.6	10.7	1.1	1
Sn	White	c, II	0	12.3	0	1
Sn^{++}		aq	$-$ 2.39	$-$ 5.9	$-$ 6.275	2
Sn^{4+}		aq			0.65	2
SnO		c	$-$ 68.4	13.5	$-$ 61.5	1
SnO_2		c			$-$ 123.2	3
$HSnO_2^-$		aq			$-$ 98.	2
$Sn(OH)_2$		c	$-$ 138.3	23.1	$-$ 117.6	1
SnF_6^{--}		aq	$-$ 474.7	0.	$-$ 420.	2
$Sn(OH)_4$		c	$-$ 270.5	29.	$-$ 227.5	2
$Sn(OH)_6^{--}$		aq			$-$ 310.5	2
SnS		c	$-$ 18.6	23.6	$-$ 19.7	1
$Sn(SO_4)_2$		c	$-$ 393.4	37.1	$-$ 346.8	2
$SnCl_2$		c	$-$ 83.6	29.3	$-$ 72.2	2
$Sn(OH)^+$		aq			$-$ 60.6	3
$Sn_2O_3^-$		aq			$-$ 141.1	3
SnO_3^{--}		aq			$-$ 137.42	3
SnH_4		g			99.	3

Titanium

Formula	Description	State	ΔH_f° (kcal)	S° (cal/deg)	ΔF_f° (kcal)	Source
Ti	Metal	c, II	0	7.24	0	1
Ti^{++}		aq			$-$ 75.1	2
Ti^{3+}		aq			$-$ 83.6	2
TiO_2	Rutile	c			$-$ 212.3	3
TiO_2	Hydrated	c	$-$ 207.		$-$ 196.3	2
$TiO(OH)_2$		c			$-$ 253.	2
TiO^{++}		aq			$-$ 138.	2
Ti_2O_3		c, II			$-$ 342.3	3
Ti_3O_5		c			$-$ 553.1	3
$FeTiO_3$		c	$-$ 288.5	25.3	$-$ 268.9	1
$HTiO_3^-$		aq			$-$ 228.5	3
TiO_2^+		aq			$-$ 111.7	3
TiO		c			$-$ 116.9	3
$Ti(OH)_3$		c			$-$ 250.9	3

Tungsten

Formula	Description	State	ΔH_f° (kcal)	S° (cal/deg)	ΔF_f° (kcal)	Source
W	Metal	c	0	8.0	0	1
WO_2		c	$-$ 136.3	17.	$-$ 124.4	2
WO_3	Yellow	c	$-$ 200.84	19.90	$-$ 182.47	1
W_2O_5		c	$-$ 337.9	34.	$-$ 360.9	2
WO_4^-		aq	$-$ 266.6	15.	$-$ 220.	2
WS_2		c	$-$ 46.3	23.	$-$ 46.2	1

Uranium

Formula	Description	State	$\Delta H_f°$ (kcal)	$S°$ (cal/deg)	$\Delta F_f°$ (kcal)	Source
U	Metal	c, III	0	12.03	0	1
U^{3+}		aq	−123.0	−30.	−124.4	1
U^{4+}		aq	−146.7	−78.	−138.4	1
UO_2		c			−246.6	3
UO_2^+		aq	−247.4	12.	−237.6	1
UO_2^{++}		aq	−250.4	−17.	−236.4	1
UO_3		c			−273.0	3
$UO_3 \cdot H_2O$		c	−375.4	33.	−343.	2
$U(OH)^{3+}$		aq	−204.1	−30.	−193.5	1
$U(OH)_3$		c			−263.2	2
$U(OH)_4$		c			−351.6	2
Na_2UO_4		c	−501.	47.	−475.	2
UO_2SO_4		aq	−467.3	−13.	−413.7	1
$UO_2(CO_3)_3^{4-}$		aq			−640.0	11
$UO_2(CO_3)_2(H_2O)_2^{2-}$		aq			−622.0	11
UO_2CO_3		c			−377.0	11
$UO_2(OH)_2 \cdot H_2O$		c			−437.0	11

Vanadium

Formula	Description	State	$\Delta H_f°$ (kcal)	$S°$ (cal/deg)	$\Delta F_f°$ (kcal)	Source
V	Metal	c	0		0	12
V^{++}		aq			54.2	12
V^{3+}		aq			−60.1	12
VO^{++}		aq			−109.0	12
VO_2^+		aq			−142.6	12
V_2O_2		c			−189.0	12
V_2O_3		c			−271.0	12
V_2O_4		c			−318.0	12
V_2O_5	Aged ppt.				−344.0	12
V_2O_5	Fresh ppt.				−342.0	12
$V(OH)_3$	Ppt.				−218.0	12
$V(OH)^{++}$		aq			−112.8	12
$VO(OH)_2$	Ppt.				−213.6	12
NH_4VO_3		c			−221.8	12
$V_4O_9^{2-}$		aq			−665.3	12
$H_2V_{10}O_{28}^{4-}$		aq			−1875.2	12
$HV_{10}O_{28}^{5-}$		aq			−1875.3	12
$V_{10}O_{28}^{6-}$		aq			−1862.4	12

Ytterbium

Formula	Description	State	$\Delta H_f°$ (kcal)	$S°$ (cal/deg)	$\Delta F_f°$ (kcal)	Source
Yb		c	0	14.45	0	2
$Yb(OH)_3$		c			−301.7	2
Yb^{++}		aq			−129.0	1
Yb^{3+}		aq	−160.6	−45.4	−156.8	2

Yttrium

Formula	Description	State	ΔH_f° (kcal)	S° (cal/deg)	ΔF_f° (kcal)	Source
Y	Metal	c	0	11.3	0	2
Y^{3+}		aq	−168.0	−48.	−164.1	2
$Y(OH)_3$		c	−339.5	23.	−307.1	2
Y_2O_3	Cubic	c			−402.0	3

Zinc

Formula	Description	State	ΔH_f° (kcal)	S° (cal/deg)	ΔF_f° (kcal)	Source
Zn	Metal	c	0	9.95	0	1
Zn^{++}		aq	−36.43	−25.45	−35.184	1
ZnO	Orthorhombic	c			−76.88	3
ZnO_2^{--}		aq			−93.03	2
$Zn(NH_3)_4^{++}$		aq			−73.5	2
$Zn(OH)^+$		aq			−78.7	3
$Zn(OH)_2$	ε, Orthorhombic	c			−133.63	3
$Zn(OH)_2$	γ, White	c			−133.31	3
$Zn(OH)_2$	β, Orthorhombic	c			−133.13	3
$Zn(OH)_2$	α	c			−131.93	3
$Zn(OH)_2$		amorph.			−131.85	3
ZnO	"Active"	c			−75.69	3
ZnS	Sphalerite	c, II	−48.5	13.8	−47.4	1
ZnS	Wurtzite	c	−45.3	13.8	−44.2	2
ZnS	Pptd.	c	−44.3(?)		−43.2(?)	2
$ZnSO_4$		c	−233.88	29.8	−208.31	1
$ZnSO_4 \cdot H_2O$		c	−310.6	34.9	−269.9	1
$ZnSO_4 \cdot 6H_2O$		c	−663.3	86.8	−555.0	1
$ZnSO_4 \cdot 7H_2O$		c	−735.1	92.4	−611.9	1
ZnSe		c	−34.	22.3	−34.7	2
$ZnSiO_3$		c	−294.6	21.4	−274.8	2
$ZnCO_3$		c	−194.2	19.7	−174.8	1
$HZnO_2^-$		aq			−110.9	3
$ZnCl_2$		c	−99.40	25.9	−88.255	1
$ZnBr_2$		c	−78.17	32.84	−74.142	1

Zirconium

Formula	Description	State	ΔH_f° (kcal)	S° (cal/deg)	ΔF_f° (kcal)	Source
Zr	Metal	c, II	0	9.18	0	1
Zr^{4+}		aq			−142.0	3
ZrO^{++}		aq			−201.5	3
ZrO_2		c			−247.7	3
$ZrO(OH)_2$		c	−338.0	22.	−311.5	2
$Zr(OH)_4$		c	−411.2	31.	−370.	2
$HZrO_3^-$		aq			−287.7	2

References

1. Rossini, F. D., D. D. Wagman, W. H. Evans, Samuel Levine, and Irving Jaffe, Selected values of chemical thermodynamic properties: *Natl. Bur. Standards Circ. 500*, U.S. Dept. Commerce (1952).
2. Latimer, W. M., *Oxidation Potentials*, 2nd edition. New York, Prentice-Hall, 1952.
3. Technical Report 684, Enthalpies libre de formation standards, à 25 °C. Centre Belge d'Étude de la Corrosion, Brussels, 1960.
4. Stephenson, C. C., personal communication.
5. Siever, R., The silica budget in the sedimentary cycle: *Am. Mineral.*, *42*, 826 (1957).
6. Estimated by R.M.G. from data in Reference 5 and the dissociation constant of H_4SiO_4.
7. Calculated by R.M.G. and C.L.C. from dissociation constants listed in Table 4.1, Chapter 4.
8. Garrels, R. M., Some free energy values from geologic relations: *Am. Mineral.*, *42*, 789 (1957).
9. Silman, J. R., Ph.D. thesis, Department of Geology, Harvard University, 1958.
10. Kelley, K. K., The thermodynamic properties of sulfur and its inorganic compounds: *U.S. Bureau of Mines Bull. 406* (1937).
11. Bullwinkel, E. P., The chemistry of uranium in carbonate solutions: U.S. Atomic Energy Comm., Raw Materials Division, RMO-2614 (1954).
12. Evans, H. T., Jr., and R. M. Garrels, Thermodynamic equilibria of vanadium in aqueous systems as applied to the interpretation of the Colorado Plateau ore deposits: *Geochim. et Cosmochim. Acta*, *15*, 131 (1958).
13. Bricker, O. P., Ph.D. thesis, Department of Geology, Harvard University, 1964.
14. Garrels, R. M., M. E. Thompson, and R. Siever, Stability of some carbonates at 25 °C and 1 atmosphere total pressure: *Am. J. Sci.*, *258*, 402 (1960).
15. Estimated from data in Kennedy, G. C., Phase relations in the system Al_2O_3-H_2O at high temperatures and pressures: *Am. J. Sci.*, *257*, 563 (1959).
16. Polzer, W., personal communication, 1961.
17. Saegusa, F., *Science Reports, Tōhoku Univ., 1st series*, *34*, 104 (1950).
18. Garrels, R. M., and M. E. Thompson, unpublished study.

APPENDIX 3

Crystal Radii of Some Ions

The radii listed below[1,2] are for ions in six-fold coordination (NaCl structure); the values are given in Angstrom units, $1\text{Å} = 10^{-8}$ cm. The ions in each horizontal row (in the body of the table) are isoelectronic.

Li^+ 0.60	Be^{++} 0.31	B^{3+} 0.20	H^- 2.08	
Na^+ 0.95	Mg^{++} 0.65	Al^{3+} 0.50	F^- 1.36	O^{--} 1.40
K^+ (1) 1.33	Ca^{++} (2) 0.99	Sc^{3+} (3) 0.81	Cl^- 1.81	S^{--} 1.84
Cu^+ 0.96	Zn^{++} 0.74	Ga^{3+} 0.62		
Rb^+ 1.48	Sr^{++} 1.13	Y^{3+} 0.93	Br^- 1.95	Se^{--} 1.98
Ag^+ 1.26	Cd^{++} 0.97 (4)	In^{3+} 0.81		
Cs^+ 1.69	Ba^{++} 1.35	La^{3+} (5) 1.15	I^- 2.16	Te^{--} 2.21
Au^+ (6) 1.37	Hg^{++} 1.10	Tl^{3+} 0.95		

(1) NH_4^+, 1.48 Å.
(2) Mn^{++}, 0.80; Fe^{++}, 0.76; Co^{++}, 0.74; Ni^{++}, 0.72 Å.
(3) Ti^{3+}, 0.74; V^{3+}, 0.74; Cr^{3+}, 0.69; Mn^{3+}, 0.66; Fe^{3+}, 0.64 Å.
(4) Sn^{++}, 1.12 Å.
(5) Trivalent rare earth ions 1.11 Å (Ce^{3+}) to 0.93 Å (Lu^{3+}).
(6) Tl^+, 1.40.

[1] Data from Linus Pauling, *The Nature of the Chemical Bond*, 3rd edition. Cornell University Press, 1960, chap. 13.
[2] Values for ions having oxidation states above 3+ are not given in the present table, although these are listed in Pauling, *op. cit.*

APPENDIX 4

Pressure of Saturated Water Vapour at Various Temperatures[a]

(Pressure in mm Hg; temperature in °C)

Temp	1	2	3	4	5	6	7	8	9	10
P_{H_2O}	4.92	5.29	5.68	6.10	6.54	7.01	7.51	8.04	8.60	9.20
Temp	11	12	13	14	15	16	17	18	19	20
P_{H_2O}	9.84	10.52	11.22	11.98	12.78	13.62	14.53	15.46	16.46	17.53
Temp	0	1	2	3	4	5	6	7	8	9
20	17.53	18.65	19.82	21.05	22.37	23.75	25.21	26.74	28.32	30.03
30	31.82	33.70	35.69	37.71	39.15	42.20	44.60	47.04	49.70	52.45
40	55.30	58.35	61.50	64.85	68.30	71.90	75.65	79.55	83.00	88.00
50	92.50	97.25	102.1	107.1	113.0	118.0	123.9	129.9	136.2	142.6
60	149.4	156.3	163.9	171.7	179.4	187.6	196.1	205.0	214.1	223.8
70	308.5	243.2	252.2	265.9	275.2	289.1	301.5	314.2	327.3	340.9
80	355.2	369.7	384.8	400.6	416.5	439.8	450.8	468.6	487.0	506.0
90	525.5	546.5	567.0	588.5	610.8	634.0	658.0	682.0	707.0	733.0
100	767.0	786.5	815.5	845.0	875.1	906.0	937.8	970.5	1004.2	1038.8

[a] From *Smithsonian Physical Tables*, 9th revised edition, prepared by W. E. Forsythe. Smithsonian Institution, Washington, D.C., 1954, p. 600. Tables covering temperatures from −60 °C (for ice) to 370 °C, and the range 0 to 25 °C in 0.1° intervals, are given in this reference.

APPENDIX 5

Reference Buffer Solutions[a]

Recommended Standard Values of pH_s

t °C	A Tetraoxalate	B Tartrate	C Phthalate	D Phosphate	E Phosphate	F Borax	G Calcium Hydroxide
0	1.666	—	4.003	6.984	7.534	9.464	13.423
5	1.668	—	3.999	6.951	7.500	9.395	13.207
10	1.670	—	3.998	6.923	7.472	9.332	13.003
15	1.672	—	3.999	6.900	7.448	9.276	12.810
20	1.675	—	4.002	6.881	7.429	9.225	12.627
25	1.679	3.557	4.008	6.865	7.413	9.180	12.454
30	1.683	3.552	4.015	6.853	7.400	9.139	12.289
35	1.688	3.549	4.024	6.844	7.389	9.102	12.133
38	1.691	3.548	4.030	6.840	7.384	9.081	12.043
40	1.694	3.547	4.035	6.838	7.380	9.068	11.984
45	1.700	3.547	4.047	6.834	7.373	9.038	11.841
50	1.707	3.549	4.060	6.833	7.367	9.011	11.705
55	1.715	3.554	4.075	6.834	—	8.985	11.574
60	1.723	3.560	4.091	6.836	—	8.962	11.449
70	1.743	3.580	4.126	6.845	—	8.921	—
80	1.766	3.609	4.164	6.859	—	8.885	—
90	1.792	3.650	4.205	6.877	—	8.850	—
95	1.806	3.674	4.227	6.886	—	8.833	—

[a] From Roger G. Bates, revised standard values for pH measurements from 0 to 95 °C: *J. Res. Natl. Bur. Std.*, 66A, 179 (1962). The five solutions of pH 3.5 to 9.5 are considered to be *primary* standards for use in calibrating glass electrodes; the tetraoxalate and calcium hydroxide solutions are *secondary* standards. For precise results, various precautions must be observed in the use of the reference solutions; these precautions are discussed in detail in the paper by Bates.

APPENDIX 6

Table of Atomic Weights—1961

(Based on Carbon-12)

Element	Symbol	Atomic No.	Atomic Weight
Actinium	Ac	89	
Aluminum	Al	13	26.9815
Americium	Am	95	
Antimony	Sb	51	121.75
Argon	Ar	18	39.948
Arsenic	As	33	74.9216
Astatine	At	85	
Barium	Ba	56	137.34
Berkelium	Bk	97	
Beryllium	Be	4	9.0122
Bismuth	Bi	83	208.980
Boron	B	5	10.811[a]
Bromine	Br	35	79.909[b]
Cadmium	Cd	48	112.40
Calcium	Ca	20	40.08
Californium	Cf	98	
Carbon	C	6	12.01115[a]
Cerium	Ce	58	140.12
Cesium	Cs	55	132.905
Chlorine	Cl	17	35.453[b]
Chromium	Cr	24	51.996[b]
Cobalt	Co	27	58.9332
Copper	Cu	29	63.54
Curium	Cm	96	
Dysprosium	Dy	66	162.50
Einsteinium	Es	99	
Erbium	Er	68	167.26
Europium	Eu	63	151.96
Fermium	Fm	100	
Fluorine	F	9	18.9984
Francium	Fr	87	
Gadolinium	Gd	64	157.25
Gallium	Ga	31	69.72
Germanium	Ge	32	72.59
Gold	Au	79	196.967
Hafnium	Hf	72	178.49

Table of Atomic Weights—1961 (continued)

Element	Symbol	Atomic No.	Atomic Weight
Helium	He	2	4.0026
Holmium	Ho	67	164.930
Hydrogen	H	1	1.00797[a]
Indium	In	49	114.82
Iodine	I	53	126.9044
Iridium	Ir	77	192.2
Iron	Fe	26	55.847[b]
Krypton	Kr	36	83.80
Lanthanum	La	57	138.91
Lead	Pb	82	207.19
Lithium	Li	3	6.939
Lutetium	Lu	71	174.97
Magnesium	Mg	12	24.312
Manganese	Mn	25	54.9380
Mendelevium	Md	101	
Mercury	Hg	80	200.59
Molybdenum	Mo	42	95.94
Neodymium	Nd	60	144.24
Neon	Ne	10	20.183
Neptunium	Np	93	
Nickel	Ni	28	58.71
Niobium	Nb	41	92.906
Nitrogen	N	7	14.0067
Nobelium	No	102	
Osmium	Os	76	190.2
Oxygen	O	8	15.9994[a]
Palladium	Pd	46	106.4
Phosphorus	P	15	30.9738
Platinum	Pt	78	195.09
Plutonium	Pu	94	
Polonium	Po	84	
Potassium	K	19	39.102
Praseodymium	Pr	59	140.907
Promethium	Pm	61	
Protactinium	Pa	91	
Radium	Ra	88	
Radon	Rn	86	
Rhenium	Re	75	186.2
Rhodium	Rh	45	102.905
Rubidium	Rb	37	85.47
Ruthenium	Ru	44	101.07
Samarium	Sm	62	150.35
Scandium	Sc	21	44.956
Selenium	Se	34	78.96
Silicon	Si	14	28.086[a]
Silver	Ag	47	107.870[b]
Sodium	Na	11	22.9898
Strontium	Sr	38	87.62
Sulfur	S	16	32.064[a]
Tantalum	Ta	73	180.948
Technetium	Tc	43	
Tellurium	Te	52	127.60
Terbium	Tb	65	158.924

Table of Atomic Weights—1961 (continued)

Element	Symbol	Atomic No.	Atomic Weight
Thallium	Tl	81	204.37
Thorium	Th	90	232.038
Thulium	Tm	69	168.934
Tin	Sn	50	118.69
Titanium	Ti	22	47.90
Tungsten	W	74	183.85
Uranium	U	92	238.03
Vanadium	V	23	50.942
Xenon	Xe	54	131.30
Ytterbium	Yb	70	173.04
Yttrium	Y	39	88.905
Zinc	Zn	30	65.37
Zirconium	Zr	40	91.22

[a] The atomic weight varies because of natural variations in the isotopic composition of the element. The observed ranges are boron, ± 0.003; carbon, ± 0.00005; hydrogen, ± 0.00001; oxygen, ± 0.0001; silicon, ± 0.001; sulfur, ± 0.003.

[b] The atomic weight is believed to have an experimental uncertainty of the following magnitude: bromine, ± 0.002; chlorine, ± 0.001; chromium, ± 0.001; iron, ± 0.003; silver, ± 0.003. For other elements, the last digit given is believed to be reliable to ± 0.5.

SOURCE: Table adopted by the International Union of Pure and Applied Chemistry; the atomic weights are based on the exact number 12 as the assigned atomic (nuclidic) mass of the principal isotope of carbon.

INDEX OF NAMES

Abd El Wahed, A. M., 264
Acree, S. F., 61
Ahrens, L. H., 100, 115
Anderson, J., 162, 170, 239, 240, 241
Anthony, J., 244, 245, 246, 247
Appleman, D. E., 335

Baas Becking, L. G. M., 143, 379, 380, 382
Babcock, R. F., 303
Back, W., 84
Barnes, H. L., 398, 399
Barnes, I., 247, 248, 249, 250
Barton, P., 397, 398
Bates, R. G., 61, 122, 126, 143, 432
Bear, F. E., 136
Belinskaya, F. A., 303
Bergman, W. E., 140
Berner, R. A., 130
Bethke, P. M., 334
Birch, F., 334
Bjerrum, J., 95, 100, 120
Blumer, M., 266
Bonner, F. T., 89, 91
Bower, C. A., 302
Bradley, A. F., 141
Bray, U. B., 104
Brenet, J., 264
Brewer, L., 61, 72, 315, 333, 350
Bricker, O. P., 429
Brinkley, S. R., Jr., 346, 347
Bullwinkel, E. P., 429
Buswell, A. M., 89

Carpenter, A. B., 376
Carrol, D., 271, 384
Casby, J. U., 290, 295, 302
Castano, J. R., 384
Cater, D. B., 141
Charlot, G., 266
Chave, K. E., 91
Chipman, J., 38
Christ, C. L., 253, 254, 255, 256, 278, 429
Clark, S. P., Jr., 335, 399
Clark, W. M., 122, 172
Clifford, A. F., 97, 98

Cloke, P. L., 394
Cobble, J. W., 331
Cohen, B., 172
Coleman, R. G., 394
Conway, B. E., 54
Crawford, J. G., 57
Criss, C. M., 331
Cropper, W. H., 384

Darken, L. S., 19, 23, 71, 171
Davies, C. W., 94, 95, 120
Deffeyes, K. S., 91
Delahay, P., 266
Deleveaux, M., 394
Deltombe, E., 263, 264, 265
Denbigh, K. G., 72

Eisenman, G., 279, 280, 290, 295, 298, 302
Ekstrand, O. R., 366
Ellington, A. C., 358
Elliott, J. F., 38
Elving, P. J., 377
Emery, K. O., 397
Eugster, H. P., 349, 399
Evans, H. T., Jr., 389, 390, 429
Evans, W. H., 429

Failey, C. F., 68, 69
Fairbridge, R., 91
Fedotov, N. A., 302
Fischer, R. B., 303
Fischer, R. P., 388
Fleming, R. H., 73, 92
Forsythe, W. E., 431
Fournie, R., 142
Franck, E. U., 116
Friedman, C. L., 302
Friedman, S. M., 302
Frondel, C., 394
Fyfe, W. S., 350

Gamson, B. W., 25
Garrels, R. M., 81, 83, 91, 93, 100, 101, 107, 113, 130, 169, 231, 232, 233, 253, 254, 255, 256, 278, 293, 299, 302, 366, 384, 389, 390, 391, 393, 394, 395, 396, 397, 429
Gaucher, E., 242, 243, 244
Germanov, A. I., 139
Glasstone, S., 19, 122, 143
Gold, V., 122
Goldsmith, J. R., 49
Goremykin, V. E., 302
Gorham, E., 130
Greenwald, I., 99
Gregor, H. P., 300, 303
Grody, W., 96
Gurry, R. W., 19, 23, 71, 171

Hale, D. K., 302
Hamer, W. J., 61, 94, 120
Harker, R. I., 49
Harned, H. S., 54, 66, 69, 72, 89, 91
Hauser, E. A., 129
Heard, H. C., 49
Helgeson, H. C., 105, 115, 117, 119, 120, 366, 399, 400
Hem, J. D., 384
Hemley, J. J., 364, 365, 368, 369, 370, 398
Hildebrand, J. H., 44, 46
Hill, V. G., 358
Holland, H. D., 327, 328, 329, 330, 400
Hostetler, P. B., 99, 253, 254, 255, 256, 391, 393
Hougen, O. A., 25
Huber, N. K., 384

Isard, J. O., 302
Ito, J., 394
Ives, D. J. G., 122, 127, 141, 142, 143

Jaffe, I., 429
James, H. L., 383
Jamieson, J. D., 302
Janz, G. J., 122, 127, 141, 142, 143
Jeffes, J. H. E., 323, 324, 325, 327
Johnson, M. W., 73, 92
Johnston, J., 104

437

INDEX OF NAMES

Jolas, F., 264
Jones, B., 108

Kanwisher, J., 130
Kaplan, I. R., 143, 379, 382
Kelley, K. K., 309, 310, 311, 315, 317, 429
Kelley, W. C., 394
Kelley, W. P., 302
Kennedy, G. C., 166, 335, 429
Kern, D. B., 76, 91
King, E. G., 315
Klotz, I. M., 54, 61, 62, 64, 72, 307, 308, 350
Kobe, K. A., 69, 89
Kolthoff, I. M., 19, 377
Krauskopf, K. B., 257, 385, 386, 387, 397, 399
Krumbein, W. C., 396, 397
Kryukov, P. A., 302
Kullerud, G., 398

Laity, R. W., 142
Larson, T. E., 89
Latimer, W. M., 19, 54, 59, 72, 429
Laureys, J., 264
Lee, G. F., 385
LePeintre, M., 142
Levine, S., 429
Lewis, G. N., 56, 58, 61, 72, 314
Lietzke, M. H., 142
Linn, K., 251, 252, 253, 362, 363, 364
Lisitsin, A. K., 139

MacInnes, D. A., 58
McIntyre, W., 234, 235, 236, 237, 238
McKinstry, H. E., 166
Magee, G. M., 264
Mahieu, C., 142
Maier, C. G., 309
Manheim, F., 138
Manov, G. G., 61
Maraghini, M., 265
Marchandise, H., 385
Markam, A. A., 69, 89
Marshall, C. E., 140, 302, 303
Mason, B., 172
Materova, E. A., 303
Mattock, G., 122, 126, 130, 143
Merkle, F. G., 136
Merwin, H. E., 339
Meyer, C., 365, 398
Moore, D., 143, 379, 382
Moore, G. W., 141
Moussard, A. M., 264

Nakashima, M., 302
Natarajan, R., 166, 168, 169, 229, 231, 232, 233

Nikolskii, B. P., 302, 303
Nygren, H. D., 141

Orville, P. M., 366, 367, 368, 369, 370
Owen, B. B., 54, 66, 69, 72, 346, 347

Parsons, R., 54, 121
Pauling, L., 98, 430
Peshekonova, N. V., 302, 303
Phillips, J., 257, 258
Pitman, A. L., 266
Pitzer, K. S., 61, 72, 315, 333, 350
Polzer, W., 429
Posnjak, F. D., 339
Pourbaix, M. J. N., 19, 172, 263, 264, 265, 266

Randall, M., 36, 56, 58, 61, 68, 69, 72, 314
Rankama, K., 115
Reed, C. E., 129
Reitzel, J., 164, 165
Revelle, R., 91
Richardson, F. D., 323, 324, 325, 327
Richter, D. H., 365, 398
Rittenberg, S. C., 397
Robertson, C. E., 141
Robie, R. A., 315, 321, 334
Robinson, R. A., 50, 54, 62, 64, 65, 72
Ross, M., 372
Rossini, F. D., 321, 429
Rudin, D. O., 290, 295, 302

Saegusa, F., 429
Sahama, T. G., 115
Salmon, J. E., 302
Salstrom, E. J., 46
Sandell, E. B., 19
Sato, M., 293, 302, 380
Scerbina, V. V., 172
Schachtschabel, P., 302
Schairer, J. F., 334
Schmalz, R. F., 338, 339
Schmets, J., 263
Schmitt, H. H., 171, 266, 370, 371, 373, 374, 375, 376, 377
Schoeller, H., 92
Scholes, S. R., 89, 91
Schonhorn, H., 300, 303
Schwarzenbach, G., 95, 100, 120
Scott, R. L., 44
Serebrennikov, V. S., 139
Severinghaus, J. W., 141
Shul'ts, M. M., 302, 303
Siever, R., 81, 83, 91, 107, 130, 131, 360, 429

Sillén, L. G., 95, 100, 120, 266, 377
Silman, J. R., 373, 374, 375, 429
Silver, I. A., 141
Skinner, B. J., 335
Skougstad, M. W., 384
Sollner, K., 303
Sosnick, B., 36
Spicer, H. C., 334
Starkey, R. L., 137, 172
Stephenson, C. C., 45, 429
Stokes, R. H., 50, 54, 62, 64, 65, 72
Stumm, W., 385
Sverdrup, H. U., 73, 92
Swift, E. H., 135

Theodorovitch, G. I., 397
Thomas, W. J. O., 96
Thomson, J. B., Jr., 319, 321, 332, 341, 350
Thompson, M. E., 91, 93, 100, 101, 107, 113, 293, 302, 429
Truesdell, A. H., 293, 302, 371
Turner, F. J., 350
Turner, R. C., 92
Tuttle, O. F., 49

Valensi, G., 213, 265
Vandervelden, F., 265
Van Eijnsbergen, J., 265
Van Muylder, J., 264, 265, 266
Van Rysselberghe, P., 265, 266
Verhoogen, J., 350
Volkov, G. A., 139

Wagman, D. D., 429
Walker, A. C., 104
Walton, H. F., 269, 270, 302
Watson, K. M., 25
Wattecamps, P., 264
Weeks, A. D., 253, 254, 255, 256
Weir, C. E., 335
Weyl, P. K., 91, 131
White, D. A., 384
Wight, K. M., 137, 172
Wiklander, L., 302
Winchester, J., 184
Withers, G., 323, 324
Wones, D. R., 349, 399
Woolmer, R. F., 122

Yoder, H. S., Jr., 335

Zawidski, J. v., 30, 34
Zen, E'an, 346
Zobell, C. E., 135, 143, 172
Zoubov, N. de, 264, 265, 266

INDEX OF SUBJECTS

Activities of dissolved species, relation to solubility, 54–56, 262–263
two-electrode measurement of, 298, 299
Activity, of cation, effect of complexing anion on, table, 114, 115
  definition, 31
  electrolyte solute, 51–54
  fugacity relations, 31 ff.
  gas, 22–28
  H_2O in water under pressure, 346
  individual ion, 53
    measurement of, 139–141, 268, 281 ff.
  of ions, from cation electrodes, 281 ff.
  liquid, 6
  OH^-, in water under pressure, 346
  pressure dependence, 35
  for Fe-O_2-Fe_3O_4 equilibrium, 343–344
  relations in ion-exchange theory, 273–274
  solid, 5, 6
  solids, corrections to, 47–49
  and standard state, 5–6
  symbols for, 6
  temperature dependence, 37
  water, in electrolyte solutions, 64–67
    at various pressures, table, 36
Activity coefficient, Debye-Hückel theory, 61–63
  electrolyte solute, 53–54
  gas, 22–26
  individual ion, 53–54, 58–60, 63–64
  and ionic strength, 60
  mean ionic, 53, 59
  mean salt method, 60
  molecular species, 67–70
    relation to solubility, 67
    temperature dependence, 69
  practical, 42
  rational, 38
  of species in sea water, 102–104
Activity coefficients, AgBr in melts, 46
  $CaCO_3$ in siderite solid-solution, 49
  Cd in Cd-Pb, 39
  CO_3^{2-}, 104
  HCO_3^-, 104
  individual ions, values of, 63
  mean, values of, 59
  molecular species, values of, 68, 69
  species in sea water, table, 103
Activity products, and electronegativities, 97–99
  and radii of cations, 99

AgBr, activity coefficients, in LiBr and in KBr, 46
Alabandite, 163, 244
Albite-analcite equilibrium, 363
Albite, authigenic, 362
  transformation to paragonite, 365
Al^{3+}, in equilibrium with $Al_2O_3 \cdot 3H_2O$, 354
$Al_2O_3 \cdot H_2O$, stability relative to $Al_2O_3 \cdot 3H_2O$, 10
AlO_2^-, in equilibrium with $Al_2O_3 \cdot 3H_2O$, 354
Al_2O_3-SiO_2-H_2O system, 352 ff.
Aluminum species, ΔF_f°, ΔH_f°, and S° values, table, 403
Analcite-albite equilibrium, 363
Anglesite, 10
Anhydrite, 10
Anion exchange, 267
Antimony species, ΔF_f°, ΔH_f°, and S° values, table, 404
Aragonite, 376
Arkosic sediments, waters in, 361
Arsenic species, ΔF_f°, ΔH_f°, and S° values, table, 404
Atmosphere, CO_2 pressure in, 11
Atomic numbers, table, 433–435
Atomic weights, table, 433–435
Au-H_2O-O_2-Cl_2, Eh-pH diagram, 257, 258
  discussion of, 257 ff.
AuS^-, 258
Azurite, 155

Bacteria, catalytic activity on decomposition of water, 178
  effect on Eh and pH, 383
Barium species, ΔF_f°, ΔH_f°, and S° values, table, 405
Bauxite formation, 352, 358
Beryllium species, ΔF_f°, ΔH_f°, and S° values, table, 405
Bicarbonate ion, activity coefficient, 104
  ionization constant, 76, 89
  temperature dependence, 89
Bismuth species, ΔF_f°, ΔH_f°, and S° values, table, 406
Boehmite, 10
Boehmite-diapore reaction, 354
Bornite, 166
Boron species, ΔF_f°, ΔH_f°, and S° values, table, 406

439

Brine, calculation of concentration in, 4
  from Deep Springs Lake, Calif., analysis of, 109
  complex ion distribution in, 109
Bromine species, ΔF_f°, ΔH_f°, and S° values, table, 407
Buffer solutions, pH values, table, 432

Ca^{++}, activity coefficient, 63
$CaCO_3$, ionization constant, 76, 89
  temperature dependence, 89
  in pure water, 77–81
  in water, first open to CO_2, then isolated, 86–89
  with fixed P_{CO_2}, 81–83
  with fixed P_{CO_2} and pH, 85–86
  with fixed ΣCO_2 and pH, 83–85
  reactions with water, 77
  undersaturation in Lake Earl, 84
Cadmium species, ΔF_f°, ΔH_f°, and S° values, table, 407
Calcite, activity product corrected for composition, 47–49
  conditions governing solubility, 75–76
  ionization constant, 76, 89
  temperature dependence, 89
  solubility, in electrolyte solutions, 107
  in sea water, 107
  See also $CaCO_3$
Calcite-siderite equilibrium, 49
Calcite-water equilibrium, change with pressure, 346–347
Calcium species, ΔF_f°, ΔH_f°, and S° values, table, 408
Calomel electrode, 127
CaO-MgO-CO_2-H_2O system, metastable relations in, 376
  stable relations in, 376
Carbonate equilibria, effect of solid solution on, 90–91
  effect of temperature on, 89
Carbonate ion, activity coefficients, 104
Carbonate-rich lakes, 370
Carbonate saturometer, 131–132
Carbon dioxide, equilibrium pressure for various compounds, see Partial pressure diagrams
Carbon dioxide, pressure in earth's atmosphere, 11
Carbonic acid, ionization constants, 76, 89
  temperature dependence, 89
Carbon species, ΔF_f°, ΔH_f°, and S° values, table, 409
Carnotite, 253, 388
Cation electrodes, calibration of, 296–298
  equations, 284, 288, 290, 291, 294
  derivation of, 281 ff.
  Eisenman's empirical, 290
  half-cell reactions in, 283
  for monovalent and divalent cations, 293
  for more than two monovalent cations, 291
  regular solution theory in, 289–290
  for single cation, 288
  for two monovalent cations, 290
  glass, effect of compositional changes on, 294–296
  as membranes, 281 ff.
  relation to cation exchangers, 281 ff.
  stearate, 300–301
  types, 285
    alkali metal-ion sensitive, 294–296
    alkaline earth-ion sensitive, 299–301
Cation exchange, capacity, clay minerals, 271
  constants, table, 270
  discussion, 267 ff.
  equations, 270, 275, 276
    activity relations in, 273–274
    regular solution theory in, 273–274
  and Law of Mass Action, 272 ff.
  non-equivalent bonding sites in, 277
  substrate, 272
Cation exchangers, selectivity, 278–281
  constant, 278–280
  and ΔF_r°, 280
Cd, activity coefficients in Cd-Pb, 39
CdI_4^{--}, dissociation constant of, 114
Cerium species, ΔF_f°, ΔH_f°, and S° values, table, 409
Cerussite, 10
Cesium species, ΔF_f°, ΔH_f°, and S° values, table, 409
Chalcedony, 362
Chalcocite, and native sulfur, 241
  non-stoichiometry, 160
  precipitation of, 241
  secondary enrichment of, 230
Chalcopyrite, 166
Chemical model, sea water, 100–106
Chemical sediments, Eh-pH diagram, 396
Chert, 228
Chloride complexes, of gold, 257, 258
  of lead, 114
  of transition, and post-transition elements, 111
Chlorine species, ΔF_f°, ΔH_f°, and S° values, table, 410
Chlorite, 363, 364
Chlorite-mica-kaolinite equilibria, 364
Chlorite-mica-phlogophite equilibria, 364
Chromium species, ΔF_f°, ΔH_f°, and S° values, table, 410
Cl^-, activity coefficient, 63
Clapeyron equation, 337
  and P-T equilibrium curve, 341
Clay minerals, as cation exchangers, 267 ff.
CO_2, activity coefficients, in NaCl solution, 69
  in KCl solution, 69
  and solubility, 69
  in Eh-pH diagrams for iron, 197
  and Eh, pH, S, 222–225
  pressure in earth's atmosphere, 11
CO_3^{--}, activity coefficient, 104
CO_3^{--}, Eh-pH diagram, 203
Cobalt species, ΔF_f°, ΔH_f°, and S° values, table, 410–411

INDEX OF SUBJECTS 441

Co-H_2O-O_2-CO_2, Eh-pH diagram, 248, 249
 discussion of, 247
Co-H_2O-O_2-CO_2-S, Eh-pH diagram, 250
 discussion of, 247
Colorado Plateau, see Uranium ores
Colorado Plateau uranium, 388 ff.
Complex formation, effect on activity of cation, table, 114, 115
Complex ions, in brine, 109
 discussion of, 51, 93 ff.
 dissociation constants, 96, 98, 114, 119
 effect of solubility, 113
 effect of temperature on, 115–120
 of minor species in natural waters, 93, 110–115
 in sea water, 113 ff.
 uranium, 253–256, 391, 393
 vanadium, 389
 See also Ion pairs
Complexing by Cl^-, with gold, 257, 258
 with lead, 114
 with transition and post-transition elements, 111
Compressibility coefficient, volume, 333
Concentration cells, 140
Concentration units, 3–5
Constants, table, 402
Conversion factors, table, 402
Copper, separation from iron, 230
Copper carbonates, effects of P_{CO_2} and P_{O_2} on, 156
 and effect of P_{S_2} on, 162
 solubility relations, 375
 stability relations, discussion, 373 ff.
 diagrams, 374, 375
Copper compounds, O_2 and CO_2 pressures in equilibrium with, 155–157
 and S_2 pressure in equilibrium with, 162
Copper ores, secondary enrichment of, 230
Copper oxides, effects of P_{CO_2} and P_{O_2} on, 156
 and effect of P_{S_2}, 162
Copper species, ΔF_f°, ΔH_f°, and S° values, table, 411
Corrosion of pipes, 137
Corundum, 308, 314
 conversion to gibbsite, ΔH_r°, 308
 ΔS_r°, 315
Covellite, 160
Crestmore, Calif., Ca,Mg carbonates, 376
Critical constants, 24
Critical pressure, values, 24
Critical temperature, values, 24
Crystal radii, table, 430
$CuCl_2$, mean activity coefficient, 59
$CuCO_3^\circ{}_{aq}$, dissociation constant, 114
$Cu(CO_3)_2^{--}$, dissociation constant, 114
Cu-CO_2-O_2-H_2O system, 154–157
 partial pressure diagram, 156
Cu-CO_2-O_2-S-H_2O, Eh-pH diagram, 240
 discussion of, 239, 240, 241

Cu-CO_2-O_2-S_2-H_2O system, 162
 partial pressure diagram, 162
Cu-Cu^{++} electrode, 16
Cu-Fe-O_2-S_2 system, 166–170
 partial pressure diagrams, 167, 169
Cu-Fe-O_2-S-H_2O, Eh-pH diagram, 231, 232
 discussion of, 230, 233
Cu-Fe-S, composition diagram, 168
CuO-CO_2-H_2O system, discussion, 373 ff.
 solubility relations, 375
 stability diagrams, 374, 375
Cu-O_2-H_2O, Eh-pH diagram, 239
 discussion of, 239, 240, 241
Cu-O_2-S_2 system, 158–161
 partial pressure diagram, 160
Cuprite, 155
CuS_5^-, dissociation constant, 114
$CuSO_4$, mean activity coefficient, 59

Debye-Hückel, equations, for individual ion activity coefficients, 61, 62
 effective diameters of ions in, table, 62
 for mean activity coefficients, 62
 values of constants in, table, 61
 theory, 61
Deep Springs Lake, Calif., brine from, 108
Diaspore, metastable persistence, 355
Digenite, 159
Dilute solution, activity coefficients in, 54
 and Henry's law, 40
 non-electrolyte, 39–41
 and practical activity coefficient, 42
Dissociation constants, values of, 70, 96, 98, 114, 119
 complex ions, 114, 213, 215
 ion pairs, 96, 98
 temperature dependence, 119
 water, 70
 See also Equilibrium constant, Ionization Constant
Dissolved species, reactions involving, 12–13
Dolomite, 376

Eh, and corrosion of pipes, 137
 definition of, 17, 132
 electrodes, 135
 equations, 17, 132, 134
 half-cell reactions, 17, 132, 134
 measurements, accuracy in, 136–139
 precautions in, 136
 recording of, 136
 readings, effect of atmosphere on, 136–137
 and irreversible oxidation potential, 136
 in natural media, 136–139
Eh-pH, diagrams, composite, 188
 effect of pressure on, 261
 effect of temperature on, 260–261
 history of development, 172–173
 relation to partial pressure diagrams, 229
 equations, relation to partial pressure, 178, 229

Eh-pH—*continued*
  values in natural aqueous environments, 380
    mine waters, 381, 382
Eh-pH diagrams, Au-O_2-S-Cl_2-H_2O, at fixed ΣCl and ΣS, 257, 258
  chemical sediments, 396
  CO_3^{--}, 203
  Co-CO_2-O_2-H_2O, at fixed ΣCO_2, 248, 249
  Co-CO_2-O_2-S-H_2O, at fixed ΣCO_2 and ΣS, 250
  Cu-Fe-H_2O-O_2-S, 231, 232
  Cu-H_2O-O_2, 239
  Cu-H_2O-O_2-CO_2-S, at fixed P_{CO_2} and ΣS, 240
  Fe^{++}, 192
  Fe^{3+}, 187
  Fe-Cu-H_2O-O_2-S, 231, 232
  Fe-H_2O-O_2, 180, 181, 183, 195
  Fe-H_2O-O_2-CO_2, at fixed ΣCO_2, 205, 208, 209
    at fixed P_{CO_2}, 199, 200
    as function of P_{CO_2}, 198
    and showing iron hydroxides and siderite, 210
  Fe-H_2O-O_2-S, at fixed ΣCO_2 and ΣS, 224
    at fixed ΣS, 218, 221, 223
    as function of P_{S_2}, 212
  Fe-H_2O-O_2-SiO_2, 227
  Fe-H_2O-O_2-SiO_2-CO_2-S, 228
  Fe(OH)$^+$, 193
  Fe(OH)$_2^+$, 191
  Fe(OH)$^{++}$, 190
  FeO_2H^- ($HFeO_2^-$), 194
  Manganese, compounds, 386, 387
    hydroxides, 387
  Metal sulfides, at fixed ΣCO_2 and ΣS, 395
  Mn-H_2O-O_2-CO_2, at fixed ΣCO_2, 242
  Mn-H_2O-O_2-CO_2-S, at fixed P_{CO_2} and ΣS, 243
  Ni-H_2O-O_2-CO_2, at fixed P_{CO_2}, 246
  Ni-H_2O-O_2-S, at fixed ΣS, 244, 245
  Ore deposition, 398
  Pb-H_2O-O_2-CO_2, at fixed P_{CO_2}, 234
    at fixed ΣCO_2, 235
  Pb-H_2O-O_2-CO_2-S, at fixed P_{CO_2} and ΣS, 237
    at fixed ΣCO_2 and ΣS, 238
  Pb-H_2O-O_2-S, at fixed ΣS, 236
  S-H_2O-O_2, 214, 217
    at fixed ΣS, 217
  Sulfur species in water, 214, 217
  U-H_2O-O_2-CO_2, as function of P_{CO_2}, 254
    as function of ΣCO_2, 255
  U-H_2O-O_2-CO_2-K-V, 256, 393
  Vanadium compounds and minerals, 389, 390
  Water stability, 176
  W-H_2O-O_2, 251
  W-H_2O-O_2-S, at fixed ΣS, 252
Electrode, Ag-AgBr, 141
  Ag-AgCl, 140
  Ag-AgI, 141
  anion, 140–141
  calomel, 127
  Cu-Cu^+, 16
  Eh, 133
  pH, 123
  for high pressures, 141–142
  for high temperatures, 141–142
  for P_{CO_2}, 141
  stearate, 300–301
  See also Cation electrodes
Electrolyte, solute, associated, 50
  Henry's law for, 51–52
  nonassociated, 50
  symmetrical, 52
  unsymmetrical, 54
  solutions, complex ions in, 51 ff.
  ion pairs in, 51 ff.
Electromotive force, *see* EMF
Electronegativities, and activity products, 97–99
Elements, table, 433–435
EMF, change with activity of ion, 284
  change with pH, 126
  and free-energy change of reaction, 16
  measurements, individual ion activity, 139–141, 268, 281
    gas pressure, 141
    at high pressures, 141–142
    at high temperatures, 141–142
  for oxidation-reduction system, 134
Enthalpy, and free-energy function, 316 ff.
  of formation, *see* Heat of formation
  of reaction, *see* Heat of reaction
  standard states for, 308
  temperature dependence, 308–309
    coefficient of, 308
    empirical equations for, 309
    graphical evaluation, 309
Entropy, absolute, 314
  definition, 313
  gram-atomic, 320
  of reaction, definition, 314
    from absolute entropies, 314
    for conversion of corundum to gibbsite, 314
    as temperature coefficient of ΔF_r°, 319
  of rock-forming minerals, 319
  sillimanite values, table, 310
  standard states for, 308
  temperature dependence, 313–314
Entropy, tables, 403–428
  See also under element name
Environment, of marine sediments, 394, 396, 397
  of ore deposition, 397–400
Equilibrium, and free-energy change of reaction, 7, 343
Equilibrium constant, definition of, 7, 342
  and E°, 16
  effect of pressure on, 343
  and standard free energy of reaction, 8, 342, 343
    calculations for various types of reactions, 9–13

Equilibrium constant—*continued*
  See also Dissociation constant, Ionization constant
Expansion coefficient, volume, 333

Facies of iron ores, 383
Faraday, 16, 402
Fe^{++}, 184 ff.
  Eh-pH diagram of, 192
Fe^{3+}, 185 ff.
  Eh-pH diagram of, 187
  in equilibrium with hematite, 185
  limited role in nature, 188
$Fe\text{-}Cu\text{-}H_2O\text{-}O_2\text{-}S$, Eh-pH diagram, 231, 232
$Fe\text{-}Cu\text{-}O_2\text{-}S_2$ system, 166–170
  partial pressure diagram, 167, 169
Fe-Cu-S, composition diagram, 168
$Fe\text{-}H_2O\text{-}O_2$, composite Eh-pH diagram of, 195
  Eh-pH diagrams, 180, 181
    of iron hydroxides, 183
  Eh-pH equations, 178–179
    for iron hydroxides, 182–183
  partial pressure diagram, 149
$Fe\text{-}H_2O\text{-}O_2\text{-}CO_2$, Eh-pH diagram, calculations for, 197 ff.
  construction, 204
  at fixed P_{CO_2}, 199
  as function of P_{CO_2}, 198
  with ionic species, 200
  of iron hydroxides and siderite, 210
  showing dissolved species, 205
  at $\Sigma CO_2 = 10^{-2}$, 208, 209
  Eh-pH relations, for calculations at fixed ΣCO_2, 201 ff.
    among solids, 207
    for determination of metastable relations, 207
$Fe\text{-}H_2O\text{-}O_2\text{-}CO_2\text{-}S\text{-}SiO_2$, Eh-pH diagram, 228
$Fe\text{-}H_2O\text{-}O_2\text{-}S$, Eh-pH diagram, at fixed ΣS, 218, 221, 223
  at fixed ΣCO_2 and ΣS, 224
  Eh-pH-P_{S_2} diagram, 212
$Fe\text{-}H_2O\text{-}O_2\text{-}SiO_2$, Eh-pH diagram, 227
$Fe\text{-}H_2O\text{-}O_2\text{-}S_2$ system, 157–158
  partial pressure diagram, 159
Feldspar-mica-kaolinite relations, discussion, 359 ff.
  equations, 360, 361
  stability diagram, 361
Feldspar-phlogopite-mica equilibria, 364
Feldspar, reaction to form kaolinite, 360
  reaction to form mica, 360
$Fe\text{-}Mn\text{-}O_2\text{-}CO_2\text{-}H_2O$ system, 166
  partial pressure diagram, 165
$Fe\text{-}Mn\text{-}O_2\text{-}S_2\text{-}H_2O$ system, 163
  partial pressure diagram, 164
FeO, disproportionation, 148
$Fe\text{-}O_2$, 146–148
  partial pressure diagram, 148
  at elevated temperatures, 327, 328
$Fe(OH)^+$, Eh-pH diagram of, 193
$Fe(OH)^{++}$, Eh-pH diagram of, 190
$Fe(OH)_2^+$, Eh-pH diagram of, 191
FeO_2H^- ($HFeO_2^-$), Eh-pH diagram of, 194
Fe_2O_3, Eh-pH equations relating ionic species to, 189
  reaction with H^+, 13, 185
  reaction with water, 12
Fe_3O_4, Eh-pH equations relating ionic species to, 189
$Fe\text{-}O_2\text{-}CO_2$ system, 11–12, 151–154
  partial pressure diagram, 152
$Fe\text{-}O_2\text{-}CO_2\text{-}H_2O$, partial pressure diagram, 153
$Fe\text{-}O_2\text{-}CO_2\text{-}S_2$ system, 161
  partial pressure diagram, 161
$Fe\text{-}O_2\text{-}Fe_3O_4$, effect of pressure on equilibrium, 343–344
$Fe\text{-}O_2\text{-}H_2$, composition diagram, 150
  partial pressure of oxygen diagram, 149
$Fe\text{-}O_2\text{-}S_2$, partial pressure diagram, at elevated temperatures, 330
$Fe\text{-}S_2$, partial pressure diagram, at elevated temperatures, 330
Ferric ion, 185 ff.
  in equilibrium with hematite, 185
  limited role in nature, 188
Ferrous-ferric couple, 14, 133
  Eh equations, 133, 134
  EMF cell, 133
  EMF equations, 133, 134
  free-energy relations, 14
Ferrous hydroxide, metastability in water, 182
Ferrous ionic species in solution, 184
Ferrous metasilicate, 226
Fluorine species, ΔF_f°, ΔH_f°, and S° values, table, 412
Formal potential, 135
Formality, 3–4
Formulas of compounds, 2
Free energy, and chemical equilibrium, 7, 8
  data, precautions in use of, 259–260
  and EMF, equation, 16
  standard, see Standard free energy
  test of validity, 154
  values, and phase composition, 158
Free-energy change of reaction, and EMF, 16
  equation for, 8, 343
  and equilibrium, 7, 8, 343
  and equilibrium quotient, 8, 343
  hematite-goethite reaction, calculations for, 338 ff.
  pressure coefficient, 332
  pressure dependence, examples, 332, 333, 334, 335, 336
    for rock-forming minerals, 332 ff.
    for sulfur, monoclinic-rhombic equilibrium, 334, 335
  temperature coefficient, 319
  temperature dependence, analytical representation, 323 ff.
    and free-energy function, 316 ff.
    graphical representation, 323 ff.
    oxide-oxide reactions, 324

Free-energy change of reaction—*continued*
of reactions involving rock-forming minerals, 320 ff.
of reactions involving S_{2g}, 325
of various types of reactions, 319 ff.
temperature-pressure dependence, 337
See also Standard free energy
Free-energy function, definition, 315
examples of use of, 316–318
values of, table, 316
Free energy of formation, tables of species, aluminum, 403; antimony, 404; arsenic, 404; barium, 405; beryllium, 405; bismuth, 406; boron, 406; bromine, 407; cadmium, 407; calcium, 408; carbon, 409; cerium, 409; chlorine, 410; chromium, 410; cobalt, 410–411; copper, 411; fluorine, 412; germanium, 412; gold, 412; hydrogen, 412; iodine, 413; iron, 413; lanthanum, 414; lead, 414; lithium, 415; magnesium, 416; manganese, 415; mercury, 416; molybdenum, 417; nickel, 417; niobium, 417; nitrogen, 418; oxygen, 418; palladium, 418; phosphorus, 419; platinum, 419; potassium, 420; rubidium, 420; scandium, 420; selenium, 421; silicon, 421; silver, 422; sodium, 422–423; strontium, 423; sulfur, 424; tantalum, 424; tellurium, 425; thorium, 425; tin, 426; titanium, 426; tungsten, 426; uranium, 427; vanadium, 427; ytterbium, 427; yttrium, 428; zinc, 428; zirconium, 428
Fugacity, activity relations, 31–35
definition of, 31
of gas, 33–35
pressure dependence, 35
temperature dependence, 37
of water, at various pressures, table, 36

Gas constant (R), numerical value, 402
Gas, ideal, *see* Gas, perfect
perfect, activity, 22
fugacity, 33–35
law of, 21
real, activity of, 22–26
activity coefficient of, 22–26
critical constants for, table, 24
critical pressure, 24
critical temperature, 24
discussion of, 21–22
fugacity of, 33–35
Gas mixtures, ideal solution, activity in, 27–28
definition of, 27
fugacity in, 33–35
perfect, activity in, 26
definition of, 26
fugacity in, 33–35
Germanium, 394
Germanium species, $\Delta F_f°$, $\Delta H_f°$, and $S°$ values, table, 412

Gibbsite, 10, 308, 314
effect of pH change on, 353, 354
equilibrium with water, 353, 354
formation from corundum, $\Delta H_r°$, 308
$\Delta S_r°$, 315
reaction to form diaspore, 355
reaction to form mica, 361
stability diagram, 354
Gibbsite-kaolinite relations, *see* Kaolinite-gibbsite relations
Glass electrode, *see* Cation electrodes
Goethite, relation to hematite, 154
Gold, chloride complexes, 258
Gold species, $\Delta F_f°$, $\Delta H_f°$, and $S°$ values, table, 412
Gypsum, equilibrium in sea water, 106

H^+, activity coefficient, 63
H_2, activity coefficients in NaCl solution, 68
Half-cells, addition of, 14
conventions for writing, 132
definition of, 14
and Eh, 17, 132, 134
hydrogen, 14, 15
potential of, 16
sign of, 132
and standard free energy of reaction, 14
Half-reactions, *see* Half-cells
HCO_3^-, activity coefficients, 104
ionization constant, 76, 89
H_2CO_3, activity coefficients, 69
ionization constant, 76, 89
Heat capacity, definition of, 308
empirical equation for, 309
gram-atomic, definition of, 319
of rock-forming minerals, 319
Heat content, sillimanite values, table, 310
See also Enthalpy
Heat of formation, definition, 308
heat of reaction from, 308
rock-forming minerals, table, 321
Heat of formation, tables, 403–428, *see* under element name
Heat of reaction, for conversion of gibbsite to corundum, 308
definition of, 307
and free-energy function, 316 ff.
from heats of formation, 308
temperature dependence, 310–311
coefficient, 310
empirical equation for, 311
of formation of H_2S, 311–313
from tabulated data, 313
Hematite, equations of reactions to form ionic species, 189
ferric ion in equilibrium with, 12, 186, 189
reaction with H^+, 13, 185
reaction with water, 12
relation to goethite, 154
relation to pyrite in SO_4^{--} field, 219
solubility of, 197
Hematite-goethite-water, P-T diagram, 339

Hematite-goethite-water—*continued*
 calculation of thermochemical quantities from, 338–341
Hematite-magnetite, Eh-pH equation, effect of temperature on, 260, 261
 effect of pressure on, 261
Henry's Law, 39–41
 electrolyte solute, 51–52
$HFeO_2^-$, Eh-pH diagram, 194
HS^-, dissociation constant of, 213
H_2S, activity coefficients in NaCl solution, 68
 dissociation constant of, 213
 temperature dependence of ΔH_r°, 311–313
HSO_4^-, dissociation constant of, 215
Huntite, 376
Hydrocerrussite, as indicator of environment, 233
Hydrogen, ion, free energy, 7
 half-cell, 14–15
 partial pressure of, in equilibrium with water, 145
 production of, by bacteria, 383
Hydrogen species, ΔF_f°, ΔH_f°, and S° values, table, 412
Hydromagnesite, 376
Hydrothermal alteration, felsic rocks, 365
Hydrothermal ores, Eh-pH environment, 398

Ideal gas, *see* Gas, perfect
Ideal solution, definition of, 31 ff.
 non-electrolyte, 31 ff.
 solid, 44
Iodine species, ΔF_f°, ΔH_f°, and S° values, table, 413
Ion exchange, *see* Cation exchange
Ion pairs, discussion of, 94 ff.
 dissociation constants, measurement of, 94
 values for, 96, 98
 Debye-Hückel theory in relation to, 94
 in natural waters, 94
 in sea water, 113 ff.
 See also Complex ions
Ion product, change with pressure, 346–347
Ionic activity coefficient, *see* Activity coefficient, individual ion
Ionic equilibria, pressure dependence, 345–347
 temperature dependence, 331
Ionic partial molal volume, 346
Ionic radii, and solid solution, 43
 table, 430
Ionic strength, definition of, 56
 calculation of, in natural waters, 57–58
Ionization constant, $CaCO_3$, 76, 89
 HCO_3^-, 76, 89
 H_2CO_3, 76, 89
 See also Dissociation constant, Equilibrium constant
Individual ion activity coefficient, *see* Activity coefficient, individual ion

Iron, separation from copper, 230
Iron carbonates, effects of P_{CO_2} and P_{O_2} on, 152, 153
 and effect of P_{S_2} on, 161
Iron compounds, metastability of, 154
 oxygen and sulfur pressures in equilibrium with, 157–158
 and carbon dioxide pressure in equilibrium with, 161
Iron hydroxides, Eh-pH equations, 182, 183
Iron-magnetite, metastability in water, 179
Iron ores, environment of, 384
 facies of, 383
Iron oxides, carbon dioxide and oxygen pressures in equilibrium with, 152, 153
 and sulfur pressure in equilibrium with, 161
 Eh-pH equations relating, 178, 179
 oxygen pressures in equilibrium with, 146–148
 oxygen and sulfur pressures in equilibrium with, 159
 and carbon dioxide pressure in equilibrium with, 161
 relation to manganese oxides, 163
 and water stability, 149
Iron silicates, 225 ff.
 and iron oxides, 226
 and magnetite, 226
 and silica, 226
Iron species, ΔF_f°, ΔH_f°, and S° values, table, 413
Iron species, in solution, equations for reactions, 189
Iron sulfides, oxygen and sulfur pressures in equilibrium with, 159
 and carbon dioxide pressure in equilibrium with, 161
Irreversible oxidation potential, 136, 380

K^+, activity coefficient, 63
Kaolinite, reaction to form feldspar, 360
 reaction with water, 353, 355
 soil, behavior, 357, 358
 stability diagram, 356
 stability relations, 355
Kaolinite-gibbsite, relations, discussion, 352 ff.
 equations, 353 ff.
 stability diagrams, 357, 358
Kaolinite-feldspar-mica relations, *see* Feldspar-mica-kaolinite relations
Kaolinite-mica-chlorite equilibria, 364
KCl, mean activity coefficient, 59
K_2O-Al_2O_3-SiO_2-H_2O system, 359 ff.
 associations diagram, 359
 mineral compatibilities, 360
 stability diagram, 361
 for elevated temperatures, 365
K_2O-MgO-Al_2O_3-SiO_2-H_2O system, discussion, 363, 364, 365
 stability diagram, 364
K_2SO_4, mean activity coefficient, 59

Lakes, carbonate-rich, 370
Lanthanum species, ΔF_f°, ΔH_f°, and S° values, table, 414
Law of Mass Action, 6, 7
and equilibrium constant, 7
Lead minerals, Eh-pH relations among, 233 ff.
Lead species, ΔF_f°, ΔH_f°, and S° values, table, 414
Liquid junction potential, 124, 125
Lithium species, ΔF_f°, ΔH_f°, and S° values, table, 415

MacInnes assumption, 58
Magnesite, 376
Magnesium-calcium carbonates, metastable relations in, 376
stable relations in, 376
Magnesium species, ΔF_f°, ΔH_f°, and S° values, table, 416
Magnetite, association with pyrite and pyrrhotite, 225
Eh-pH relation to pyrite, 219
equations of reactions to form ionic species, 189
facies, 228
formation from FeO, 148
and iron silicate, 226
relation to pyrite in SO_4^{--} field, 219
Magnetite-hematite, Eh-pH equation, effect of temperature on, 260, 261
effect of pressure on, 261
Malachite, 155
Manganese, compounds, Eh-pH diagram, 386, 387
deposits, 385, 386, 387, 388
hydroxides, Eh-pH diagram, 387
Manganese oxides, relation to iron oxides, 163
Manganese species, ΔF_f°, ΔH_f°, and S° values, table, 415
Marcasite, oxidation by ferric salts, 222
relation to pyrite, 222
Marine sediments, Eh-pH relations, 394, 396, 397
Mass Action, Law of, 6–7
and equilibrium constant, 7
Mean activity coefficient, see Activity coefficient, mean
Mean salt method, for individual ion activity coefficients, 60
Membrane, semipermeable, 268
Membrane electrodes, see Cation electrodes
Mercury species, ΔF_f°, ΔH_f°, and S° values, table, 416
Metal sulfides, Eh-pH diagram, 395
Metastable relations, among Ca, Mg carbonates, 376
Metastability, of amorphous silica, 360
of diaspore, 355
of ferrous hydroxide in water, 182
of iron compounds, 152, 153, 154
of iron hydroxides relative to oxides, 182, 183
of iron-magnetite in water, 179, 180
of kaolinite-feldspar, 362
of MnO_4^-, 241
of siderite at earth's surface, 201
of SO_4^{--}, 383
of wegscheiderite, 372, 373
Mica, reaction to form feldspar, 360
reaction to form gibbsite, 361
Mica-feldspar-kaolinite relations, see Feldspar-mica-kaolinite relations
Mica-feldspar-phlogopite equilibria, 364
Mica-kaolinite-chlorite equilibria, 364
Mica-phlogopite-chlorite equilibria, 364
Mine water, Eh and pH, 381, 382
Mn-Fe-H_2O-O_2-CO_2 system, 166
partial pressure diagram, 165
Mn-Fe-H_2O-O_2-S_2 system, 163
partial pressure diagram, 164
Mn-H_2O-O_2-CO_2, Eh-pH diagram, 242
discussion of, 241 ff.
Mn-H_2O-O_2-CO_2-S, Eh-pH diagram, 243
discussion of, 241 ff.
MnO_4^-, metastability of, 241
Molal volume, 334
temperature dependence, 335
Molality, 3–4
Molarity, 3–4
Mole fraction, definition, 4, 5
and partial pressure, 26
and rational activity coefficient, 38
Molecular species, activity coefficients, 67 ff.
Molybdenum species, ΔF_f°, ΔH_f°, and S° values, table, 417
Montmorillonite, 262, 263
Montroseite, 389

N_2, activity coefficients in NaCl solution, 68
Na^+, activity coefficient, 63
NaCl, mean activity coefficient, 59
Nahcolite, equilibrium with trona, 372
Na_2O-Al_2O_3-SiO_2-H_2O system, discussion, 362–363
stability diagram, 363
at elevated temperatures, discussion, 365
stability diagram, 365
Na_2O-CO_2-H_2O system, activity of water in, 370, 371, 372
discussion of, 370, 371
equations for, 371, 372
phase diagrams, 371, 372
Na_2O-K_2O-Al_2O_3-SiO_2-H_2O system, discussion, 366 ff.
equations involved in, 367, 368
stability diagrams, 366, 369
Na_2SO_4, mean activity coefficient, 59
Natron, 370, 371
Natural environments, Eh and pH, 380
NH_3, activity coefficients in NaCl solution, 68
Nickel species, ΔF_f°, ΔH_f°, and S° values, table, 417
Ni-H_2O-O_2-CO_2, Eh-pH diagram, 246
discussion of, 244 ff.
Ni-H_2O-O_2-S, Eh-pH diagram, 244, 245
discussion of, 244 ff.

Niobium species, ΔF_f°, ΔH_f°, and S° values, table, 417
Nitrogen species, ΔF_f°, ΔH_f°, and S° values, table, 418
NO_3^-, activity coefficient, 63
Nonassociated electrolyte, 50
Non-electrolyte solutions, 29
Numbers, atomic, table, 433–435

O_2, activity coefficients in NaCl solution, 68
Ore deposition, Eh-pH diagram, 398
 environment of, 397 ff.
Ore-forming fluid, 400
Organic matter, effect on Eh and pH, 382, 383
Oxidation of sulfide ores, 394, 395
Oxidation potential, see Eh
Oxygen, equilibrium pressure for various compounds, see Partial pressure diagrams
 partial pressure in equilibrium with water, 145
 pressure dependence of fugacity in $Fe-O_2-Fe_3O_4$, 343–344
Oxygen species, ΔF_f°, ΔH_f°, and S° values, table, 418

Palladium species, ΔF_f°, ΔH_f°, and S° values, table, 418
Paragenetic relations, in $CaO-MgO-CO_2-H_2O$ system, 376
Paragonite, transformation to albite, 365
Partial molal volume 35, 344
 ionic, 346
Partial pressure, symbol for, 5
 and mole fraction, 26
 activity relations, 27 ff.
Partial pressure diagrams, $Cu-Fe-O_2-S_2$, 167, 169
 $Cu-CO_2-O_2-H_2O$, 156
 $Cu-CO_2-O_2-S_2-H_2O$, 162
 $Cu-O_2-S_2$, 160
 at elevated temperatures, 327–331
 $Fe-CO_2-O_2$, 152
 $Fe-CO_2-O_2-H_2O$, 153
 $Fe-CO_2-O_2-S_2$, 161
 $Fe-Mn-CO_2-O_2-H_2O$, 165
 $Fe-Mn-O_2-S_2-H_2O$, 164
 $Fe-O_2$, 148
 at elevated temperatures, 327, 328
 $Fe-O_2-H_2O$, 149
 $Fe-O_2-S_2$, at elevated temperatures, 330
 $Fe-O_2-S_2-H_2O$, 159
 $Fe-S_2$, at elevated temperatures, 330
 limitations of, 170–171
 O_2 and H_2 in equilibrium with water, 145
 $Pb-O_2-SO_2-CO_2$, 170
 relation to Eh-pH diagrams, 229
 sphalerite-depositing ore solutions, 399
$PbCl^+$, dissociation constant, 114
$Pb-CO_2O_2-H_2O$, Eh-pH diagram at fixed P_{CO_2}, 234
 at fixed ΣCO_2, 235
 discussion of Eh-pH diagrams, 233, 236, 238

$Pb-CO_2-O_2-S-H_2O$, Eh-pH diagram at fixed P_{CO_2} and ΣS, 237
 at fixed ΣCO_2 and ΣS, 238
 discussion of Eh-pH diagrams, 233 ff.
$Pb-H_2O-O_2-S$, Eh-pH diagram at fixed ΣS, 236
 discussion of, 233 ff.
$Pb-O_2-SO_2-CO_2$, partial pressure diagram, 170
PbO, red-yellow stability relations, 9
PbO_2, as indicator of environment, 233
$Pb_3(OH)_2(CO_3)_2$, as indicator of environment, 233
PbS, and native sulfur, 236
pH, definition, in terms of hydrogen ion activity, 13, 123
 operational, 124
 electrode, Ag-AgCl, 142
 equation for, 294
 glass, 123, 125–126
 for high pH values, 126
 metal-metal oxide, 141
 measurement, accuracy of, 125
 and liquid-junction potential, 124, 125
 precautions in, 128–129
 readings, in natural media, 129–131
 and effect of atmosphere on, 129–131
 reference buffer solutions, 124
 calibration of glass electrode with, 126
 table, 432
 reference electrodes, 126–128
Phlogophite-feldspar-mica equilibria, 363, 364
Phlogophite-mica-chlorite equilibria, 364
Phosphorus species, ΔF_f°, ΔH_f°, and S° values, table, 419
Platinum species, ΔF_f°, ΔH_f°, and S° values, table, 419
Plattnerite, as indicator of environment, 233
pNa, definition, 268
Polymorphs, criterion for equilibrium, 9, 326
Potassium chloride, for use as standard in obtaining single-ion activity coefficients, 58, 140
Potassium species, ΔF_f°, ΔH_f°, and S° values, table, 420
Potential, formal, 135
 irreversible oxidation, 136, 380
 liquid junction, 124, 125
Practical activity coefficient, 42
Pressure, effect on Eh-pH diagrams, 261
Pressure units, conversion factors, 402
P-T curve, and the Clapeyron equation, 341
 for hematite-geothite-water, 339
 from thermochemical data, 341
Pyrite, association with magnetite and pyrrhotite, 222
 Eh-pH relation to magnetite, 218
 oxidation by ferric salts, 222
 relation to Fe^{++}, in HS^- field, 222
 in HSO_4^- field, 220
 in SO_4^{--} field, 220

Pyrite—*continued*
  relation to hematite in HSO_4^- field, 220
  in SO_4^{--} field, 219
  relation to magnetite in SO_4^{--} field, 220
  relation to marcasite, 222
Pyrochroite, 241
Pyrrhotite, association with pyrite and magnetite, 222
  coexistence with water, 158, 216, 217, 219
  oxidation to pyrite, 216
  stability relative to other iron compounds, 216

Radii of cations, and activity products, 99
  of ions in crystals, table, 430
Raoult's law, 29–31
Rational activity coefficient, 38
Reaction quotient (Q), and electromotive force, 16
  and free-energy change of reaction, 8, 343
Reduced pressure, 24
Reduced temperature, 24
Reference electrodes, calomel, 127
  Ag-AgCl, 140
Regular solution, activity coefficients from, 44
  for AgBr, 46
  applied to ion-exchange theory, 273–274
  constant, values of, 45
  equations for, 44–47
Rhodochrosite vs. siderite, stability fields, 241
Rubidium species, ΔF_f°, ΔH_f°, and S° values, table, 420

S^{--}, 215
Salt bridge, 127, 128
Salting-out effect, 68
Saturometer, carbonate, 131–132
Scandium species, ΔF_f°, ΔH_f°, and S° values, table, 420
Schoepite, 253
Sea water, activity coefficients of species in, 102–104
  table, 103
  chemical model of, 100–106
  composition of average surface, table, 101
  distribution of species in, 104–106
  as equilibrating medium, 364
  ion pairs in, table, 106
  ionic strength, 101
  location in Eh-pH diagram, 396
  relative solubilities in, 115
Sedimentary deposits, Eh-pH diagram, 396
Sediments, marine, pH, 130
Selectivity, in cation-exchangers, 278–281
Selenium, 394
Selenium species ΔF_f°, ΔH_f°, and S° values, table, 421
Semipermeable membrane, 268
Shale, as semipermeable membrane, 268
$S-H_2O-O_2$, Eh-pH diagram, 214, 217
  at fixed ΣS, 217

Siderite, 10, 11, 151
  and CO_3^{--}, 202 ff.
  control by open and closed systems, 206
  as criterion of reducing conditions, 225
  Eh-pH relation to magnetite, 197
  facies, 228
  metastability at earth's surface, 201
Siderite-calcite equilibrium, 49
Siderite-magnetite, in iron ores, 206
Silica, and iron mineral stability, 226
  and iron silicates, 225 ff.
  reaction with gibbsite to form mica, 361
  reaction with kaolinite to form feldspar, 360
  reaction with mica to form feldspar, 360
  solubility controls, 360
Silicate facies of iron, 228
Silicates of iron, 225 ff.
Silicon species, ΔF_f°, ΔH_f°, and S° values, table, 421
Sillimanite, entropy values, table, 310
  heat content values, table, 310
Silver-silver chloride electrode, 140
Silver species, ΔF_f°, ΔH_f°, and S° values, table, 422
SO_4^{--}, activity coefficient, 63
  Eh-pH relations, 215
  metastability, 383
Sodium species, ΔF_f°, ΔH_f°, S° values, table, 422–423
Soil, kaolinite, 357
  pH, 130
  water-logged, Eh and pH of, 137
Solid solution, effect of ionic radii on, 43
  and temperature, 43
Solids, reactions involving, 9–10
Solubility, and activities of dissolved species, 54–56, 262–263
  calculation from activities, 54–56
  and complex ions, 113
  of gases in electrolyte solutions, 67 ff.
  solids in solids, 42 ff.
Solutions, dilute, 39–41
  electrolyte, 50–51
  gas, 26–28
  ideal, 31
  non-electrolyte, 29
  regular, 44–47
  solid, 42–44
Sphalerite, ore solutions depositing, 399
$S-S_2$ relations, 163
Stability diagrams, valid only for species considered, 154, 158
Stability field of solid, definition for Eh-pH diagrams, 188
Standard free energy, of formation, 7 ff.
  compound, 7
  element, 7
  hydrogen ion, aqueous, 7
  symbol for, 7
  values, tables, 403–428
  of reaction, calculations for various types of reactions, 9–13
  equations for, 8, 342, 343

INDEX OF SUBJECTS

Standard free energy—*continued*
  and equilibrium constant, 8, 342, 343
  and half-cells, 14
  relation to EMF, 16
Standard state, adsorbed ion, 289
  and activity, 5–6
  of electrolyte solute, 52–54
  enthalpy, 308
  entropy, 308
  free energy, 7
Stearate electrodes, 300–301
Strontium species, ΔF_f°, ΔH_f°, and S° values, table, 423
Subscripts, use of in formulas, 2, 5, 401
Sulfide ores, oxidation of, 394
Sulfur, effect on iron minerals, 222–225
  and Eh, pH, CO_2 changes, 222–225
  Eh-pH field of native, 216
  rhombic-monoclinic equilibrium, 334, 335
  vapor, equilibrium pressure for various compounds, *see* Partial pressure diagrams containing S_2
Sulfur dioxide, equilibrium pressure for lead compounds, 170
Sulfur species, ΔF_f°, ΔH_f°, and S° values, table, 424
Sulfur species, in water, Eh-pH diagram, 214
  at fixed ΣS, 217
  equilibria relating, 213–216
  relations at ΣS fixed, 215–216
  stable ions, 213
Superscripts, 401
Surface waters, Eh and pH, 380
Symbols, atomic, table, 433–435
Symbols, table, 401
Symmetrical solution, *see* Regular solution

Tantalum species, ΔF_f°, ΔH_f°, and S° values, table, 424
Tellurium species, ΔF_f°, ΔH_f°, S° values, table, 425
Temperature, centigrade and Kelvin scales relation, 2, 402
  effect on Eh-pH diagram, 260, 261
Tenorite, 155
Thermodynamic data, references to, 429
Thermonatrite, equilibrium with trona, 371
Third law, calculation of ΔF_r° using, 307, 315
  formation of gibbsite from corundum, 315
Third law of thermodynamics, statement of, 314
Thorium species, ΔF_f°, ΔH_f°, and S° values, table, 425
Tin species, ΔF_f°, ΔH_f°, and S° values, table, 426
Titanium species, ΔF_f°, ΔH_f°, and S° values, table, 426
Transference numbers, and cation electrodes, 284

Trona, equilibrium with nahcolite, 372
  equilibrium with thermonatrite, 371
Tungsten species, ΔF_f°, ΔH_f°, and S° values, table, 426

U-H_2O-O_2-CO_2, Eh-pH diagram, as function of P_{CO_2}, 254
  as function of ΣCO_2, 255
  discussion of, 253
U-H_2O-O_2-CO_2-K-V, Eh-pH diagram, 256
Underground waters, control of Eh and pH, 382, 383
Units, table, 402
$UO_2(CO_3)_3^{4-}$, dissociation constant, 114
Uraninite, 253, 388
Uranium, carbonate complexes, 253–256, 391, 393
  deposits, 388 ff.
  ores, predicted oxidation sequence, 392
  stability limits of minerals, 391
Uranium species, ΔF_f°, ΔH_f°, and S° values, table, 427
U-V-K-O_2-H_2O-CO_2 system, Eh-pH diagram, 393

Vanadium, compounds, Eh-pH diagram, 389
  deposits, 388 ff.
  minerals, Eh-pH diagram, 390
Vanadium species, ΔF_f°, ΔH_f°, and S° values, table, 427
van't Hoff, equation, 348
  applied to annite-sanidine-magnetite-water, 349
Vapor pressure, water, at various temperatures, table, 431
V-H_2O-O_2, Eh-pH diagrams, 389, 390

Water, activity, in electrolyte solutions, 64–67
  and fugacity values, at various pressures, table, 36
  in ideal solution under pressure, 37
  in KCl, NaCl, $CaCl_2$ solutions, values, 65
  measurement from lowering of vapor pressure, 65, 66
  and Raoult's law, 65, 66
  temperature dependence, 66
  controls of Eh and pH, 380, 381, 382, 383
  dissociation constant, table, 70, 71
  dissociation into H_2 and O_2, 144
  ground, Eh and pH, 139
  ionic equilibria, effect of pressure on, 346
  partial pressure diagram, 145
  pressure of saturated vapor, at various temperatures, table, 431
  stability, Eh-pH diagram, 176
  stagnant, brackish, Eh and pH, 138
Waters, in arkosic sediments, 361

Weathering products, kaolinite-feldspar in, 362
Weathering of rocks, and sea-water content, 113 ff.
Wegscheiderite, 372, 373
W-H_2O-O_2, Eh-pH diagram, 251
  discussion of, 247 ff.
W-H_2O-O_2-S, Eh-pH diagram, 252
  discussion of, 247 ff.
Wüstite, 413

Ytterbium species, ΔF_f°, ΔH_f°, and S° values, table, 427
Yttrium species, ΔF_f°, ΔH_f°, and S° values, table, 428

Zinc species, ΔF_f°, ΔH_f°, and S° values, table, 428
Zirconium species, ΔF_f°, ΔH_f°, and S° values, table, 428
ZoBell solution, 135